# PHASE SPACE PICTURE OF QUANTUM MECHANICS

*Group Theoretical Approach*

Lecture Notes in Physics Series — Vol. 40

# PHASE SPACE PICTURE OF QUANTUM MECHANICS

*Group Theoretical Approach*


**Y. S. Kim**
Department of Physics and Astronomy
University of Maryland
College Park, Maryland 20742, USA

**M. E. Noz**
Department of Radiology
New York University
New York 10016, USA



World Scientific
Singapore • New Jersey • London • Hong Kong

*Published by*

World Scientific Publishing Co. Pte. Ltd.
5 Toh Tuck Link, Singapore 596224
*USA office:* 27 Warren Street, Suite 401-402, Hackensack, NJ 07601
*UK office:* 57 Shelton Street, Covent Garden, London WC2H 9HE

**British Library Cataloguing-in-Publication Data**
A catalogue record for this book is available from the British Library.

**World Scientific Lecture Notes in Physics — Vol. 40**
**PHASE SPACE PICTURE OF QUANTUM MECHANICS**
**Group Theoretical Approach**



ISBN-13 978-981-02-0360-3
ISBN-10 981-02-0360-8
ISBN-13 978-981-02-0361-0 (pbk)
ISBN-10 981-02-0361-6 (pbk)

# PREFACE

Quantum mechanics can take different forms. The Schrödinger picture of quantum mechanics is very useful in atomic and nuclear physics. The Heisenberg picture is the basic language for the covariant formulation of quantum field theory. Is there then any need for a new picture of quantum mechanics? This depends on whether there are branches of physics where the Schrödinger or Heisenberg picture is less than fully effective.

Quantum optics and relativistic bound-state problems are relatively new fields. In quantum optics, we deal with creation and annihilation of photons and linear superposition of multiphoton states. It is possible to construct the mathematics of harmonic oscillators in the Schrödinger picture to describe the photon's states. However, the mathematics becomes complicated when we attempt to describe generalized coherent states often called the squeezed states. Is there a language simpler than the Schrödinger picture?

Quantum field theory accommodates both the uncertainty principle and special relativity. However, it is less than fully effective in describing bound-state problems or localized probability distributions. It is possible to construct models of relativistic hadrons consisting of quarks starting from the Schrödinger picture of quantum mechanics. The question then is whether it is possible to formulate the uncertainty relations in a covariant manner (Dirac 1927).

The phase-space picture of quantum mechanics provides the answer to these questions. Starting from the Schrödinger wave function, it is possible to construct a distribution function, often called the Wigner function, in phase space in terms of the c-number position and momentum variables. In this picture, it is possible to perform canonical transformations as in the case of classical mechanics. This will bring us a deeper understanding of the uncertainty principle.

This phase-space picture of quantum mechanics is not new. The earliest application of the Wigner phase-space distribution function was made in quantum corrections to thermodynamics in 1932 (Wigner 1932a). Since then, the Wigner function has been discussed in many branches of physics including statistical mechanics, nuclear physics, atomic and molecular physics, and foundations of physics. However, it is difficult to see the advantage of using the Wigner function over the existing method in those traditional branches of physics.

In this book, we discuss applications of the Wigner function in quantum optics and the relativistic quark model which are relatively new subjects in physics and which still need a basic scientific language. From the mathematical point of view, the Wigner function for the ground-state harmonic oscillator is the basic language for these new branches of physics. However, its symmetry properties constitute the most interesting aspect of this new scientific language.

Indeed, the symmetry property of the Wigner function in phase space is that of the Lorentz group. The Lorentz group is known to be a difficult subject to mathematicians, because it is a non-compact group. To physicists, group theory is a difficult subject when its representations have no physical applications. However, the situation is quite the opposite when the representation can extract physical implications.

In this book, we discuss the physical consequences of the symmetries of the Wigner function in phase space. This book is written for those scientists and students who wish to study the basic principles of the phase-space picture of quantum mechanics and physical applications of the Wigner distribution functions. This book will also serve a useful purpose for those who simply wish to study the physical applications of the Lorentz group.

We are indebted to Professor Eugene P. Wigner for encouraging us to formulate a group theoretical approach to the phase-space picture of quantum mechanics. Professor Wigner suggested the use of the light-cone coordinate system for the covariant formulation of the Wigner function. Indeed, Chapter 10 of this book is based on Professor Wigner's ideas. He suggested the possibility that the work of Inonu and Wigner (1953) on group contractions be extended to study the space-time geometry of relativistic particles (Kim and Wigner 1987a and 1990a). He also suggested the use of the concept of entropy when the measurement process is less than complete in a relativistic system (Kim and Wigner 1990c).

While this book was being written, we received helpful comments and suggestions from many of our colleagues, including K. Cho, D. Han, C. H. Kim, M. Kruger, P. McGrath, H. S. Pilloff, L. Rana, Y. H. Shih, J. Soln, C. Van Hine, and W. W. Zachary.

September 1990
YSK and MEN

# INTRODUCTION

The concept of phase space arises naturally from the Hamiltonian formulation of classical mechanics, and plays an important role in the transition from classical physics to quantum theory. However, in quantum mechanics, the position and momentum variables cannot be measured simultaneously. In the Schrödinger picture, the wave function is written as a function of either the position or the momentum variable, but not of both. For this reason, in quantum mechanics, the density matrix (Von Neumann 1927 and 1955) replaces phase space as a device for describing the density of states. It therefore appears that phase space is not a useful concept in quantum mechanics. We disagree. The role of phase space in quantum mechanics has not yet been fully explored.

Starting from the density matrix, is it possible to develop an algorithm of quantum mechanics based on phase space? This question has been raised repeatedly since the publication in 1932 of Wigner's paper on the quantum correction for thermodynamic equilibrium (Wigner 1932a). Since it is not possible to measure simultaneously position and momentum without error, it is meaningless to define a point in phase space. However, this does not prevent us from defining an area element in phase space whose size is not smaller than Planck's constant. Since the measurement problem is stated in terms of the least possible value of the product of the uncertainties in the position and momentum, it is of interest to see how the uncertainty product can be stated in phase space.

The basic advantage of this phase-space picture of quantum mechanics is that it is possible to perform canonical transformations, just as in classical mechanics. The purpose of this book is to study the physical consequences derivable from canonical transformations in quantum mechanics. Using these transformations, we can compare quantum mechanics with classical physics in terms of many illustrative examples. In addition, the phase-space picture of quantum mechanics is becoming a new scientific language for modern optics which is a rapidly expanding field. Furthermore, the Lorentz transformation in a given direction of boost is a canonical transformation in the light-cone coordinate system. This allows us to state the uncertainty relation in a Lorentz-invariant manner.

There are still many questions concerning the uncertainty relations for which answers are not well known. For instance, in the Schrödinger picture, the free-particle wave packet becomes widespread, and the uncertainty product increases as time progresses or regresses. Is it possible to state the uncertainty relation in terms of the quantity which remains constant? Can phase space provide an answer to this question? The answer to this question is YES. In the phase space-picture, the uncertainty is define in terms of the area which the Wigner distribution function occupies. The spread of a wave packet is an area-preserving canonical transformation in the phase-space picture of quantum mechanics.

Quantum optics is a rapidly expanding subject, and it is increasingly clear that coherent and squeezed states of light will play a major role in a new understanding of the uncertainty principle, and will provide innovations in high-technology industrial applications. These optical states are minimum-uncertainty states, and transformations among these state are therefore canonical transformations. Indeed, the phase-space picture of quantum mechanics is the natural language for these relatively new quantum states.

Most physicists these days learn classical mechanics from Goldstein's textbook (Goldstein 1980). However, Goldstein's book does not emphasize the importance of linear canonical transformations, which are discussed in more advanced books (Arnold 1978, Abraham and Marsden 1978, Guilemin and Sternberg 1984). In this book, we shall discuss the group of linear canonical transformations in phase space which is the inhomogeneous symplectic group (Han *et al.* 1988). For a single pair of canonically conjugate variables, the group is the inhomogeneous symplectic group ISp(2), and it is ISp(4) for two pairs of conjugate variables.

If we do not take into account translations in phase space, the symmetry groups become those of homogeneous symplectic transformations. The groups Sp(2) and Sp(4) are locally isomorphic to the (2 + 1)-dimensional and (3 + 2)-dimensional Lorentz groups. Thus the study of the symmetries in phase space requires the study of Lorentz transformations.

The Lorentz transformation is one of the most fundamental transformations in physics, and this subject can be formulated in terms of the inhomogeneous Lorentz group (Wigner 1939). Since this group governs the fundamental space-time symmetries of elementary particles, there are many papers and books on this subject (Kim and Noz 1986). In this book, we treat Lorentz transformations as canonical transformations.

One of the persisting question in modern physics is whether the uncertainty relations can be Lorentz-transformed. Does Planck's constant remain invariant under Lorentz transformations? Is localization of the probability distribution a Lorentz-invariant concept? It is very difficult to answer these questions in the Heisenberg or Schrödinger picture of quantum mechanics. The basic limitation of these pictures is that they do not tell us how the uncertainty relations appear to observers in different Lorentz frames. The question of whether quantum mechanics can be made consistent with special relativity has been and still is the central issue of modern physics.

We shall address this question within the framework of the phase-space picture of quantum mechanics. It is interesting to note that the Lorentz boost in a given direction is a canonical transformation in phase space using the light-cone variables. This allows us to state the uncertainty relations in a Lorentz-invariant manner. Feynman's parton picture (Feynman 1969) and the nucleon form factors are discussed as illustrative examples.

In the first two Chapters, we discuss the forms of classical mechanics and quantum mechanics useful for the formulation of the Wigner phase-space picture of quantum mechanics, which is discussed in detail in Chapters 3 and 4. Chapters 5 and 6 are for the applications of the Wigner function to coherent and squeezed states of light. It is seen in these chapters that the study of the Wigner function requires the knowledge of the Lorentz group.

In Chapters 7 and 8, we present a detailed discussion of the physical representations of the inhomogeneous Lorentz group or the Poincaré group which governs the fundamental space-time symmetries of relativistic particles. By constructing the representation based on harmonic oscillators, we study the phase-space picture of relativistic extended particles. Chapters 9 contains a detailed discussion of experimental observation of Lorentz-squeezed hadrons. Finally, in Chapter 10, we discuss some fundamental issues in space-time symmetries of relativistic system, including the unification of space-time symmetries of massive and massless particles and the entropy increase due to the incompleteness in measurements.

Since we are combining the Wigner function with group theory, we have reprinted in the Appendix Wigner's 1932 paper on the Wigner function as well as his 1939 paper on the representations of the inhomogeneous Lorentz group. The study of phase space requires a knowledge of harmonic oscillators. P.A.M. Dirac was interested in constructing representations of the Lorentz group based on four dimensional harmonic oscillators. We have therefore included Dirac's 1945 paper on the Lorentz group and his 1963 paper on the de Sitter group.

There are many other interesting subjects which can be studied within the framework of the phase-space picture of quantum mechanics but are not discussed in this book. However, there are now a number of review articles (Wigner 1971, O'Connell 1983, Carruthers and Zachariasen 1983, Hillery *et al.* 1984, Balazs and Jennings 1984, Littlejohn 1986) containing applications of the Wigner phase-space distribution function to various branches of modern physics. The scope of this book is limited to the simplest form of the Wigner function with maximum symmetry applicable to the branches of physics in which the phase-space picture is definitely superior to other forms of quantum mechanics.

# Contents

# PHASE SPACE PICTURE OF QUANTUM MECHANICS

*Group Theoretical Approach*

# Chapter 1

# PHASE SPACE IN CLASSICAL MECHANICS

The concept of phase space originates from classical mechanics in its Hamiltonian formalism, in which a given dynamical system depends on a number of independent coordinate variables and the same number of conjugate momentum variables. The Cartesian space consisting of these $2n$ coordinate variables is called phase space (Goldstein 1980).

This phase-space formalism is the starting point for the modern approach to classical mechanics (Arnold 1978, Abraham and Marsden 1978), including nonlinear dynamics and chaos. Traditionally, the phase-space formalism of classical mechanics plays the role of a bridge between classical mechanics and quantum mechanics. In this Chapter, we shall study the properties of classical phase space which will be shared by the phase-space formulation of quantum mechanics.

Of particular interest are linear canonical transformations which correspond to unitary transformations in the Schrödinger picture of quantum mechanics. The mathematics for linear canonical transformations is that of the symplectic group which is relatively new in physics (Weyl 1946). The linear transformations of the $n$ pairs of canonical variables is governed by the group $Sp(2n)$ (Gilmore 1974, Guillemin and Sternberg 1984). In this book, we will be primarily concerned with physical problems requiring one and two pairs of canonical variables. With this point in mind, we shall start this section with the Hamiltonian formulation of classical mechanics.

## 1.1 Hamiltonian Form of Classical Mechanics

Classical mechanics starts with Newton's second law stating that force is proportional to acceleration. There are several reformulations of this law such as the Lagrangian and Hamiltonian formalisms. The Lagrangian form is useful when we do not wish to consider constraint forces. It plays the key role in quantum field theory. It is also serves as the bridge between Newton's second law and the Hamiltonian formalism.

If there are $n$ independent coordinates $q_1, q_2, \cdots, q_n$ in a given dynamical system, the Lagrangian is a function of these coordinates and their time derivatives, as well as the time variable:

$$L = L(q_1, q_2, \cdots, q_n; \dot{q}_1, \dot{q}_2, \cdots, \dot{q}_n; t). \tag{1.1}$$

From this, the momentum variable conjugate to $q$ is defined as

$$p_i = \frac{\partial L}{\partial \dot{q}_i}. \tag{1.2}$$

For each $i$, the equation of motion is

$$\dot{p}_i = \frac{\partial L}{\partial q_i}. \tag{1.3}$$

This is the Lagrangian form of the equation of motion.

The Hamiltonian is defined as

$$H = \sum_i \dot{q}_i p_i - L. \tag{1.4}$$

Then, from the Lagrangian equations of motion,

$$\delta H = \sum_i (\dot{q}_i \delta p_i - \dot{p}_i \delta q_i) + \frac{\partial H}{\partial t} \delta t. \tag{1.5}$$

Thus, for each $i$, we can write the Hamiltonian equation of motion as

$$\dot{q}_i = \frac{\partial H}{\partial p_i}, \quad \dot{p}_i = -\frac{\partial H}{\partial q_i}. \tag{1.6}$$

Now the Hamiltonian can be regarded as a function of $q_1, q_2, \cdots, q_n$ and $p_1, p_2, \cdots, p_n$.

As far as the time dependence is concerned, from the definition of the Hamiltonian given in Eq. 1.4,

$$\frac{\partial L}{\partial t} = -\frac{\partial H}{\partial t}. \tag{1.7}$$

The total derivative of the Hamiltonian is

$$\frac{dH}{dt} = \sum_i \left( \frac{\partial H}{\partial q_i} \dot{q}_i + \frac{\partial H}{\partial p_i} \dot{p}_i \right) + \frac{\partial H}{\partial t}. \tag{1.8}$$

As a consequence of the equations of motion, the quantity in parenthesis vanishes, and

$$\frac{dH}{dt} = \frac{\partial H}{\partial t} = -\frac{\partial L}{\partial t}. \tag{1.9}$$

Thus, if the Lagrangian does not depend on time explicitly, the Hamiltonian is a constant of motion.

Let us consider some examples. The Hamiltonian for a free particle is naturally $H = p^2/2m$, and the Hamiltonian for $n$ particles is

$$H = \sum_{i=1}^{n} p_i^2/2m_i, \tag{1.10}$$

which is the total energy. The equations of motion lead to $\dot{p}_i = 0$ for every $i$.

The Hamiltonian for a charged particle in an electromagnetic field generated by the vector potential $\mathbf{A}$ and the scalar potential $\phi$ is

$$H = \frac{1}{2m}\left(\mathbf{p} - \frac{e}{c}\mathbf{A}\right)^2 + e\phi, \tag{1.11}$$

where $e$ is the charge of the particle. In this case, the momentum vector $\mathbf{p}$ is not the mass times the velocity, but the quantity $\frac{1}{m}\left(\mathbf{p} - \frac{e}{c}\mathbf{A}\right)$ is the velocity. Since the magnetic field does not change the magnitude of velocity, the above Hamiltonian is the total energy of the system. This form of the Hamiltonian has been discussed extensively in standard textbooks on classical and quantum mechanics (Goldstein 1980, Schiff 1968). Since the potential can be gauge-transformed, the above Hamiltonian is not invariant under gauge transformations. However, the resulting equations of motion are invariant under gauge transformations.

For example, for a particle in a constant magnetic field along the $z$ direction, $\mathbf{B} = \hat{e}_z B$, the vector potential can be written as

$$\mathbf{A} = \hat{e}_y x B, \quad \text{or} \quad \mathbf{A} = \frac{1}{2}\left(\hat{e}_y x - \hat{e}_x y\right) B. \tag{1.12}$$

The difference between the two potentials is $\frac{1}{2}\left(\hat{e}_y x + \hat{e}_x y\right) B$. This is the gradient of the scalar function $xyB/2$. These two potentials give the same set of equations of motion.

For the first choice of $\mathbf{A}$, the Hamiltonian can be written as

$$H = \frac{1}{2m}\left\{p_x^2 + (p_y - exB/c)^2\right\}. \tag{1.13}$$

Thus, according to the Hamiltonian equations of motion,

$$\dot{x} = \left(\frac{1}{m}\right) p_x, \quad \dot{y} = \frac{1}{m} p_y - \left(\frac{eB}{mc}\right) x,$$
$$\dot{p}_x = \left(\frac{eB}{mc}\right)\left(p_y - \frac{exB}{c}\right), \quad \dot{p}_y = 0. \tag{1.14}$$

From the first two equations,

$$p_x = m\dot{x}, \qquad p_y = m\dot{y} + exB/c. \tag{1.15}$$

The substitution of these into Eq. 1.14 leads to the familiar set of Newton's equations for the circular orbit with the cyclotron frequency $eB/mc$. The Hamiltonian equations of motion will take a different form for the second vector potential in Eq. 1.12, but they will lead to the same set of Newton's equations.

The Hamiltonian for the one-dimensional harmonic oscillator is

$$H = \left(\frac{1}{2m}\right) p^2 + \left(\frac{K}{2}\right) x^2. \tag{1.16}$$

According to the equations of motion, $p = m\dot{x}$, and $Kx = -\dot{p}$. This result is well known. This form of the Hamiltonian plays the central role in modern optics and relativistic quantum mechanics, and will be discussed extensively in this book.

Let us consider the Galilei transformation of this system, where the coordinate is transformed as

$$x' = x + vt. \tag{1.17}$$

This means that the above harmonic oscillator is on a truck which moves with the velocity $v$. Then to the observer on the ground, the Hamiltonian will be

$$H = \left(\frac{1}{2m}\right) p'^2 + \left(\frac{K}{2}\right) (x' - vt)^2. \tag{1.18}$$

The equations of motion are

$$\dot{p}' = K(x' - vt), \quad \dot{x}' = p'/m. \tag{1.19}$$

This leads to the conclusion that the acceleration in the truck frame is the same as that on the ground. Furthermore, the Lagrangian can be written as

$$L = \left(\frac{m}{2}\right) \dot{x}'^2 - \left(\frac{K}{2}\right) (x' - vt)^2, \tag{1.20}$$

which leads to $p' = m\dot{x}' = m(\dot{x} + v)$. Thus

$$p' = p + mv. \tag{1.21}$$

Thus, in terms of $x$ and $p$, the Hamiltonian becomes

$$H = \left(\frac{1}{2m}\right) (p + mv)^2 + \left(\frac{K}{2}\right) x^2. \tag{1.22}$$

This is consistent with what we expect from the Galilei transformation.

## 1.2 Trajectories in Phase Space

It is quite clear that, in the Hamiltonian formalism, the dynamical system of $n$ degrees of freedom is described by $n$ coordinate variables $q_1, q_2, \cdots, q_n$, and their conjugate momenta $p_1, p_2, \cdots, p_n$. It is then possible to consider a 2n-dimensional space spanned by $n$ coordinate and $n$ momentum variables. This space is called phase space.

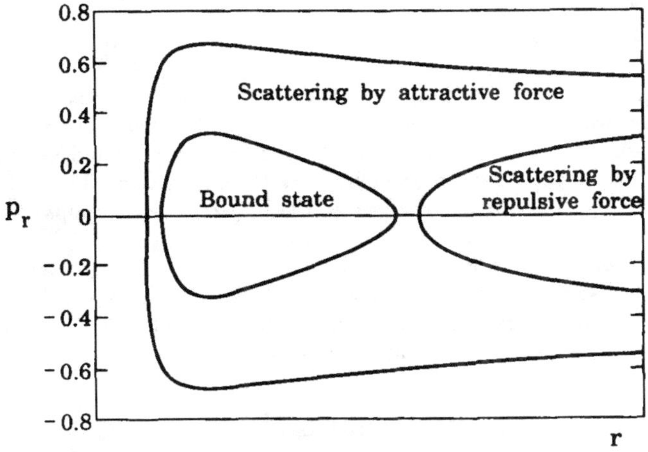


Figure 1.1: Phase-space description of the Kepler problem. If the total energy is negative, the orbit is closed. If the energy is positive, the orbit is open. The particle comes in with a negative momentum and goes out with a positive momentum. This figure gives an interesting description of the transition from negative to positive energy. If $k$ is negative, the potential is repulsive, and the momentum takes its maximum value at an infinite distance.

If there is one degree of freedom, then a two-dimensional phase space consisting of $x$ and $p$ can completely determine the dynamical system. For one free particle with a given momentum $p$, the particle trajectory is a line parallel to the $x$ axis with a fixed value of $p$. For the one-dimensional harmonic oscillator, the Hamiltonian is the total energy, and the trajectory is an ellipse in the phase space of $x$ and $p$.

Let us next consider a simple pendulum whose Hamiltonian is (Goldstein 1980)

$$H = p_\theta^2/2I + mg\ell(1 - cos\theta). \tag{1.23}$$

Here again the Hamiltonian is the total energy. The angular momentum is given by $p_\theta$. If the energy is very small, the small-angle approximation can be made, and thus the energy becomes

$$E = p_\theta^2/2I + (mg\ell/2)\theta^2. \tag{1.24}$$

Thus the trajectory in the phase space of $\theta$ and $p_\theta$ is an ellipse. If however, the energy is much larger than $mg\ell$, then the potential energy terms can be ignored, and the energy becomes that of a free particle.

It is possible to give a phase space description of the Kepler problem. The total energy can be written in terms of the radial momentum and the effective potential as (Goldstein 1980):

$$E = p_r^2/2m + \ell^2/2mr^2 - k/r, \tag{1.25}$$

where $\ell$ is the total angular momentum. If $k$ is positive, the system can have a bound state. Then the trajectory in the phase space of $r$ and $p_r$ is closed for a negative value of $E$. If the energy is positive, the trajectory is open as is indicated in Figure 1.1. It is interesting to see the transition from negative to positive values of energy in Figure 1.1.

Let us go back to the one-dimensional harmonic oscillator whose trajectory is an ellipse in the two-dimensional phase space. If the Galilei transformation is applied as in the case of Section 1.1, the trajectory in $x'$ and $p'$ would be described by Eq. 1.18. Here the particle traces an ellipse while moving along the $x$ direction with velocity $v$. In the phase space of $x$ and $p$, the trajectory is an ellipse whose center is translated from the origin to $x = 0$ and $p = -mv$.

Indeed, many interesting physical phenomena can be interpreted in terms of coordinate transformations in phase space. Transformations which preserve the Hamiltonian form of equations of motion are called canonical transformations. We shall study these transformations in Section 1.3.

## 1.3 Canonical Transformations

Classical mechanics starts with Newton's second law stating that force is proportional to acceleration. There are several reformulations of this law such as the Lagrangian and Hamiltonian formalisms. The Hamiltonian formulation plays the role of a bridge between classical mechanics and several branches of modern physics. In this formalism, the Hamiltonian is a function of coordinate variables $q_1, q_2, \cdots, q_n$ and their respective conjugate momentum variables $p_1, p_2, \cdots, p_n$. Let us rewrite here the Hamiltonian equations given in Eq. 1.6:

$$\dot{q}_1 = \frac{\partial H}{\partial p_i}, \qquad \dot{p}_i = -\frac{\partial H}{\partial q_i}. \tag{1.26}$$

We can write these two equations in one form by introducing a 2n-dimensional phase space whose coordinate variables are

$$(\eta_1, \eta_2, \cdots, \eta_{n+1}, \eta_{n+2}, \cdots, \eta_{2n}) = (q_1, q_2, \cdots, q_n, p_1, p_2, \cdots, p_n). \tag{1.27}$$

Then the Hamiltonian equations of motion can be written as

$$\dot{\eta}_i = J_{ij}\frac{\partial H}{\partial \eta_j}, \tag{1.28}$$

where $J$ is a 2n-by-2n matrix of the form (Arnold 1978, Abraham and Marsden 1978, Goldstein 1980, Kim and Noz 1983)

$$J = \begin{pmatrix} 0 & I \\ -I & 0 \end{pmatrix}. \tag{1.29}$$

and $I$ is the n-by-n identity matrix.

For one pair of canonical variables, $J$ takes the form

$$J = \begin{pmatrix} 0 & 1 \\ -1 & 0 \end{pmatrix}. \tag{1.30}$$

For two pairs of canonical variables in the four-dimensional phase space, $J$ is

$$J = \begin{pmatrix} 0 & 0 & 1 & 0 \\ 0 & 0 & 0 & 1 \\ -1 & 0 & 0 & 0 \\ 0 & -1 & 0 & 0 \end{pmatrix}. \tag{1.31}$$

The transformations which leave the form of Hamilton's equations invariant are called canonical transformations (Goldstein 1980). Let us consider the transformation of variables from $\eta_i$ to $\xi_i$. Then we are led to consider the matrix $M_{ij}$, defined as

$$M_{ij} = \frac{\partial}{\partial \eta_j}\xi_i. \tag{1.32}$$

Then

$$\dot{\xi}_i = M_{ij}J_{jk}M_{\ell k}\frac{\partial H}{\partial \xi_\ell}. \tag{1.33}$$

Thus, in matrix notation, if

$$MJ\tilde{M} = J, \tag{1.34}$$

then the system of Hamilton's equations remains invariant under the coordinate transformation from $\eta_i$ to $\xi_i$ . Indeed, the above equation is the condition that the transformation be canonical. The group of linear transformations which satisfy this condition is called the symplectic group (Weyl 1946, Guillemin and Sternberg 1984). For convenience, we shall call the above condition the symplectic condition.

For one pair of canonical variables, transformations consist of rotations and squeezes. The rotation is represented by the antisymmetric matrix (Kim and Noz 1983):

$$R(\phi) = \begin{pmatrix} \cos(\phi/2) & -\sin(\phi/2) \\ \sin(\phi/2) & \cos(\phi/2) \end{pmatrix}. \tag{1.35}$$

This matrix satisfies the symplectic condition (Kim and Noz 1983):

$$R(\phi)JR(-\phi) = J, \tag{1.36}$$

with $J$ of Eq. 1.30. In addition, the symmetric form of the squeeze matrix:

$$S(0,\lambda) = \begin{pmatrix} e^{\lambda/2} & 0 \\ 0 & e^{-\lambda/2} \end{pmatrix}. \tag{1.37}$$

satisfies

$$S(0,\lambda)JS(0,\lambda) = J. \tag{1.38}$$

Then the squeeze along the direction which makes an angle of $\phi/2$ with the $q$ axis is (Kim and Noz 1983)

$$\begin{aligned} S(\phi,\lambda) &= R(\phi)S(\lambda,0)R(-\phi) \\ &= \begin{pmatrix} \cosh(\lambda/2) + (\cos\phi)\sinh(\lambda/2) & (\sin\phi)\sinh(\lambda/2) \\ (\sin\phi)\sinh(\lambda/2) & \cosh(\lambda/2) - (\cos\phi)\sinh(\lambda/2) \end{pmatrix}. \end{aligned} \tag{1.39}$$

This matrix also satisfies the symplectic condition (Kim and Noz 1983):

$$S(\phi, \lambda) J \tilde{S}(\phi, \lambda) = J. \tag{1.40}$$

As we shall see in Chapter 4, the symmetry of linear canonical transformations is governed by the group $Sp(2)$ which is locally isomorphic to the (2+1)-dimensional Lorentz group.

For the four-dimensional phase space consisting of two pairs of canonical variables, the geometry is much more complicated. However, a partial view of the full geometry can still be helpful. It is not difficult to check the symplectic condition $RJ\tilde{R} = J$ for the following form of $R$.

$$R(\phi) = \begin{pmatrix} \cos(\phi/2) & -\sin(\phi/2) & 0 & 0 \\ \sin(\phi/2) & \cos(\phi/2) & 0 & 0 \\ 0 & 0 & \cos(\phi/2) & -\sin(\phi/2) \\ 0 & 0 & \sin(\phi/2) & \cos(\phi/2) \end{pmatrix}, \tag{1.41}$$

with $J$ of Eq. 1.31. The squeeze matrix

$$S(0, \lambda) = \begin{pmatrix} e^{\lambda/2} & 0 & 0 & 0 \\ 0 & e^{-\lambda/2} & 0 & 0 \\ 0 & 0 & e^{-\lambda/2} & 0 \\ 0 & 0 & 0 & e^{\lambda/2} \end{pmatrix} \tag{1.42}$$

also satisfies the condition:

$$S(0, \lambda) J \tilde{S}(0, \lambda) = J. \tag{1.43}$$

It should be noted that the four-by-four matrix with diagonal elements $e^{\lambda/2}, e^{-\lambda/2}, e^{\lambda/2}, e^{-\lambda/2}$, which is clearly different from that of Eq. 1.42 does not satisfy the above condition. It is important to note that if the $x_1x_2$ coordinate system is squeezed along the $x_1$ direction, then the $p_1p_2$ coordinate system is squeezed along the $p_2$ direction.

In addition, there are also useful nonlinear canonical transformations. The most important transformation is the transformation of one pair of canonical variables from the Cartesian to polar coordinate system. For this purpose, we are led to consider

$$Q = \frac{1}{2}\left(q^2 + p^2\right), \quad P = \tan^{-1}\left(\frac{p}{q}\right). \tag{1.44}$$

For one pair of variables, the symplectic condition is equivalent to

$$\begin{vmatrix} \dfrac{\partial Q}{\partial q} & \dfrac{\partial P}{\partial q} \\ \\ \dfrac{\partial Q}{\partial p} & \dfrac{\partial P}{\partial p} \end{vmatrix} = 1. \tag{1.45}$$

Indeed, the transformation of Eq. 1.44 satisfies the above condition. This canonical transformation will play an important role in interpreting the uncertainty principle in the Fock space governing the creation and annihilation of particles.

## 1.4 Coupled Harmonic Oscillators

The coupled oscillator problem is a standard item in the established physics curriculum. Let us consider a system of coupled oscillators described in Figure 1.2.

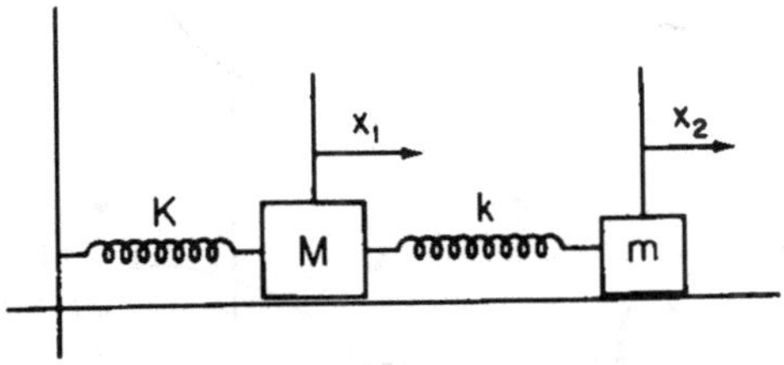


Figure 1.2: Coupled oscillators of two different masses. $x_1$ and $x_2$ measure displacements from the equilibrium positions.

The Hamiltonian for this system is

$$H = \frac{1}{2}\left\{\frac{1}{M}(p_1)^2 + \frac{1}{m}(p_2)^2 + K(x_1)^2 + k(x_1 - x_2)^2\right\}. \tag{1.46}$$

In the above expression, the kinetic energy $T$ is

$$T = \frac{1}{2}\left\{\frac{1}{M}(p_1)^2 + \frac{1}{m}(p_2)^2\right\}, \tag{1.47}$$

while the potential energy is

$$V = \frac{1}{2}(K + k)(x_1)^2 - k(x_1 x_2) + \frac{1}{2}k(x_2)^2 .$$

The problem of solving the coupled oscillator problem is to simultaneously diagonalize the kinetic and potential energy terms by choosing a suitable linear combinations of the coordinate variables $x_1$ and $x_2$ (Goldstein 1980, Aravind 1989, Rana and Kim 1989).

In the above expression, the kinetic energy is already decoupled. We can decouple the potential energy by making a rotation of the coordinate system spanned by $x_1$ and $x_2$. However, if we rotate the $x_1x_2$ coordinate system, the kinetic energy will acquire a coupling term. Therefore, it is not possible to decouple the system by rotation alone.

In order to gain a better understanding of the problem, let us plot both the kinetic and potential energy terms in the $p_1p_2$ and $x_1x_2$ coordinate systems. Both the kinetic and potential energy terms are represented by ellipses in their respective coordinate systems. This is illustrated in Figure 1.3.

We can see that any attempt to align the ellipse by rotation in Figure 1.3(b) will result in disalignment of the ellipse in Figure 1.3(a). It is not possible to align

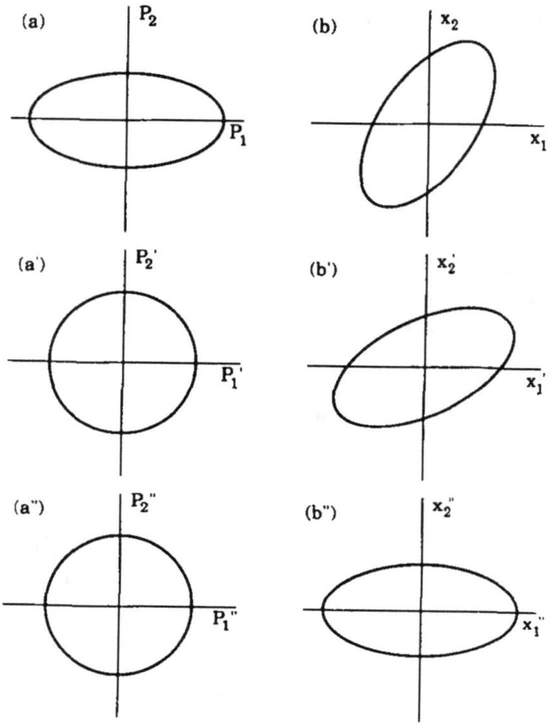


Figure 1.3: Kinetic and potential energies. Both the kinetic and potential energies take quadratic forms which can be described by ellipses. Figures 1.3(a) and 1.3(b) represent the quadratic forms in $(p_1, p_2)$ and $(x_1, x_2)$ for the kinetic energy and potential energy respectively. The kinetic energy ellipse is aligned along its principal axes, while the potential energy is not. It is not possible to align both ellipses simultaneously. It is possible, however, to change the scales of the $(p_1, p_2)$ and $(x_1, x_2)$ coordinates by squeezing to transform the kinetic ellipse into a circle in the $(p_1', p_2')$ coordinate system. Then the $(x_1', x_2')$ coordinate system undergoes the same scale transformation. Figures 1.3(a′) and 1.3(b′) indicate these new coordinate systems. It is now possible to rotate the $(x_1', x_2')$ coordinate system to align the potential energy ellipse without affecting the kinetic energy counterpart. The coordinates $(p_1'', p_2'')$ and $(x_1'', x_2'')$ represent the completely decoupled system indicated by Figures 1.3 (a″) and 1.3 (b″).

both ellipses simultaneously along their principal axes. However, we can consider the possibility of deforming the ellipse of Figure 1.3(a) into a circle first, resulting in Figure 1.3(a′). Figure 1.3(b) will then become Figure 1.3(b′). Then rotating the ellipse of Figure 1.3(b′) will leave invariant the circle of Figure 1.3(a′). We can then rotate Figure 1.3(b′) to Figure 1.3(b″) to align the potential energy ellipse along its principal axes. By doing this, we would be able to decouple the quadratic form of the potential energy without introducing any coupling in the kinetic energy.

Let us make the following coordinate transformations:

$$\begin{pmatrix} x_1' \\ x_2' \\ p_1' \\ p_1' \end{pmatrix} = \begin{pmatrix} e^{\lambda/2} & 0 & 0 & 0 \\ 0 & e^{-\lambda/2} & 0 & 0 \\ 0 & 0 & e^{-\lambda/2} & 0 \\ 0 & 0 & 0 & e^{\lambda/2} \end{pmatrix} \begin{pmatrix} x_1 \\ x_2 \\ p_1 \\ p_1 \end{pmatrix}, \tag{1.48}$$

where $e^{\lambda} = \left(\frac{M}{m}\right)^{1/2}$. The above matrix performs a scale transformation. This particular form of scale transformation is often called a "squeeze" in the current literature (Ekert *et al.* 1989, Kim and Li 1989, Rana and Kim 1989). In terms of the new coordinate variables,

$$T = \frac{1}{2}\left(\frac{1}{mM}\right)^{1/2}\left((p_1')^2 + (p_2')^2\right),$$

$$V = \frac{1}{2}(K+k)\sqrt{m/M}\,(x_1')^2 - k\,(x_1'x_2') + \frac{1}{2}k\sqrt{M/m}\,(x_2')^2 . \tag{1.49}$$

Indeed, the kinetic energy term is invariant under rotations. We can decouple the potential energy by making the rotation:

$$\begin{pmatrix} x_1'' \\ x_2'' \\ p_1'' \\ p_1'' \end{pmatrix} = \begin{pmatrix} \cos(\phi/2) & -\sin(\phi/2) & 0 & 0 \\ \sin(\phi/2) & \cos(\phi/2) & 0 & 0 \\ 0 & 0 & \cos(\phi/2) & -\sin(\phi/2) \\ 0 & 0 & \sin(\phi/2) & \cos(\phi/2) \end{pmatrix} \begin{pmatrix} x'_1 \\ x'_2 \\ p'_1 \\ p'_1 \end{pmatrix}. \tag{1.50}$$

Under this transformation,

$$T = \frac{1}{2}\left(\frac{1}{mM}\right)^{1/2}\left((p_1'')^2 + (p_2'')^2\right). \tag{1.51}$$

On the other hand, $V$ becomes

$$V = \frac{\alpha}{2}(x_1'')^2 + \frac{\beta}{2}(x_2'')^2 + \gamma\,(x_1''x_2''), \tag{1.52}$$

with

$$\alpha = \frac{(k+K)\sqrt{m/M} + k\sqrt{M/m}}{2} + k\,(\sin\phi) + \left((K+k)\sqrt{m/M} - k\sqrt{M/m}\right)\frac{\cos\phi}{2},$$

$$\beta = \frac{(k+K)\sqrt{m/M} + k\sqrt{M/m}}{2} - k\,(\sin\phi) - \left((K+k)\sqrt{m/M} - k\sqrt{M/m}\right)\frac{\cos\phi}{2},$$

$$\gamma = -k\,(\cos\phi) + \left((K+k)\sqrt{m/M} - k\sqrt{M/m}\right)\frac{\sin\phi}{2}.$$

In order to decouple the quadratic form for the potential energy, we can choose that value of angle $\phi$ which will cause the coefficient $\gamma$ to vanish. The angle is then given by

$$\tan\phi = \frac{2\sqrt{mM}}{m\,(1+K/k) - M}. \tag{1.53}$$

The coefficients $\alpha$ and $\beta$ can be determined from this particular value of the angle $\phi$.

Since $\gamma = 0$ in the expression of Eq. 1.52, both the kinetic and potential energy terms are decoupled. The problem of finding the frequencies of the normal modes is now trivial. The frequency for the $x_1''$ and $x_2''$ modes are

$$\omega_1 = \left(\alpha/\sqrt{mM}\right)^{1/2}, \quad \omega_2 = \left(\beta/\sqrt{mM}\right)^{1/2}, \tag{1.54}$$

respectively, and the solutions are

$$x_1'' = A_1 \cos(\omega_1 t + \phi_1), \quad x_2'' = A_2 \cos(\omega_2 t + \phi_2). \tag{1.55}$$

The constants $A_1, A_2, \phi_1$, and $\phi_2$ are to be determined from the initial conditions.

From these solutions, we can obtain $x_1$ and $x_2$ by combining the transformations given in Eq. 1.48 and Eq. 1.50:

$$\begin{pmatrix} x_1 \\ x_2 \end{pmatrix} = \begin{pmatrix} (m/M)^{1/4} & 0 \\ 0 & (M/m)^{1/4} \end{pmatrix} \begin{pmatrix} \cos(\phi/2) & \sin(\phi/2) \\ -\sin(\phi/2) & \cos(\phi/2) \end{pmatrix} \begin{pmatrix} x_1'' \\ x_2'' \end{pmatrix},$$

$$\begin{pmatrix} p_1 \\ p_2 \end{pmatrix} = \begin{pmatrix} (M/m)^{1/4} & 0 \\ 0 & (m/M)^{1/4} \end{pmatrix} \begin{pmatrix} \cos(\phi/2) & \sin(\phi/2) \\ -\sin(\phi/2) & \cos(\phi/2) \end{pmatrix} \begin{pmatrix} p_1'' \\ p_2'' \end{pmatrix}. \tag{1.56}$$

The above analysis shows that it is not possible to decouple the coupled oscillator system with rotation alone. We need, in addition, the scale transformation:

$$T = \frac{1}{2}\left(M(\dot{x}_1)^2 + m(\dot{x}_2)^2\right), \tag{1.57}$$

which can be scale transformed to

$$T = \frac{1}{2}\sqrt{mM}\left((\dot{x}_1')^2 + (\dot{x}_2')^2\right). \tag{1.58}$$

This expression is invariant under rotations in the $\dot{x}_1\dot{x}_2$ coordinate system. As a consequence, the kinetic energy can be written as

$$T = \frac{1}{2}\sqrt{mM}\left((\dot{x}_1'')^2 + (\dot{x}_2'')^2\right). \tag{1.59}$$

while the potential energy is decoupled. As we shall see in later Chapters, the coupled oscillator problem is the foundation for two-mode squeezed states and special relativity. The mathematics of the coupled-oscillator problem will play an important role throughout this book.

## 1.5 Group of Linear Canonical Transformations in Four-Dimensional Phase Space

In this book, we shall be mostly concerned with two- and four-dimensional phase spaces. For the two-dimensional case, the group of linear canonical transformations is locally isomorphic to the (2 + 1)-dimensional Lorentz group. This is simple enough to study when the problem arises in later chapters. However, the story is quite different for four-dimensional phase space.

For linear canonical transformations involving two-pairs of canonical variables, we must work with the group of four-by-four real matrices satisfying the symplectic condition of Eq. 1.34. This group is the four-dimensional symplectic group or $Sp(4)$. There are many physical applications of this group.

It is more convenient to discuss this group in terms of its generators, defined as

$$M = e^{-i\alpha G}, \tag{1.60}$$

where $G$ represents a set of pure imaginary four-by-four matrices. The symplectic condition of Eq. 1.34 dictates that $G$ be symmetric and anticommute with $J$ or be antisymmetric and commute with $J$.

In terms of the Pauli spin matrices and the two-by-two identity matrix, we can construct the following four antisymmetric matrices which commute with $J$ of Eq. 1.31.

$$J_1 = \frac{i}{2}\begin{pmatrix} 0 & \sigma_1 \\ -\sigma_1 & 0 \end{pmatrix}, \quad J_2 = \frac{1}{2}\begin{pmatrix} \sigma_2 & 0 \\ 0 & \sigma_2 \end{pmatrix},$$

$$J_3 = \frac{i}{2}\begin{pmatrix} 0 & \sigma_3 \\ -\sigma_3 & 0 \end{pmatrix}, \quad J_0 = \frac{i}{2}\begin{pmatrix} 0 & I \\ -I & 0 \end{pmatrix}. \tag{1.61}$$

The following six symmetric generators anticommute with $J$.

$$K_1 = +\frac{i}{2}\begin{pmatrix} 0 & \sigma_3 \\ \sigma_3 & 0 \end{pmatrix}, \quad K_2 = +\frac{i}{2}\begin{pmatrix} I & 0 \\ 0 & -I \end{pmatrix}, \quad K_3 = -\frac{i}{2}\begin{pmatrix} 0 & \sigma_1 \\ \sigma_1 & 0 \end{pmatrix},$$

$$Q_1 = -\frac{i}{2}\begin{pmatrix} \sigma_3 & 0 \\ 0 & -\sigma_3 \end{pmatrix}, \quad Q_2 = +\frac{i}{2}\begin{pmatrix} 0 & I \\ I & 0 \end{pmatrix}, \quad Q_3 = +\frac{i}{2}\begin{pmatrix} \sigma_1 & 0 \\ 0 & -\sigma_1 \end{pmatrix}. \tag{1.62}$$

These generators satisfy the commutation relations:

$$[J_i, J_j] = +i\epsilon_{ijk}J_k, \quad [J_i, J_0] = 0,$$

$$[J_i, K_j] = +i\epsilon_{ijk}K_k, \quad [J_i, Q_j] = +i\epsilon_{ijk}Q_k,$$

$$[K_i, K_j] = [Q_i, Q_j] = -i\epsilon_{ijk}J_k, \tag{1.63}$$

$$[K_i, Q_j] = +i\delta_{ij}J_0,$$

$$[K_i, J_0] = +iQ_i, \quad [Q_i, J_0] = -iK_i.$$

The group of homogeneous linear transformations with this closed set of generators is called the symplectic group $Sp(4)$ (Arnold 1978, Guillemin and Sternberg 1984, Han *et al.* 1990b).

This group is becoming increasingly important in physics. It is locally isomorphic to the $(3+2)$-dimensional Lorentz group which is called the $O(3,2)$ de Sitter group (Dirac 1963). The de Sitter group was originally introduced to physics for the purpose of studying the curvature of the universe (Friedmann 1922 and 1924, Robertson 1933), and is still one of the important space time symmetry groups in relativity and elementary particle physics. Recently, it was observed that Dirac's two-oscillator representation of $O(3,2)$ is the fundamental scientific language for twomode squeezed states in quantum optics (Han *et al.* 1989 and 1990b). Needless to say, canonical transformations still play a very important role in classical mechanics. The modern version of classical mechanics is based on phase space (Arnold 1978, and Abraham and Marsden 1978).

## 1.6 Poisson Brackets

In the Hamiltonian formulation of classical mechanics, there is a conjugate momentum variable for each position variable. Let us use $q_i$ and $p_i$ to denote one pair of these variables. For one pair of variables, the Poisson bracket is defined to be

$$[U,V]_{q,p} = \sum_i \left(\frac{\partial U}{\partial q_i}\frac{\partial V}{\partial p_i} - \frac{\partial U}{\partial p_i}\frac{\partial V}{q_i}\right), \tag{1.64}$$

where $U$ and $V$ are functions of the variables $q$ and $p$. In terms of the phase-space variables given in Section 1.3, this expression can be written as

$$[U,V]_{q,p} = \sum_{ij} J_{ij}\frac{\partial U}{\partial \eta_i}\frac{\partial V}{\partial \eta_j}. \tag{1.65}$$

It is thus quite clear that the Poisson bracket is invariant under canonical transformations. With this understanding we can delete the subscripts $q$ and $p$ from the bracket.

From the definition,

$$[q_i,q_j] = 0, \quad [p_i,p_j] = 0, \quad [q_i,p_j] = \delta_{ij}. \tag{1.66}$$

These Poisson brackets form the bridge between classical and quantum mechanics (Dirac 1958). The Poisson bracket plays an interesting role in taking time derivatives. Let A be a function of canonical variables. Then

$$\frac{dA}{dt} = \sum_i \left(\frac{\partial A}{\partial q_i}\dot{q}_i + \frac{\partial A}{\partial p_i}\dot{p}_i\right) + \frac{\partial A}{\partial t}. \tag{1.67}$$

Then, from the Hamiltonian equations of motion, the above form can be written as

$$\frac{dA}{dt} = [A,H] + \frac{\partial A}{\partial t}. \tag{1.68}$$

This becomes Heisenberg's equation of motion in quantum mechanics (Schiff 1968, Merzbacher 1970).

Let us go back to classical mechanics. In terms of Poisson brackets, Hamilton's equation of motion can be written as

$$\dot{q} = [q, H], \quad \dot{p} = [p, H]. \tag{1.69}$$

The Poisson bracket of the Hamiltonian with the Hamiltonian should vanish. The form of time derivative in Eq. 1.68 is consistent with the timederivative of the Hamiltonian given in Eq. 1.9.

## 1.7 Distributions in Phase Space

In the Hamiltonian formalism of classical mechanics, one particle at a given instance corresponds to a point in phase space. As time progresses, the particle traces a trajectory in phase space. If there are $N$ particles, there will be $N$ points and $N$ trajectories. If $N$ is large, we can treat the problem statistically.

For a given volume element in phase space $\Delta q_1 \Delta q_2 \cdots \Delta q_n \Delta p_1 \Delta p_2 \cdots \Delta p_n$, we can consider the probability distribution function $f(q_1, q_2, \cdots, q_n, p_1, p_2, \cdots, p_n, t)$ such that the number of particles within this volume element is

$$\Delta N = N f(q_1, q_2, \cdots, q_n, p_1, p_2, \cdots, p_n, t) \Delta q_1 \Delta q_2 \cdots \Delta q_n \Delta p_1 \Delta p_2 \cdots \Delta p_n. \tag{1.70}$$

Since the total number $N$ is distributed all over phase space, the distribution function is normalized:

$$\int f(q_1, q_2, \cdots, q_n, p_1, p_2, \cdots, p_n, t) dq_1 dq_2 \cdots dq_n dp_1 dp_2 \cdots dp_n = 1. \tag{1.71}$$

If the particles are allowed to move in the three-dimensional Euclidean world, $N$ is 3, and the dimension of phase space is 6. However, it is possible that we are interested in distributions in one pair of canonical variables, say $q_1$ and $p_1$. Then the probability distribution in this two-dimensional phase space is

$$f(q_1, p_2, t) = \int f(q_1, q_2, \cdots, q_n, p_1, p_2, \cdots, p_n, t) dq_2 \cdots dq_n dp_1 dp_2 \cdots dp_n. \tag{1.72}$$

It is also possible to define the distribution over the single variable $q_1$ or $p_1$ as

$$f(q_1, t) = \int f(q_1, p_1, t) dp_1, \quad f(p_1, t) = \int f(q_1, p_1, t) dq_1. \tag{1.73}$$

For a given pair of $f(q_1, t)$ and $f(p_1, t)$, the phase space distribution function $f(q_1, p_1, t)$ does not have to be unique (Feynman 1973), because it is always possible to construct a function $g(q_1, p_1, t)$ such that

$$\int g(q_1, p_1, t) dq_1 dp_1 = 0, \tag{1.74}$$

and add it to $f(q_1, p_1, t)$. As we shall see in Chapter 3, this enables us to define a quantum phase space which is drastically different from classical phase space.

Let us consider the time dependence of this distribution function. The time derivative is

$$\frac{df}{dt} = \sum_i \left( \frac{\partial f}{\partial q_i}\dot{q}_i + \frac{\partial f}{\partial p_i}\dot{p}_i \right) + \frac{\partial f}{\partial t}. \tag{1.75}$$

Here again, we can use the Poisson bracket to write

$$\frac{df}{dt} = [f, H] + \frac{\partial f}{\partial t}. \tag{1.76}$$

If the total probability is conserved, $\frac{df}{dt} = 0$, and

$$[f, H] + \frac{\partial f}{\partial t} = 0. \tag{1.77}$$

This is known as Liouville's equation. The time derivative of $f$ given in Eq. 1.75 can also be written as

$$\frac{df}{dt} = \sum_i \left( \frac{\partial}{\partial q_i}(f\dot{q}_i) + \frac{\partial}{\partial p_i}(f\dot{p}_i) \right) + \frac{\partial f}{\partial t}. \tag{1.78}$$

Thus Liouville's equation is the continuity equation in phase space.

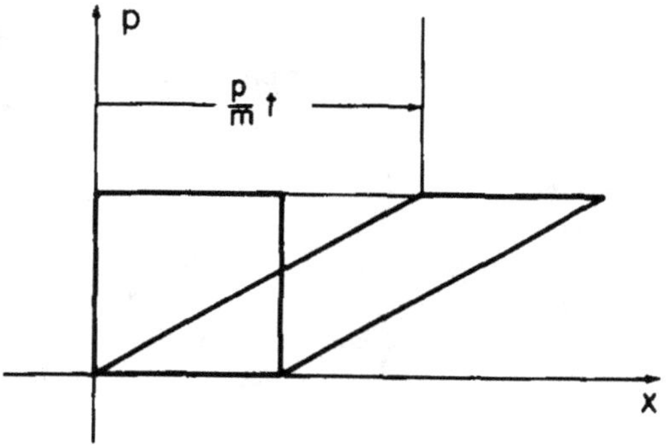


Figure 1.4: Time development of free particles in phase space. At $t = 0$, the particles are distributed in the square region specified in the graph. If all the particles have the same mass, those with larger momentum will move faster. As a consequence, the distribution will be sheared as time progresses.

For example, let us consider a large number of free particles initially distributed as

$$f(x, p, t = 0) = F(x, p), \tag{1.79}$$

at $t = 0$. Since the particles are free,

$$\frac{\partial H}{\partial x} = 0, \quad \frac{\partial H}{\partial p} = \frac{p}{m}. \tag{1.80}$$

Thus Liouville's equation for this system is

$$\frac{\partial f}{\partial t} = -\left(\frac{p}{m}\right)\frac{\partial f}{\partial x}. \tag{1.81}$$

The solution of this equation is

$$f(x,p,t) = F(x - pt/m, p). \tag{1.82}$$

This form of time development is illustrated in Figure 1.4. As we shall see in Chapter 3, this will correspond to the wave packet spread in quantum mechanics. This form of shear is also a canonical transformation.

# Chapter 2

# FORMS OF QUANTUM MECHANICS

There are several different but equivalent representations of quantum mechanics. The Schrödinger picture is very convenient for atomic and nuclear physics. The Heisenberg picture is useful in quantum field theory and in making the correspondence with classical mechanics through the Poisson bracket. The interaction representation serves a useful purpose in time-dependent perturbation theory in both quantum mechanics and quantum field theory. This representation takes advantage of both the Heisenberg and Schrödinger pictures. These different representations of quantum mechanics describe the same physics. Thus, choosing a particular representation of quantum mechanics depends on convenience.

Starting from Chapter 3, we shall discuss the phase-space picture of quantum mechanics whose algorithm is quite different from that of the Schrödinger or Heisenberg picture. The Wigner phase-space distribution function plays the central role in this picture. This distribution function is intimately related to the density matrix. We shall therefore discuss in detail the density matrix representation of quantum mechanics in this Chapter. Since there are already many textbooks putting emphasis on this representation (Pantell and Puthoff 1969, Louisell 1973, Blum 1981), we shall not give a full-fledged formalism. Instead, we shall rely on illustrative examples, including a spin 1/2 particle in a magnetic field and harmonic oscillators.

## 2.1 Schrödinger and Heisenberg Pictures

In the Schrödinger picture of quantum mechanics, the time evolution of the dynamical system is described solely by the wave functions, and the dynamical operators are time-independent. The situation is quite the opposite in the Heisenberg picture. There, the time dependence is within the operators, and the state vectors are independent of time. There are only a small number of problems which can be solved exactly in both pictures. We shall discuss here one example.

Let us consider a spin-1/2 particle at rest in a constant magnetic field. The time-dependent Schrödinger equation is

$$i\frac{\partial}{\partial t}\psi(t) = H\psi(t), \tag{2.1}$$

where the operator and the wave functions are both defined in the Schrödinger picture. If the magnetic field is in the $z$ direction, the Hamiltonian for this system is

$$H = \frac{1}{2}\begin{pmatrix} \omega & 0 \\ 0 & -\omega \end{pmatrix} \tag{2.2}$$

Then the solution is

$$\psi(t) = \begin{pmatrix} e^{-i(\omega/2)t} & a \\ e^{i(\omega/2)t} & b \end{pmatrix} \tag{2.3}$$

with the normalization condition $|\, a\,|^2 + |\, b\,|^2 = 1$ . If the spin is in the $z$ direction at $t = 0$, there is no change in the direction of spin. If on the other hand, the spin is initially along the $x$ direction with $a = b = 1/\sqrt{2}$ then the above solution can be written as

$$\psi(t) = (\cos(\omega t/2))\begin{pmatrix} 1/\sqrt{2} \\ 1/\sqrt{2} \end{pmatrix} - i\,(\sin(\omega t/2))\begin{pmatrix} 1/\sqrt{2} \\ -1/\sqrt{2} \end{pmatrix}. \tag{2.4}$$

This means that the probability of spin up along the $x$ direction is $(\cos(\omega t/2))^2$ while the probability of it being down is $(\sin(\omega t/2))^2$. We can expand the above wave function in terms of spinors along the $y$ direction:

$$\psi(t) = e^{-i\pi/4}\,(\cos(\omega t/2 + \pi/4))\begin{pmatrix} 1/\sqrt{2} \\ i/\sqrt{2} \end{pmatrix}$$

$$+\, e^{+i\pi/4}\,(\sin(\omega t/2 + \pi/4))\begin{pmatrix} 1/\sqrt{2} \\ -i/\sqrt{2} \end{pmatrix}. \tag{2.5}$$

The probability of spin being along the $y$ direction is $\sin^2(\omega t/2 + \pi/4)$. Since

$$\cos^2(\omega t/2) = \frac{1 + \cos\omega t}{2}, \quad \cos^2(\omega t/2 + \pi/4) = \frac{1 - \sin\omega t}{2},$$

$$\sin^2(\omega t/2) = \frac{1 - \cos\omega t}{2}, \quad \sin^2(\omega t/2 + \pi/4) = \frac{1 + \sin\omega t}{2}, \tag{2.6}$$

the spin precesses around the $z$ axis with the angular frequency of $\omega$. Indeed, the expectation value of $\sigma_x$ and $\sigma_y$ are

$$<\sigma_x> = \cos\omega t, \quad <\sigma_y> = \sin\omega t. \tag{2.7}$$

In the Schrödinger picture, the time dependence is strictly in the wave function, and the operators do not depend on time. On the other hand, in the Heisenberg picture, the time dependence is in the operators. If $\sigma_z$ corresponds to the spin along the $z$ direction, the operator for the spin along the $\theta\phi$ direction is

$$A = \begin{pmatrix} Z & X - iY \\ X + iY & -Z \end{pmatrix}, \tag{2.8}$$

where

$$Z = \cos\theta, \quad X = (\sin\theta)\cos\phi, \quad Y = (\sin\theta)\sin\phi.$$

The direction of the spin is a function of time. Heisenberg's equation of motion is

$$i\frac{d}{dt}(A_H(t)) = [A_H(t), H]. \tag{2.9}$$

Thus, $X$, $Y$, and $Z$ satisfy the following differential equations.

$$\frac{dZ}{dt} = 0, \quad \frac{dX}{dt} - \omega Y = 0, \quad \frac{dY}{dt} + \omega X = 0. \tag{2.10}$$

If the spin is along the $x$ direction at $t = 0$ , $Z = 0$ , $X = \cos\omega t$, $Y = -\sin\omega t$, and

$$A_H(t) = \begin{pmatrix} 0 & \exp(i\omega t) \\ \exp(-i\omega t) & 0 \end{pmatrix}. \tag{2.11}$$

This spinor corresponds to the spin precessing around the $z$ axis with angular frequency $-\omega$. The angular frequency of the rotation of this operator is in the direction opposite to that of spin precession frequency obtained in the Schrödinger picture. Is this correct? Let us examine how the Heisenberg picture is related to the Schrödinger picture. We can get the above spinor starting from the Schrödinger equation. The solution of the differential equation can be written as

$$\psi(t) = e^{-iHT}\psi(0). \tag{2.12}$$

In general, the expectation value of a given time-independent operator $A$ is

$$<A> = (\psi(t), A(0)\psi(t)), \tag{2.13}$$

with $A(0) = A$. This can then be written as

$$<A> = (\psi(0), A_H(t)\psi(0)), \tag{2.14}$$

with

$$A_H(t) = e^{iHt}A(0)e^{-iHt}, \tag{2.15}$$

where $e^{-iHt}$ is a unitary operator and carries out the time evolution from $t = 0$ to $t$. The wave function in the Heisenberg picture is $\psi(0)$ , and is independent of time. Thus

$$\psi_H = e^{iHt}\psi(t). \tag{2.16}$$

For the form of $H$ given in Eq. 2.2,

$$e^{iHt} = \begin{pmatrix} e^{i\omega t/2} & 0 \\ 0 & e^{-i\omega t/2} \end{pmatrix}. \tag{2.17}$$

Thus the operator $A_H(t)$ in the Heisenberg picture should take the form of Eq. 2.11.

It is important to note that the Heisenberg operator is quite different from the time translation of the operator in the Schrödinger picture. We note that the time evolution in the Schrödinger picture is given in Eq. 2.12. Thus, the time translation of a given operator is

$$A(t) = e^{-iHt}A(0)e^{iHt}. \tag{2.18}$$

This form is quite different from $A_H(t)$ in the Heisenberg picture given in Eq. 2.15. The difference is in the sign of $t$. This operator satisfies

$$i\frac{\partial}{\partial t}A(t) = [H, A(t)], \tag{2.19}$$

which is different from Heisenberg's equation of motion given in Eq. 2.9 . We may call the above equation Liouville's form of the Schrödinger equation. In this form,

$$A(t) = \begin{pmatrix} 0 & \exp(i\omega t) \\ \exp(i\omega t) & 0 \end{pmatrix}. \tag{2.20}$$

This form of the Schrödinger equation, while not widely known, is used very frequently in atomic physics (Blum 1981) and in quantum optics (Pantell and Puthoff 1969). The above operator is different from the time-dependent Heisenberg operator given in Eq. 2.11. The difference is in the direction of time. As we shall see, in Chapter 3, Liouville's equation serves many useful purposes in the phase-space picture of quantum mechanics.

## 2.2 Interaction Representation

The interaction representation takes advantage of the convenience of both the Schrödinger and Heisenberg pictures. In the example given in Section 2.1, the transformation from the Schrödinger to the Heisenberg picture can be performed easily because $H$ is independent and the transformation matrix $e^{iHt}$ takes a simple form. On the other hand, if we add a complicated term to this simple Hamiltonian so that the total Hamiltonian is $H + G$, the transformation is no longer simple. However, we can make a transformation with the simple portion of the total Hamiltonian.

In order to illustrate this point, let us start with the example discussed in Section 2.1. We are now adding a weak sinusoidal magnetic field along the $x$ direction. Then the total Hamiltonian is $H + G$, where $H$ is given in Eq. 2.2, and $G$ is

$$G(t) = b\begin{pmatrix} 0 & \cos\alpha t \\ \cos\alpha t & 0 \end{pmatrix}. \tag{2.21}$$

The coefficient $b$ measures the strength of the interaction. The Schrödinger equation is

$$i\frac{\partial}{\partial t}\psi(t) = (H + G)\psi(t), \tag{2.22}$$

where $H$ is a time-independent Hamiltonian. We can now consider the wave function in the interaction representation defined as

$$\psi_I(t) = e^{iHt}\psi_S(t) \quad \text{or} \quad \psi_S(t) = e^{-iHt}\psi_I(t). \tag{2.23}$$

Then $\psi_I(t)$ satisfies the equation

$$i\frac{\partial}{\partial t}\psi_I(t) = G_I(t)\psi(t). \tag{2.24}$$

with

$$G_I(t) = e^{iHt}G(t)e^{-iHt}.$$

The wave function $\psi_I(t)$ and the operator $G_I(t)$ are called the wave function and the Hamiltonian in the interaction representation. The differential equation of Eq. 2.24 is simpler than that of Eq. 2.22 particularly when $b$ of Eq. 2.21 is small. Indeed, this form of quantum mechanics is the starting point for relativistic quantum field theory and quantum electrodynamics leading to Feynman diagrams (Itzykson and Zuber 1984). This representation is important in quantum optics where the radiation and absorption of a photon is the main subject.

In order to understand the essence of this representation, let us go back to the problem of a spinning particle in the magnetic field. From Eq. 2.17 and Eq. 2.21, $G_I(t)$ becomes

$$G_I(t) = b(\cos\alpha t)e^{iHt}\sigma_x e^{-iHt}$$

$$= b\begin{pmatrix} 0 & e^{i\omega t}\cos\alpha t \\ e^{-i\omega t}\cos\alpha t & 0 \end{pmatrix}. \tag{2.25}$$

The difficulty in solving the Schrödinger equation of Eq. 2.24 arises from the fact that $G_I(t_1)$ and $G_I(t_2)$ do not commute with each other for different values of $t_1$ and $t_2$ . For this reason, time ordering is needed when we solve the equation by iteration. The solution takes the form

$$\psi_I(t) = \left\{\sum_{n=1}^{\infty}((-i)^n/n!)\,P\int_0^t\cdots\int_0^t G_I(t_1)\cdots G_I(t_n)dt_1\cdots dt_n\right\}\psi_I(0). \tag{2.26}$$

where P is the time-ordering operator which dictates $G_I(t_i)G_I(t_j)$ in the integrand be ordered in such a way that $t_i > t_j$. If $b$ is sufficiently small, it is valid to take a few lowest order terms in the above series. However, if $b$ is large, the series is not useful for calculating measurable numbers (Shirley 1963).

There are many other approaches to this problem. The rotating wave approximation is one of them. If we write

$$\psi_I(t) = \begin{pmatrix} C_1(t) \\ C_2(t) \end{pmatrix}, \tag{2.27}$$

then Eq. 2.24 can be written as (Louden 1973)

$$\dot{C}_1(t) = -ib\left(e^{i\omega t}\cos\alpha t\right)C_2(t),$$

$$\dot{C}_2(t) = -ib\left(e^{-i\omega t}\cos\alpha t\right)C_1(t). \quad (2.28)$$

These equations appear very simple, but cannot be solved analytically. It is a simple matter these days to solve these equations numerically. This does not prevent us from studying the properties of these equations. First of all, it is easy to prove that the total probability is conserved:

$$\mid C_1(t)\mid^2 + \mid C_2(t)\mid^2 = 1. \quad (2.29)$$

Next, it should be noted that the difference between the two frequencies $(\omega - \alpha)$ can be much smaller than the sum $(\omega + \alpha)$. Thus, if we are primarily interested in the frequency region where $(\omega - \alpha)$ is small, the high frequency component of $(\omega + \alpha)$ can be ignored. Thus we can write $G_I(t)$ as

$$G_I(t) = \frac{b}{2}\begin{pmatrix} 0 & e^{i(\omega-\alpha)t} \\ e^{-i(\omega-\alpha)t} & 0 \end{pmatrix}. \quad (2.30)$$

Then the differential equations in Eq. 2.28 become

$$\dot{C}_1(t) = -i\frac{b}{2}e^{i(\omega-\alpha)t}C_2(t),$$

$$\dot{C}_2(t) = -i\frac{b}{2}e^{-i(\omega-\alpha)t}C_1(t). \quad (2.31)$$

It is possible to decouple the above differential equations. If we impose the initial condition that $C_1(0) = 0$ and $\mid C_2(0)\mid = 1$. Then the solutions are

$$C_1(t) = \left(\frac{b}{\Omega}\right)e^{i(\omega-\alpha)t/2}\sin(\Omega t/2),$$

$$C_2(t) = ie^{-i(\omega-\alpha)t/2}\left\{\cos(\Omega t/2) + i\left(\frac{\omega-\alpha}{\Omega}\right)\sin(\Omega t/2)\right\}, \quad (2.32)$$

with

$$\Omega = \left((\omega-\alpha)^2 + b^2\right)^{1/2}.$$

Thus

$$\mid C_1(t)\mid^2 = \left(\frac{b}{\Omega}\right)^2\sin^2(\Omega t/2),$$

$$\mid C_2(t)\mid^2 = \cos^2(\Omega t/2) + \left(\frac{\omega-\alpha}{\Omega}\right)^2\sin^2(\Omega t/2). \quad (2.33)$$

This set of solutions satisfies the normalization condition of Eq. 2.29.

## 2.3 Density-Matrix Formulation of Quantum Mechanics

In quantum mechanics, measurable quantities are associated with probability, rather than the probability amplitude. In Sections 2.1 and 2.2 we are eventually interested in calculating $| C_1(t) |^2$ and $| C_2(t) |^2$ from $C_1(t)$ and $C_2(t)$. We may therefore consider the two-by-two matrix defined as

$$\rho_{ij}(t) = C_i(t)C_j^*(t), \tag{2.34}$$

or

$$\rho(t) = \begin{pmatrix} C_1(t)C_1^*(t) & C_1(t)C_2^*(t) \\ C_2(t)C_1^*(t) & C_2(t)C_2^*(t) \end{pmatrix}.$$

This is called the density matrix for the system. In this particular case, the density matrix is formulated in the interaction representation. The conservation of probability given in Eq. 2.29 can be stated as

$$Tr(\rho) = 1, \quad \rho^2 = \rho. \tag{2.35}$$

The density matrix of Eq. 2.34 is Hermitian and can therefore be diagonalized. Thus, if diagonalized, the density matrix should take the form

$$\rho_+ = \begin{pmatrix} 1 & 0 \\ 0 & 0 \end{pmatrix} \quad \text{or} \quad \rho_- = \begin{pmatrix} 0 & 0 \\ 0 & 1 \end{pmatrix}, \tag{2.36}$$

which correspond to the spin-up and spin-down states respectively.

Next, we are interested in whether it is possible to set up the equation of motion solely in terms of the density matrix without resorting to wave functions. Indeed, the time derivative of $\rho$ is

$$\dot{\rho}_{ij}(t) = \dot{C}_i(t)C_j^*(t) + C_i(t)\dot{C}_j^*(t). \tag{2.37}$$

Thus, for the system discussed in Section 2.2, the differential equations of Eq. 2.28 for $C_i(t)$ lead to

$$\dot{\rho}_{11}(t) = -\dot{\rho}_{22}(t) = ib(\cos\alpha t)\left(e^{-i\omega t}\rho_{12} - e^{+i\omega t}\rho_{21}\right),$$

$$\dot{\rho}_{12}(t) = \dot{\rho}_{21}^*(t) = ib(\cos\alpha t)e^{i\omega t}\left(\rho_{11} - \rho_{22}\right). \tag{2.38}$$

It is thus possible to write the equation of motion in terms of the density matrix. If we use for $\rho(t)$ the two-by-two matrix as is defined in Eq. 2.34, the above set of equations can be written as

$$i\frac{\partial}{\partial t}\rho(t) = [G_I(t), \rho(t)], \tag{2.39}$$

where $G_I(t)$ is given in Eq. 2.25. This is the Liouville equation for the density matrix in the interaction representation.

Solving this differential equation would be as difficult as the case in the interaction representation discussed in Section 2.2. However, it should be possible to obtain a solution once the rotating wave approximation is made. In this approximation,

$$\dot{\rho}_{11}(t) = -\dot{\rho}_{22}(t) = i\frac{b}{2}\left(e^{-i(\omega-\alpha)t}\rho_{12} - e^{+i(\omega-\alpha)t}\rho_{21},\right),$$

$$\dot{\rho}_{12}(t) = \dot{\rho}^*_{21}(t) = i\frac{b}{2}e^{i(\omega-\alpha)t}(\rho_{11} - \rho_{22}). \tag{2.40}$$

These are called the optical Bloch equations. In solving these equations, we can start with the parametrization

$$\rho_{11}(t) = a_{11}e^{\lambda t}, \quad \rho_{22} = a_{22}e^{\lambda t},$$

$$\rho_{12}(t) = a_{12}e^{i(\omega-\alpha)t}e^{\lambda t},$$

$$\rho_{21}(t) = a_{21}e^{-i(\omega-\alpha)t}e^{\lambda t}. \tag{2.41}$$

Then

$$\lambda a_{11} = -\lambda a_{22} = i\frac{b}{2}(a_{12} - a_{21}),$$

$$(\lambda + i(\omega-\alpha))\, a_{12} = i\frac{b}{2}(a_{11} - a_{22}),$$

$$(\lambda - i(\omega-\alpha))\, a_{21} = -i\frac{b}{2}(a_{11} - a_{22}). \tag{2.42}$$

This set of equations can be written in the matrix form

$$\begin{pmatrix} \lambda & 0 & -ib/2 & ib/2 \\ 0 & \lambda & ib/2 & -ib/2 \\ -ib/2 & ib/2 & \lambda + i(\omega-\alpha) & 0 \\ ib/2 & -ib/2 & 0 & \lambda - i(\omega-\alpha) \end{pmatrix} \begin{pmatrix} a_{11} \\ a_{22} \\ a_{12} \\ a_{21} \end{pmatrix} = 0. \tag{2.43}$$

The resulting equation for $\lambda$ is

$$\lambda^2\left(\lambda^2 + (\omega-\alpha)^2 + b^2\right) = 0. \tag{2.44}$$

The solutions of this equation are

$$\lambda_1 = 0, \quad \lambda_2 = i\Omega, \quad \lambda_3 = -i\Omega, \tag{2.45}$$

and each element of $\rho(t)$ may be written as

$$\rho_{ij}(t) = a^{(1)}_{ij} + a^{(2)}_{ij}\exp(i\Omega t) + a^{(3)}_{ij}\exp(-i\Omega t). \tag{2.46}$$

Thus, if we start with the initial condition $\rho_{11}(0) = \rho_{12}(0) = \rho_{21}(0)$ and $\rho_{22}(0) = 1$ , then $\rho_{11}(t)$ may take the form

$$\rho_{11}(t) = \frac{K}{2}(1 - \cos \Omega t), \tag{2.47}$$

where $K$ is a constant smaller than one. To complete the solution, we have to determine this constant using the optical Bloch equations of Eq. 2.40. Since $\rho_{11}(t) + \rho_{22}(t) = 1$,

$$\rho_{22}(t) = 1 - \frac{K}{2} + \left(\frac{K}{2}\right) \cos \Omega t. \tag{2.48}$$

Furthermore,

$$\dot{\rho}_{12}(t) = i\frac{b}{2}e^{i(\omega-\alpha)t}(K - 1 - K(\cos \Omega t)). \tag{2.49}$$

This equation is integrable. The result of integration is

$$\rho_{12}(t) = \left(\frac{b}{2}\right) e^{i(\omega-\alpha)t} \left\{ \left(\frac{K-1}{\omega - \alpha}\right) \right.$$

$$\left. + (K/b^2)((\omega - \alpha)\cos \Omega t - i\Omega(\sin \omega t)) \right\}. \tag{2.50}$$

From the initial condition $\rho_{12}(0) = 0$, $K$ is determined to be $(b/\omega)^2$. We can substitute this solution and its complex conjugate for $\rho_{21}$ into the first equation of Eq. 2.40 for $\dot{\rho}_{11}$ to confirm the validity of the solution. The result is the same as the one derived in Section 2.2, and will be discussed further in Section 2.6

Thus far, we used the density matrix formalism of quantum mechanics in order to reproduce the result which is available in other representations. Thus, there is no compelling reason to choose this particular representation while abandoning others. However, the density matrix becomes an indispensable tool when ensemble averages are taken (Von Neumann 1927). For instance, we can consider a system consisting of a statistical mixture of the states with two different initial conditions.

## 2.4 Mixed States

Not all quantum states are in pure states. The purpose of this section is to discuss a concrete example of a non-pure or mixed state. For a spin-1/2 particle, the eigenspinors for the $\theta\phi$ direction are

$$u_+(\theta,\phi) = \begin{pmatrix} e^{-i\phi/2}\cos(\phi/2) \\ e^{i\phi/2}\sin(\theta/2) \end{pmatrix}, \quad u_-(\theta,\phi) = \begin{pmatrix} -e^{-i\phi/2}\sin(\theta/2) \\ e^{i\phi/2}\cos(\theta/2) \end{pmatrix} \tag{2.51}$$

for the positive and negative directions respectively. The polarization vector $\mathbf{P}$ is defined as

$$P_i = (u_+(\theta,\phi))^\dagger \sigma_i u_+(\theta\phi), \tag{2.52}$$

which will hereafter be written as $<\theta\phi \mid \sigma_i \mid \theta\phi>$. Then

$$P_x = (\sin\theta)\cos\phi, \quad P_y = (\sin\theta)\sin\phi, \quad P_z = \cos\theta. \tag{2.53}$$

It is clear that this polarization vector has a unit length:

$$P_x^2 + P_y^2 + P_z^2 = 1. \tag{2.54}$$

Let us now consider the case of two independently prepared groups of particles: $N_1$ particles with spin along the $\theta_1\phi_1$ direction and $N_2$ particles with spin along the $\theta_2\phi_2$ direction. Then the polarization vector is the statistical average:

$$P_i = w_1 <\theta_1\phi_1 \mid \sigma_i \mid \theta_1\phi_1> + w_2 <\theta_2\phi_2 \mid \sigma_i \mid \theta_2\phi_2>, \tag{2.55}$$

where

$$w_1 = N_1/(N_1 + N_2), \quad w_2 = N_2/(N_1 + N_2).$$

The (magnitude)$^2$ of this mixed-state polarization vector is

$$P^2 = w_1^2 + w_2^2 + 2(w_1 w_2)\cos\delta_{12}, \tag{2.56}$$

where $\delta_{12}$ is the angle between the two directions. $P^2$ is smaller than 1 and is greater than $(w_1 - w_2)^2$. If the two directions are the same, $P^2 = 1$, and, if they are opposite to each other, $P^2 = (w_1 - w_2)^2$.

If we go back to the eigenspinors of Eq. 2.51, the projection operator which selects only the spin states along the positive $\theta\phi$ direction is

$$\Lambda_+(\theta,\phi) = u_+(\theta,\phi)\,[u_+(\theta,\phi)]^\dagger, \tag{2.57}$$

which can be written in matrix form as

$$\Lambda_+(\theta,\phi) = \frac{1}{2}\begin{pmatrix} 1+\cos\theta & e^{-i\phi}\sin\theta \\ e^{i\phi}\sin\theta & 1-\cos\theta \end{pmatrix}. \tag{2.58}$$

The projection operator which selects only the spinors for the negative direction is

$$\Lambda_-(\theta,\phi) = \frac{1}{2}\begin{pmatrix} 1-\cos\theta & -e^{-i\phi}\sin\theta \\ -e^{i\phi}\sin\theta & 1+\cos\theta \end{pmatrix}. \tag{2.59}$$

The Pauli spin operator along the $\theta\phi$ direction is $\sigma(\theta,\phi) = (\sin\theta\cos\phi)\sigma_x + (\sin\theta\sin\phi)\sigma_y + (\cos\theta)\sigma_z$, and takes the form

$$\sigma(\theta,\phi) = \begin{pmatrix} \cos\theta & e^{-i\phi}\sin\theta \\ e^{i\phi}\sin\theta & -\cos\theta \end{pmatrix}. \tag{2.60}$$

In terms of this matrix, the projection operators are

$$\Lambda_+(\theta,\phi) = \frac{1}{2}(1+\sigma(\theta,\phi)), \quad \Lambda_-(\theta,\phi) = \frac{1}{2}(1-\sigma(\theta,\phi)). \tag{2.61}$$

It is clear that these operators are Hermitian and satisfy

$$\Lambda_+(\theta,\phi) + \Lambda_-(\theta,\phi) = I. \tag{2.62}$$

Furthermore, simple matrix multiplications will lead to $(\Lambda_+(\theta,\phi))^2 = \Lambda_+(\theta,\phi)$ and $(\Lambda_-(\theta,\phi))^2 = \Lambda_-(\theta,\phi)$.

The density matrix for the mixed state is defined as

$$\rho = w_1\Lambda_+(\theta,\phi) + w_2\Lambda_-(\theta,\phi), \tag{2.63}$$

Thus

$$\rho = \frac{1}{2} + \frac{1}{2}(w_1 - w_2)\,\sigma(\theta,\phi). \tag{2.64}$$

Since $\sigma(\theta,\phi)$ is traceless, $Tr(\rho) = 1$. However,

$$\rho^2 = \frac{1}{4}\left\{1 + (w_1 - w_2)^2\right\} + \frac{1}{2}(w_1 - w_2)\,\sigma(\theta,\phi). \tag{2.65}$$

Therefore $\rho^2$ is not equal to $\rho$. It is equal to $\rho$ only for a pure state when $w_1 = 1$ and $w_2 = 0$, or vice versa. The trace of $\rho^2$ is less than one for mixed states.

The density matrix of Eq. 2.64 is Hermitian and can be diagonalized by rotation. The diagonalized form of $\sigma(\theta,\phi)$ is $\sigma_z$. As a consequence, the diagonal form of the density matrix is

$$\rho_D = \frac{1}{2}\begin{pmatrix} 1 + w_1 - w_2 & 0 \\ 0 & 1 - w_1 + w_2 \end{pmatrix}. \tag{2.66}$$

If $w_1 = 1$ and $w_2 = 0$, then the system is fully polarized, and

$$\rho_D = \begin{pmatrix} 1 & 0 \\ 0 & 0 \end{pmatrix}. \tag{2.67}$$

These diagonal expressions clearly demonstrates the trace properties of the density matrix, since the trace is invariant under rotations.

Furthermore, the polarization vector can be obtained through the relation

$$P_i = Tr(\sigma_i \rho). \tag{2.68}$$

Indeed, the calculation gives

$$P_x = (w_1 - w_2)(\sin\theta)\cos\phi, \quad P_y = (w_1 - w_2)(\sin\theta)\sin\phi,$$

$$P_z = (w_1 - w_2)\cos\theta. \tag{2.69}$$

If $w_1 = 1$ and $w_2 = 0$, the system is fully polarized. If one the other hand, $w_1 = w_2 = 1/2$, then the polarization vector vanishes. The magnitude of the polarization is measured from

$$P^2 = P_x^2 + P_y^2 + P_z^2 = (w_1 - w_2)^2. \tag{2.70}$$

This expression is for a mixed states, and becomes that of Eq. 2.54 for a pure state with $(w_1 - w_2)^2 = 1$.

In this section, we have worked out a concrete example of the density matrix for a mixed state of spin-1/2 particles. We should now be able to formulate a general theory of the density matrix.

## 2.5 Density Matrix and Ensemble Average

We have thus far discussed only two-by-two density matrices. The size of the matrix can be arbitrarily large, and can even be infinite-by- infinite. The matrix index can be continuous. For example the wave function $\psi(x)$ can be regarded as a column vector with a continuous index $x$.

If a given quantum state is a linear superposition of many different eigenstates $\psi_n$,

$$\psi(x) = \sum_n C_n \psi_n(x), \tag{2.71}$$

with

$$\sum_n | C_n |^2 = 1.$$

One way to obtain $C_m$ for a particular $m$ is to use the projection operator:

$$\Lambda_m = \psi_m \left(\psi_m\right)^\dagger . \tag{2.72}$$

Then

$$\Lambda_m \psi(x) = C_m \psi_m(x). \tag{2.73}$$

with an understanding that

$$\Lambda_n \psi(x) = \psi_n(x) \left(\psi_n(x'), \psi(x')\right). \tag{2.74}$$

Thus the projection operator is a function of two variables $x$ and $x'$. It is appropriate to write it as $\Lambda_n(x, x')$ whenever necessary with $x$ and $x'$ as the continuous matrix indices. This operator is Hermitian and has the following property:

$$\Lambda_n \Lambda_m = \delta_{nm} \Lambda_m, \quad \sum_n \Lambda_n = I, \tag{2.75}$$

where $I$ is the identity operator.

The density matrix is defined as

$$\rho(x, x') = \sum_n | C_n |^2 \Lambda_n(x, x'). \tag{2.76}$$

Then, for a given operator $A(x)$, $Tr(\rho A)$ is

$$Tr(A\rho) = \sum_n | C_n |^2 \left(\psi_n, A\psi_n\right). \tag{2.77}$$

This is quite different from the expectation value

$$<A> = \sum_n \sum_m C_m^* C_n (\psi_m, A\psi_n). \tag{2.78}$$

Thus the difference between the two quantities is

$$<A> - Tr(\rho A) = \sum_{n \neq m} C_m^* C_n \left(\psi_m, A\psi_n\right). \tag{2.79}$$

When does this quantity vanish? The coefficient $C_m^* C_n$ is real and positive when $m = n$ . Otherwise, it is complex and carries a phase factor. Thus, when we take an ensemble average with random phase factors, the average of each $C_m^* C_n$ vanishes when $n \neq m$.

For example, let us discuss the one-dimensional harmonic oscillator, for which the Schrödinger equation is

$$i\frac{\partial}{\partial t}\psi(x,t) = \frac{1}{2}\left(m\omega^2 x^2 - \frac{1}{m}\left(\frac{\partial}{\partial x}\right)^2\right)\psi(x,t). \tag{2.80}$$

Then the most general form of normalized solution is

$$\psi(x,t) = e^{-i\omega t/2}\sum_n C_n e^{-in\omega t}\psi_n(x), \tag{2.81}$$

with

$$(\psi(x,t),\psi(x,t)) = \sum_n \mid C \mid^2 = 1.$$

$\psi_n(x)$ is the solution of the time-independent equation:

$$\frac{1}{2}\left(m\omega^2 x^2 - \frac{1}{m}\left(\frac{\partial}{\partial x}\right)^2\right)\psi_n(x) = \omega(n+1/2)\psi_n(x). \tag{2.82}$$

The expectation value $<A>= (\psi(x,t), A\psi(x,t))$ of an operator $A(x)$ can be written as

$$\begin{aligned} <A>= &\sum_n \mid C_n \mid^2 (\psi_n(x), A(x)\psi_n(x)) \\ &+ \sum_{n\neq m} C_m^* C_n e^{i\omega(m-n)t}\left(\psi_m(x), A(x)\psi_n(x)\right). \end{aligned} \tag{2.83}$$

If we take the time average of this quantity, the second term vanishes. If we take the ensemble average for many particles with different phases of $C_n$, the result is the same. As a consequence, the ensemble average is

$$\overline{<A>} = \sum_n \mid C_n \mid^2 (\psi_n(x), A(x)\psi_n(x)). \tag{2.84}$$

It is very convenient to treat this problem if we introduce the density matrix defined as

$$\rho(x,x') = \sum_n \mid C_n \mid^2 \psi_n(x)\psi_n^*(x'), \tag{2.85}$$

and

$$\overline{<A>} = \int dx' \int A(x',x)\rho(x,x')dx, \tag{2.86}$$

with

$$A(x',x) = \delta(x'-x)A(x).$$

The above expression is then the trace of the matrix $A(x',x)\rho(x,x')$ often written as

$$\overline{<A>} = Tr(\rho A). \tag{2.87}$$

If $C_n = \delta_{nm}$ for a given value of $m$, then we say that the system is in a pure state. Otherwise, the system is in a mixed state. Since $\sum_n | C_n |^2 = 1$, $Tr(\rho) = 1$ and $Tr(\rho^2) = \sum_n | C_n |^4$, it is 1 for a pure state and is less than 1 for mixed states.

The best known example is the harmonic oscillator in thermal equilibrium (Feynman 1973), for which the density matrix is

$$\rho_T(x,x') = \left(1 - e^{-\omega/kT}\right) \sum_n e^{-n\omega/kT} \psi_n(x)\psi_n^*(x'). \tag{2.88}$$

For this matrix, $Tr(\rho)$ is

$$Tr(\rho_T) = \int \rho_T(x,x)dx = \left(1 - e^{-\omega/kT}\right) \sum_n e^{-n\omega/kT}, \tag{2.89}$$

which becomes 1 after summation. $Tr\left((\rho_T)^2\right)$ is

$$Tr(\rho_T \rho_T) = \int \left\{ \int \rho_T(x,x')\, \rho_T(x',x)dx' \right\} dx. \tag{2.90}$$

From the expression of Eq. 2.88,

$$Tr\left((\rho_T)^2\right) = \left(1 - e^{-\omega/kT}\right)^2 \sum_n e^{-2n\omega/kT} = \tanh(\omega/2kT). \tag{2.91}$$

This is less than one, and becomes one when $T$ becomes zero. When $T = 0$, the system is in a pure state which is the ground-state of the harmonic oscillator.

## 2.6 Time Dependence of the Density Matrix

When we discussed the density matrix for the harmonic-oscillator in Section 2.5, we started with a time-dependent system, but ended up with a time-independent density matrix. Does this mean that the density matrix is intrinsically a time-independent quantity? The answer to this question is clearly NO, because the example discussed in Section 2.3 clearly leads to a time dependent density matrix.

In terms of the time-dependent wave functions, the density matrix can be written as

$$\rho(x,x',t) = \sum_n w_n \psi_n(x,t) \left(\psi_n(x',t)\right)^\dagger. \tag{2.92}$$

Then

$$i\frac{\partial}{\partial t}\rho(x,x',t)$$

$$= \sum_n w_n \left\{ \left( i\frac{\partial}{\partial t}\psi_n(x,t) \right) \left(\psi_n(x',t)\right)^\dagger - \left(\psi_n(x,t)\right) \left( i\frac{\partial}{\partial t}\psi_n(x',t) \right)^\dagger \right\}$$

$$= \sum_n w_n \left\{ \left(H\psi_n(x,t)\right) \left(\psi_n(x',t)\right)^\dagger - \left(\psi_n(x,t)\right) \left(H\psi_n(x',t)\right)^\dagger \right\}. \tag{2.93}$$

As a consequence, the density matrix in the Schrödinger picture satisfies the Liouville equation:

$$i\frac{\partial}{\partial t}\rho(x,x',t) = [H,\rho(x,x',t)]. \tag{2.94}$$

In the Schrödinger picture, the Hamiltonian is independent of time, and the solution formally takes the form

$$\rho(t) = e^{-iHt}\rho(0)e^{iHt}. \tag{2.95}$$

If the Hamiltonian commutes with $\rho(0)$, then the density matrix is independent of time.

For example, in the case of the one-dimensional harmonic oscillator discussed in Section 2.5

$$\rho(t) = e^{-iHt}\rho(0)e^{iHt} = \sum_n w_n \left(e^{-iHt}\psi_n(x,t)\right)\left(e^{-iHt}\psi_n(x',t)\right)^\dagger . \tag{2.96}$$

If the wave function $\psi_n(x,t)$ is an eigenstate of $H$, which is the case, then the density matrix is independent of time. The density matrix commutes with the Hamiltonian.

For an example of the cases where the Hamiltonian does not commute with the density matrix, let us consider the case where a spin 1/2 particle is in a constant magnetic field. The Hamiltonian for this system is given in Eq. 2.2. The wave function of Eq. 2.3 is not an energy eigenstate. The density matrix is

$$\rho(t) = \begin{pmatrix} aa^* & ab^*e^{-\omega t} \\ a^*be^{i\omega t} & bb^* \end{pmatrix}. \tag{2.97}$$

Indeed, in the Schrödinger picture, the time evolution can be written in the form of Eq. 2.18 as

$$\rho(t) = \begin{pmatrix} e^{-i\omega t/2} & 0 \\ 0 & e^{i\omega/2} \end{pmatrix}\begin{pmatrix} aa^* & ab^* \\ a^*b & bb^* \end{pmatrix}\begin{pmatrix} e^{i\omega t/2} & 0 \\ 0 & e^{-i\omega t/2} \end{pmatrix}. \tag{2.98}$$

This matrix does not commute with the Hamiltonian unless $a = 1$ and $b = 0$, or $a = 0$ and $b = 1$. In these special cases, the density matrices are

$$\begin{pmatrix} 1 & 0 \\ 0 & 0 \end{pmatrix} \quad \text{and} \quad \begin{pmatrix} 0 & 0 \\ 0 & 1 \end{pmatrix}, \tag{2.99}$$

respectively. This means that the spin is along the positive and negative directions respectively. The direction of the spin does not change in time. As a consequence, the density matrix is independent of time.

If the spin is along the $x$ direction at $t = 0$, the density matrix is

$$\rho_x(t) = \frac{1}{2}\begin{pmatrix} 1 & e^{-i\omega t} \\ e^{i\omega t} & 1 \end{pmatrix}. \tag{2.100}$$

It is possible to calculate the expectation values of $\sigma_i$ , using the formula $<\sigma_i>= Tr(\rho_x\sigma_i)$. This trace calculation will lead to $<\sigma_z>= 0, <\sigma_x>= \cos\omega t$, and $<\sigma_y>= \sin\omega t$, as are expected from the same result as in Eq. 2.7

If the spin is along the negative $x$ direction at $t = 0$,

$$\rho_{-x}(t) = \frac{1}{2}\begin{pmatrix} 1 & -e^{-i\omega t} \\ -e^{i\omega t} & 1 \end{pmatrix}. \tag{2.101}$$

Both $\rho_x(t)$ and $\rho_{-x}(t)$ are pure-state density matrices, and satisfy the condition $Tr(\rho) = 1$ and $\rho^2 = \rho$.

For the system of $N$ spin 1/2 particles, if $N/2$ particles have spin along the $x$ direction, and the other half of them are along the negative $x$ direction, then the ensemble average of the density matrix is

$$\begin{pmatrix} 1/2 & 0 \\ 0 & 1/2 \end{pmatrix}. \tag{2.102}$$

This is not a pure-state density matrix. It satisfies the trace relation $Tr(\rho) = 1$, but $Tr(\rho^2)$ is less than one.

It is possible to derive the Liouville equation in the interaction representation, as the example in Section 2.3 illustrates. Let us continue the discussion of Section 2.3 on the rotating wave approximation. If the spin is along the $z$ direction initially,

$$\begin{aligned} \rho_{11}(t) &= 1 - \rho_{22}(t) = \left(\frac{b}{\Omega}\right)^2 (\sin(\Omega t/2))^2, \\ \rho_{12}(t) &= \rho_{21}^*(t) = -i\left(\frac{b}{\Omega}\right) e^{i(\omega-\alpha)t} (\sin(\Omega t/2)) \\ &\quad \times \left\{\cos(\Omega t/2) - i\left(\frac{\omega - \alpha}{\Omega}\right)\sin(\Omega t/2)\right\}. \end{aligned} \tag{2.103}$$

If, on the other hand, the spin is along the negative $z$ axis,

$$\begin{aligned} \rho_{11}(t) &= 1 - \rho_{22}(t) = 1 - \left(\frac{b}{\Omega}\right)^2 (\sin(\Omega t/2))^2, \\ \rho_{12}(t) &= \rho_{21}^*(t) = i\left(\frac{b}{\Omega}\right) e^{i(\omega-\alpha)t} (\sin(\Omega t/2)) \\ &\quad \times \left\{\cos(\Omega t/2) - i\left(\frac{\omega - \alpha}{\Omega}\right)\sin(\Omega t/2)\right\}. \end{aligned} \tag{2.104}$$

The ensemble average of these two initial conditions is again of the diagonal form given in Eq. 2.102.

In order to study expectation values of operators, we note that the formula for the measurable quantity:

$$<A> = Tr(A\rho) \tag{2.105}$$

is invariant under unitary transformation of the operators and the density matrix. Indeed, this expression is valid in the interaction representation, as well as in the Schrödinger and Heisenberg pictures. The density matrices of Eq. 2.103 and Eq. 2.104 have been computed in the interaction representation.

If we wish to calculate the expectation value of $<\sigma_i>$, we have to use the operators in the interaction representation, in which

$$\sigma_x(t) = \begin{pmatrix} e^{i\omega t/2} & 0 \\ 0 & e^{-i\omega t2} \end{pmatrix} \begin{pmatrix} 0 & 1 \\ 1 & 0 \end{pmatrix} \begin{pmatrix} e^{-\omega t/2} & 0 \\ 0 & e^{i\omega t/2} \end{pmatrix},$$

$$\sigma_y(t) = \begin{pmatrix} e^{i\omega t/2} & 0 \\ 0 & e^{-i\omega t2} \end{pmatrix} \begin{pmatrix} 0 & -i \\ i & 0 \end{pmatrix} \begin{pmatrix} e^{-\omega t/2} & 0 \\ 0 & e^{i\omega t/2} \end{pmatrix},$$

$$\sigma_z(t) = \begin{pmatrix} e^{i\omega t/2} & 0 \\ 0 & e^{-i\omega t2} \end{pmatrix} \begin{pmatrix} 1 & 0 \\ 0 & -1 \end{pmatrix} \begin{pmatrix} e^{-\omega t/2} & 0 \\ 0 & e^{i\omega t/2} \end{pmatrix}. \quad (2.106)$$

As a consequence, in the interaction representation, the Pauli spin matrices are

$$\sigma_x(t) = \begin{pmatrix} 0 & e^{i\omega t/2} \\ e^{-i\omega t2} & 0 \end{pmatrix}, \quad \sigma_y(t) = \begin{pmatrix} 0 & -ie^{i\omega t/2} \\ ie^{-i\omega t2} & 0 \end{pmatrix},$$

$$\sigma_z(t) = \begin{pmatrix} 1 & 0 \\ 0 & -1 \end{pmatrix}. \quad (2.107)$$

The polarization vector can now be calculated from

$$P_x(t) = Tr\left(\rho(t)\sigma_x(t)\right) = 2\left\{Re\left(e^{-i\omega t}\rho_{12}(t)\right)\right\},$$

$$P_y(t) = Tr\left(\rho(t)\sigma_y(t)\right) = 2\left\{Im\left(e^{-i\omega t}\rho_{12}(t)\right)\right\},$$

$$P_z(t) = Tr\left(\rho(t)\sigma_z(t)\right) = \rho_{11}(t) - \rho_{22}(t). \quad (2.108)$$

The above calculations were carried out in the interaction representation. Since, however,

$$P_i(t) = Tr\left(\rho(t)\sigma_i(t)\right) = Tr\left(e^{-iHt}\rho(0)e^{iHt}e^{-iHt}\sigma_i(0)e^{iHt}\right). \quad (2.109)$$

The calculation in the Schrödinger picture will produce the same result. The story will be the same for the Heisenberg picture. This illustrates the fact that $P_i(t)$ is a measurable quantity, and is independent of the representation of quantum mechanics we choose for computational purposes. There is only one quantum mechanics.

If the spin is initially along the $z$ direction, we use the $\rho$ matrix given in Eq. 2.103.

$$P_x(t) = 2\left(\frac{b}{\Omega}\right)(\sin(\Omega t/2))\left\{(\cos(\Omega t/2))\sin(\alpha t)\right.$$

$$\left. + \left(\frac{\omega-\alpha}{\Omega}\right)(\sin(\Omega t/2))\cos(\alpha t)\right\},$$

$$P_y(t) = 2\left(\frac{b}{\Omega}\right)(\sin(\Omega t/2))\left\{(\cos(\Omega t/2))\cos(\alpha t)\right.$$

$$\left. - \left(\frac{\omega-\alpha}{\Omega}\right)(\sin(\Omega t/2))\sin(\alpha t)\right\},$$

$$P_z(t) = 1 - 2\left(\frac{b}{\Omega}\right)^2 (\sin(\Omega t/2))^2 . \tag{2.110}$$

The magnitude of this polarization vector is measured again from

$$(P_x(t))^2 + (P_y(t))^2 + (P_z(t))^2 . \tag{2.111}$$

For the polarization vector given in Eq. 2.110, this quantity is one and independent of time.

If we use the solution for $\rho(t)$ given in Eq. 2.104 for the spin initially along the negative $z$ direction, the polarization vector is the negative of the vector given in Eq. 2.110. This is also a pure state. If we mix the states with two different initial polarizations, as we did in Section 2.4, then

$$(P_x(t))^2 + (P_y(t))^2 + (P_z(t))^2 = (w_1 - w_2)^2 , \tag{2.112}$$

which is independent of time.

# Chapter 3

# WIGNER PHASE-SPACE DISTRIBUTION FUNCTIONS

We noted in Chapter 2 that there are several different representations of quantum mechanics. It was noted that the density matrix formalism is a convenient representation for non-pure mixed states. In this Chapter, we would like to discuss another form of quantum mechanics which is also convenient for mixed states. In this new form, both the position and momentum variables are c-numbers. It is thus possible to formulate quantum mechanics in phase space. We shall call this form the phase-space picture of quantum mechanics.

As the wave function plays the central role in the Schrödinger picture, the phase-space distribution function introduced by Wigner (1932a) is the starting point in the phase-space picture of quantum mechanics. This distribution function is widely known as the Wigner function. The Wigner function is constructed from the Schrödinger wave function through the density matrix. However, it is a function of both position and momentum variables. In quantum mechanics, it is not possible to determine the position and momentum variables simultaneously. Then how can we represent the uncertainty principle in this picture of quantum mechanics. This is one of the questions to be addressed in this Chapter.

We shall first study the basic properties of the Wigner distribution function together with illustrative examples. It is shown that the phase-space picture gives a very accurate description of the uncertainty relation for spreading wave packets. We use the one-dimensional harmonic oscillator to illustrate the Wigner function for mixed states and its connection with the density matrix. A brief review of the literature on the Wigner function is also given.

## 3.1 Basic Properties of the Wigner Phase-Space Distribution Function

If the wave function depends on $x_1, x_2, \cdots, x_n$, and $t$, then the density matrix is

$$\rho(x_1, \cdots, x_n; x_1', \cdots, x_n'; t) = \sum_k w_k \psi_k(x_1, \cdots, x_n, t)\psi_k^*(x_1', \cdots, x_n', t). \tag{3.1}$$

It is a straight-forward generalization of the single variable-density matrix discussed in Chapter 2, with

$$\sum_n w_k = 1. \tag{3.2}$$

Then the Wigner phase-space distribution function is defined as (Wigner 1932a and 1987)

$$W(x_1, \cdots, x_n; p_1, \cdots, p_n; t) = \left(\frac{1}{\pi}\right)^n \int \exp\{2i(p_1 y_1 + \cdots + p_n y_n)\}$$

$$\times \rho(x_1 - y_1, \cdots, x_n - y_n; x_1 + y_1, \cdots, x_n + y_n; t)\, dy_1 dy_2 \cdots dy_n. \tag{3.3}$$

Here, $x_i$ and $p_i$ are c-numbers, and the Wigner function is defined over the 2n-dimensional phase space. If the system is in a pure state with the wave function $\psi(x_1, \cdots, x_n, t)$,

$$W(x_1, \cdots, x_n; p_1, \cdots, p_n; t) = \left(\frac{1}{\pi}\right)^n \int \exp\{2i(p_1 y_1 + \cdots + p_n y_n)\}$$

$$\times \psi^*(x_1 + y_1, \cdots, x_n + y_n, t)\psi(x_1 - y_1, \cdots, x_n - y_n, t) dy_1 dy_2 \cdots dy_n. \tag{3.4}$$

This pure-state Wigner function is simplest when the system depends only one pair of $x$ and $p$ variables.

Thus the easiest way to study the properties of the Wigner function is to start with this simplest Wigner function. For this simple case, the Wigner function takes the form

$$W(x, p, t) = \frac{1}{\pi} \int e^{2ipy} \psi^*(x + y, t)\psi(x - y, t) dy. \tag{3.5}$$

For the time-independent wave function $\psi(x)$, the Wigner function does not depend on time. Indeed, the Wigner function of the form

$$W(x, p) = \frac{1}{\pi} \int e^{2ipy} \psi^*(x + y)\psi(x - y) dy \tag{3.6}$$

is most frequently seen in the literature.

Let us study this simplest form first, and then generalize its properties to the more complicated expressions given in Eq. 3.4.

1. If integrated over $p, W(x,p,t)$ gives the quantum probability distribution in $x$:

$$|\psi(x)|^2 = \int W(x,p)dp. \tag{3.7}$$

Similarly, if $W(x,p)$ is integrated over $x$, it gives the probability that the momentum coordinates have the values $p$:

$$|\chi(p)|^2 = \int W(x,p)dx, \tag{3.8}$$

where $\chi(p,t)$ is the momentum wave function derivable from $\psi(x)$ through

$$\chi(p) = \left(\frac{1}{\pi}\right)^{1/2} \int e^{-ip\cdot x}\psi(x)dx. \tag{3.9}$$

In the definition of the Wigner function throughout Eq. 3.3-Eq. 3.6, $p$ was introduced as a Fourier-transform parameter. From the properties of Eq. 3.7 and Eq. 3.8, we can regard it as the momentum variable. In the Wigner phase space, both $x$ and $p$ are c-numbers. This aspect is different from the case in either the Schrödinger or Heisenberg picture of quantum mechanics. As we shall see in Section 3.2, the uncertainty principle therefore has to be stated in a manner different from the conventional Schrödinger or Heisenberg picture.

2. It is possible to define the Wigner function starting from the momentum wave function. Since

$$\psi(x) = \left(\frac{1}{\pi}\right)^{1/2} \int e^{ip\cdot x}\chi(p)dp, \tag{3.10}$$

the Wigner function can also be written as

$$W(x,p) = \left(\frac{1}{\pi}\right) \int e^{-2ix\cdot z}\chi^*(p+z)\chi(p-z)dz. \tag{3.11}$$

It is possible to reproduce the probability distributions of Eq. 3.7 and Eq. 3.8 from the above expression.

3. If the system is in the state $\psi(x)$ and an observation is made as a result of which the system's state vector becomes $\phi(x)$, the probability of this result of the observation is $|(\psi,\phi)|^2$, which is the absolute square of the scalar product of the two state vectors. In terms of the Wigner functions, the transition probability can be written as

$$|(\psi,\phi)|^2 = 2\pi \int W_\psi(x,p)W_\phi(x,p)dxdp. \tag{3.12}$$

This expression is a clear proof that the Wigner function cannot be positive everywhere in phase space. If $\psi$ and $\phi$ are orthogonal, the above expression has to vanish. This cannot happen if both of the Wigner functions on the right-hand side are positive everywhere in phase space.

4. If $A(x)$ and $B(p)$ are dynamical operators depending only on $x$ and $p$ respectively, their expectation values are

$$(\psi(x), A(x)\psi(x)) = \int A(x)W(x,p)dxdp, \tag{3.13}$$

$$(\chi(p), B(p)\chi(p)) = \int B(p)W(x,p)dxdp. \tag{3.14}$$

In the Schrödinger picture, $A(x)$ and $B(p)$ do not always commute with each other. Thus the expectation value of the product $A(x)B(p)$ does not take a simple form. This problem will be discussed in Section 3.6.

5. Since both $x$ and $p$ are c-numbers in the phase-space picture, it is possible to perform canonical transformations in phase space, as in the case of distribution functions in classical phase space. However, unlike the Liouville distribution function in classical mechanics, the Wigner function is not a probability distribution function. As was noted in in the paragraph following Eq. 3.12, the Wigner function can become negative in phase space.

Properties (1) - (5) mentioned above for a single pair of canonical variables can be generalized to the Wigner function of many pairs of variables. The time-dependence of the Wigner function and the mixed-state Wigner function will be addressed in Section 3.2 and Section 3.5 respectively. Canonical transformations in phase space will be discussed extensively in Chapter 4.

## 3.2 Time Dependence of the Wigner Function

The Wigner function can be constructed from a time dependent wave function. Then it should be possible to derive a differential equation for the Wigner function starting from the time-dependent Schrödinger equation:

$$i\frac{\partial}{\partial t}\psi(x,t) = -\left(\frac{1}{2m}\right)\left(\frac{\partial}{\partial x}\right)^2 \psi(x,t) + V(x)\psi(x), \tag{3.15}$$

where $m$ is the mass of the particle, and $V(x)$ is the potential. From the definition of the Wigner function,

$$\begin{aligned}\frac{\partial}{\partial t}W(x,p,t) = \left(\frac{i}{2\pi m}\right)\int e^{2ipy}&\left\{\psi^*(x+y,t)\left[\left(\frac{\partial}{\partial x}\right)^2\psi(x-y,t)\right]\right.\\ &\left.-\left[\left(\frac{\partial}{\partial x}\right)^2\psi^*(x+y,t)\right]\psi(x-y,t)\right\}dy\\ &+\frac{i}{\pi}\int e^{2ipy}\left(V(x+y)-V(x-y)\right)\\ &\times\psi^*(x+y,t)\psi(x-y,t)dy.\end{aligned} \tag{3.16}$$

The potential factor $(V(x+y) - V(x-y))$ can be expanded as

$$(V(x+y) - V(x-y)) = \sum_{n=0}^{\infty} \frac{2}{(2n+1)!} \left\{ \left( \frac{\partial}{\partial x} \right)^{2n+1} V(x) \right\} y^{2n+1}. \tag{3.17}$$

Thus,

$$\begin{aligned} \frac{\partial}{\partial x} W(x,p,t) &= -\left( \frac{i}{2\pi m} \right) \frac{\partial}{\partial x} \int e^{2ipy} \frac{\partial}{\partial y} (\psi^*(x+y,t)\psi(x-y,t))\, dy \\ &\quad + \left( \frac{i}{\pi} \right) \sum_{n=0}^{\infty} \frac{2}{(2n+1)!} \left\{ \left( \frac{\partial}{\partial x} \right)^{2n+1} V(x) \right\} \\ &\quad \times \int e^{2ipy} y^{2n+1} \psi^*(x+y,t)\psi(x-y,t) dy. \end{aligned} \tag{3.18}$$

In the above expression, $y^{2n+1}$ and $\partial/\partial y$ can be replaced by $(i/2)\partial/\partial p$ and $2ip$ respectively. The differential equation then takes the form

$$\begin{aligned} \frac{\partial}{\partial t} W(x,p,t) &= -\left( \frac{p}{m} \right) \frac{\partial}{\partial x} W(x,p,t) \\ &\quad + \sum_{n=0}^{\infty} \left( \frac{-i}{2} \right)^{2n} \frac{1}{(2n+1)!} \left[ \left( \frac{\partial}{\partial x} \right)^{2n+1} V(x) \right] \left( \frac{\partial}{\partial p} \right)^{2n+1} W(x,p,t). \end{aligned} \tag{3.19}$$

This is an infinite-order differential equation if $V(x)$ does not take the form of a finite polynomial. On the other hand, if

$$\left( \frac{\partial}{\partial x} \right)^{3} V(x) = 0, \tag{3.20}$$

then the above equation can be reduced to the classical Liouville equation:

$$\frac{\partial}{\partial t} W(x,p,t) = \left( \frac{\partial H}{\partial x} \right) \frac{\partial}{\partial p} W(x,p,t) - \left( \frac{\partial H}{\partial p} \right) \frac{\partial}{\partial x} W(x,p,t). \tag{3.21}$$

Indeed, if the potential is of the form $V(x) = ax^2 + bx + c$, the Wigner distribution function satisfies the Liouville equation. While we can use the same mathematics for both the classical phase-space distribution and the Wigner distribution function, there is a clear difference between them. In the case of classical distribution, it is a probability distribution in phase space (Feynman 1973, Goldstein 1980). The Wigner distribution is not a probability distribution in phase space (Wigner 1932a and 1987).

If the particle is free with $V = 0$, the differential equation of Eq. 3.19 becomes

$$\frac{\partial}{\partial t} W(x,p,t) = -\left( \frac{p}{m} \right) \frac{\partial}{\partial x} W(x,p,t). \tag{3.22}$$

We shall return to this equation in Section 3.3 to study the wave-packet spread.

In the case of the harmonic oscillator with $V(x) = Kx^2/2$, the differential equation becomes

$$\frac{\partial}{\partial t}W(x,p,t) = -\left(\frac{p}{m}\right)\frac{\partial}{\partial x}W(x,p,t) + (Kx)\frac{\partial}{\partial p}W(x,p,t). \tag{3.23}$$

Since the harmonic oscillator plays the central role throughout this book, we shall study this equation in detail in Section 3.4.

If the potential is of the form $V(x) = gx$, as in the case of a particle in a constant gravitational field, the quantum Liouville equation is

$$\frac{\partial}{\partial t}W(x,p,t) = -\left(\frac{p}{m}\right)\frac{\partial}{\partial x}W(x,p,t) + g\frac{\partial}{\partial p}W(x,p,t). \tag{3.24}$$

If the Wigner distribution function is known at $t = 0$, then the solution of this differential equation is

$$W(x,p,t) = W(x - pt/m, p + gt, t = 0). \tag{3.25}$$

As time progresses, the phase-space distribution moves from $p = 0$ to $p = -gt$. At the same time, $x = 0$ moves to $pt/m$. The solution can therefore be represented by a coordinate transformation:

$$\begin{pmatrix} x' \\ p' \\ 1 \end{pmatrix} = \begin{pmatrix} 1 & t/m & 0 \\ 0 & 1 & -gt \\ 0 & 0 & 1 \end{pmatrix} \begin{pmatrix} x \\ p \\ 1 \end{pmatrix}. \tag{3.26}$$

This means that the solution of the Liouville equation for the linear potential can be represented by a linear coordinate transformation. As we shall see in later chapters as well as in Sections 3.3 and 3.4, coordinate transformation in phase space teach us many interesting lesson in physics.

## 3.3 Wave Packet Spreads

The wave packet spread for a free particle is one of the most puzzling feature of the Schrödinger picture of quantum mechanics. The spread is caused by the fact that particles have different phase velocities for different momenta. Thus, the position distribution becomes widespread while the momentum distribution remains invariant, and the uncertainty product $<\Delta x><\Delta p>$ increases as time progresses or regresses. The question then is whether there is a way of stating this uncertainty relation in a time-independent manner.

If the particle is free, the Wigner function satisfies the differential equation given in Eq. 3.22, and its solution is (Kijowski 1974, Lee 1982, Littlejohn 1986, Royer 1987, Kim and Wigner 1990b)

$$W(x,p,t) = W(x - pt/m, p, 0). \tag{3.27}$$

If the Wigner function is localized within a given region in phase space at $t = 0$, its localization region is sheared while its area remains invariant.

In the phase-space picture of quantum mechanics, both $x$ and $p$ are c-numbers. Since, however, $(\Delta x)(\Delta p)$ cannot be smaller than Planck's constant, it is possible to quantify the uncertainty relation in terms of the area within which the Wigner function is concentrated. The area of localization cannot be smaller than Planck's constant. This region may undergo deformation, but the uncertainty remains invariant as long as the area of localization is invariant under certain transformations. This region of localization is called the error box (Caves *et al.* 1981).

For example, Let us consider the spread of a Gaussian wave packet with its momentum distribution:

$$g(p) = \left(\frac{b}{\pi}\right)^{1/4} e^{-bp^2/2} \tag{3.28}$$

at $t = 0$. As time progresses, the momentum distribution function remains unchanged:

$$| g(p) |^2 = \left(\frac{b}{\pi}\right)^{1/2} e^{-bp^2}. \tag{3.29}$$

Thus, if we define $(\Delta p)$ as

$$(\Delta p) = \left\{2 \int | g(p) |^2 p^2 dp\right\}^{1/2}, \tag{3.30}$$

then $(\Delta p) = (1/b)^{1/2}$, which is invariant under time evolution.

After the construction:

$$\psi(x,t) = \left(\frac{1}{2\pi}\right)^{1/2} \int \exp\left(ipx - ip^2 t/2m\right) g(p) dp, \tag{3.31}$$

the time-dependent Schrödinger wave function becomes

$$\psi(x,t) = \left(\frac{b}{\pi}\right)^{1/4} \left(\frac{1}{b + it/m}\right)^{1/2} e^{-x^2/2(b+it/m)}. \tag{3.32}$$

The spatial probability distribution is

$$| \psi(x,t) |^2 = \left\{b/\pi \left(b^2 + (t/m)^2\right)\right\}^{1/2} \exp\left\{-bx^2/\left(b^2 + (t/m)^2\right)\right\}. \tag{3.33}$$

Thus

$$(\Delta x) = \left\{b\left(1 + (t/mb)^2\right)\right\}^{1/2}. \tag{3.34}$$

Therefore, the uncertainty product in the Schrödinger picture is

$$(\Delta x)(\Delta p) = \left(1 + (t/mb)^2\right)^{1/2}. \tag{3.35}$$

This quantity increases as $t$ progresses. This is widely known as wave packet spread in Schrödinger's form of quantum mechanics.

We are now interested in the question of whether the uncertainty can be stated in a time-invariant manner in the phase-space picture of quantum mechanics. If we construct the Wigner function for the above wave function, its form is

$$W(x,p,t) = \frac{1}{\pi}\exp\left\{-\left((x - pt/m)^2/b + bp^2\right)\right\}. \tag{3.36}$$

This distribution is concentrated within the region where the exponent is less than one in magnitude:

$$(x - pt/m)^2/b + bp^2 < 1. \tag{3.37}$$

This region is described by the tilted ellipse in Figure 3.1. At $t = 0$, the distribution function is localized within the elliptic region $x^2/b + bp^2 = 1$. The elliptic region becomes sheared as time progresses.

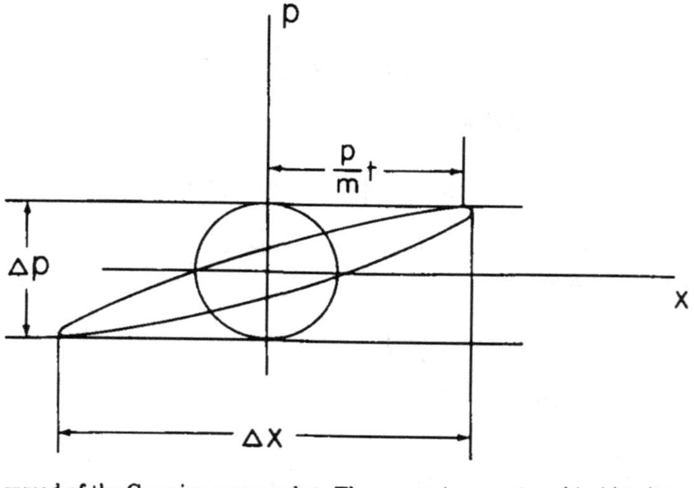


Figure 3.1: The spread of the Gaussian wave packet. The uncertainty product $(\Delta x)(\Delta p)$ increases in the Schrödinger picture as time progresses or regresses. However, the area of the localization remains constant. This area of localization is Planck's constant.

Thus we can define the magnitude of the uncertainty as the area divided by $\pi$. As time progresses, since the area remains unchanged, the uncertainty remains invariant. The area is the volume of the error box. The error box becomes sheared as time progresses, but its volume remains invariant.

Here again, the time evolution can be described in terms of a linear transformation in phase space:

$$\begin{pmatrix} x' \\ p' \end{pmatrix} = \begin{pmatrix} 1 & t/m \\ 0 & 1 \end{pmatrix}\begin{pmatrix} x \\ p \end{pmatrix}. \tag{3.38}$$

We shall discuss the mathematical property of this transformation in Chapter 4.

## 3.4 Harmonic Oscillators

The classical Hamiltonian for describing the one-dimensional harmonic oscillator is

$$H = \frac{1}{2m}p^2 + \frac{K}{2}x^2. \tag{3.39}$$

It is possible to measure $p$ and $x$ in units of $\sqrt{m}$ and $1/\sqrt{K}$ respectively. Thus, the Hamiltonian can be written as

$$H = \frac{1}{2}\left(p^2 + x^2\right). \tag{3.40}$$

In this system of units, the Schrödinger equation is

$$i\frac{\partial}{\partial t}\psi(x,t) = -\left(\frac{1}{2}\right)\left(\frac{\partial}{\partial x}\right)^2 \psi(x,t) + \left(\frac{1}{2}x^2\right)\psi(x). \tag{3.41}$$

The Liouville equation for the Wigner function takes the form

$$\frac{\partial}{\partial t}W(x,p,t) = \left(x\frac{\partial}{\partial p} - p\frac{\partial}{\partial x}\right) W(x,p,t). \tag{3.42}$$

The solution of this equation is

$$W(x,p,t) = \left\{\exp\left[\left(x\frac{\partial}{\partial p} - p\frac{\partial}{\partial x}\right)t\right]\right\} W(x,p,0). \tag{3.43}$$

This is a rotation around the origin in phase space. In matrix form, this rotation can be represented as

$$\begin{pmatrix} x' \\ p' \end{pmatrix} = \begin{pmatrix} \cos(t) & -\sin(t) \\ \sin(t) & \cos(t) \end{pmatrix} \begin{pmatrix} x \\ p \end{pmatrix}. \tag{3.44}$$

If the distribution is invariant under rotations, it is independent of time.

The time-independent Schrödinger equation is

$$-\left(\frac{1}{2}\right)\left(\frac{\partial}{\partial x}\right)^2 \psi(x,t) + \left(\frac{1}{2}x^2\right)\psi(x) = (n + 1/2)\psi(x,t), \tag{3.45}$$

and its normalized solutions are

$$\psi_n(x) = [1/(\sqrt{\pi}2^n n!)]^{1/2} H_n(x) \exp(-x^2/2). \tag{3.46}$$

where $H_n(x)$ is the Hermite polynomial of the $n^{th}$ order. This wave function is in its energy eigenstate. It is possible to evaluate the Wigner function:

$$W_n(x,p) = \frac{1}{\pi}\int \psi_n^*(x+y)\psi_n(x-y)e^{2ipy}dy. \tag{3.47}$$

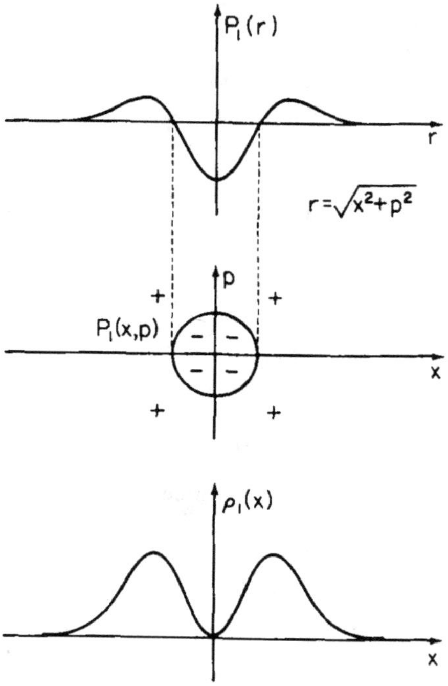


Figure 3.2: The Wigner function for the harmonic oscillator in the first excited state. It is negative at the origin, becomes positive as $(x^2+p^2)$ increases, and becomes vanishingly small as $(x^2+p^2)$ becomes very large. If the Wigner function is integrated over $p$, then it becomes the probability density in $x$, which is positive for all values of $x$.

The result of the calculation is

$$W_n(x,p) = \left(\frac{n!}{\pi}\right)\left(\exp(-r^2/2)\right)\sum_{k=0}^{\infty}(-1)^k r^{2(n-k)}/\left([(n-k)!]^2 k!\right), \tag{3.48}$$

where

$$r^2 = 2(x^2+p^2).$$

This means that $W_n(x,y)$ is a function only of $r$, and can be written as $W_n(r)$. This in fact satisfies the differential equation (Hillery *et al.* 1984)

$$-\frac{1}{2r}\left[\frac{d}{d\rho}r\left(\frac{d}{dr}W_n(r)\right)\right]+\frac{1}{2}r^2W_n(r) = (2n+1)W_n(r). \tag{3.49}$$

The solution to this equation is readily available in the literature (Schleich *et al.* 1988), and its form is

$$W_n(x,p) = ((-1)^n/\pi)\left(L_n(r^2)\right)e^{-r^2/2}, \tag{3.50}$$

where $L_n(r^2)$ is the Laguerre polynomial. This expression is invariant under rotations around the origin, and can also be written as (Schleich *et al.* 1988)

$$W_n(x,p) = \frac{1}{\pi}\left(\frac{1}{4}\right)^n e^{-(x^2+p^2)} \sum_{k=0}^{n} \left(\frac{1}{k!(n-k)!}\right) H_{2k}(\sqrt{2}x) H_{2(n-k)}(\sqrt{2}p), \tag{3.51}$$

where $H_{2k}(\sqrt{2}x)$ and $H_{2(n-k)}(\sqrt{2}p)$ are the Hermite polynomials.

For the ground-state, the Wigner function is

$$W_0(x,p) = \frac{1}{\pi}\exp\left(-(x^2+p^2)\right). \tag{3.52}$$

If $n = 1$,

$$W_1(x,p) = \frac{2}{\pi}\left(x^2+p^2-\frac{1}{2}\right)\exp\left(-(x^2+p^2)\right). \tag{3.53}$$

$W_0(x,p)$ is positive everywhere in phase space. As is illustrated in Figure 3.2, $W_1(x,p)$ is negative at the origin, but is positive for sufficiently large values of $(x^2+p^2)$. Both of them become vanishingly small for very large values of $(x^2+p^2)$. Even though $W_1(x,p)$ is negative around the origin, the probability density in $x$ is always positive.

## 3.5 Density Matrix

As was noted in Section 3.1, the transition from the Schrödinger wave function to the Wigner function takes place through the density matrix. Let us examine in this section how some of the basic properties of the density matrix become translated into the language of the Wigner function. For one pair of canonical variables, the Wigner function can be written as

$$W(x,p) = \frac{1}{\pi}\int \rho(x+y, x-y) e^{2ipy} dy. \tag{3.54}$$

From this expression, it is possible to convert the Wigner function back to the density matrix:

$$\rho(x,x') = \int W\left(\frac{x+x'}{2}, p\right) e^{-ip(x-x')} dp. \tag{3.55}$$

Using this relation, we can calculate $Tr(\rho)$ and $Tr(\rho^2)$ from the Wigner function.

From the above expression, $Tr(\rho)$ is

$$Tr(\rho) = \int \rho(x,x)dx = \int W(x,p)dxdp = 1. \tag{3.56}$$

For two different density matrices $\rho_1$ and $\rho_2$, it is possible to construct two different Wigner functions $W_1$ and $W_2$. Then,

$$\begin{aligned} Tr(\rho_1\rho_2) &= \int \left\{\int \rho_1(x,x')\rho_2(x',x)dx'\right\} dx \\ &= 2\pi \int W_1(x,p)W_2(x,p)dxdp. \end{aligned} \tag{3.57}$$

Consequently,

$$Tr(\rho^2) = \int \rho(x,x')\rho(x',x)dxdx'$$

$$= 2\pi \int (W(x,p))^2\, dxdp. \tag{3.58}$$

As a more concrete example, let us consider a one-dimensional harmonic oscillator, starting from the density matrix of Eq. 2.88. If the oscillator has a unit frequency, the density matrix takes the form

$$\rho_T(x,x') = \left(1 - e^{-1/kT}\right) \sum_n e^{-n/kT} \psi_n(x)\psi_n^*(x'). \tag{3.59}$$

Thus, according to the analysis given in Section 3.4, the Wigner function for this system is

$$W_T(x,p) = \left(1 - e^{-1/kT}\right) \sum_n e^{-n/kt} W_n(x,p), \tag{3.60}$$

where $W_n(x,p)$ is the Wigner function for the $n^{th}$ excited-state harmonic oscillator.

However, the series expansion of a function in terms of not-so-familiar functions is not very useful. We thus ask the question of whether a closed form is available for $W_T(x,p)$ or $\rho_T(x,x')$. For this purpose, let us start with the trial function (Ruiz 1974, Kim and Noz 1986):

$$f(x,x') = \left(\frac{1}{\pi}\right)^{1/2} \exp\left\{-\left(\frac{1}{4}\right)\left(e^{-2\eta}(x+x')^2 + e^{2\eta}(x-x')^2\right)\right\}, \tag{3.61}$$

and consider a linear expansion in terms of the harmonic oscillator wave functions:

$$f(x,x') = \sum_n C_{nm}\psi_n(x)\psi_m(x'). \tag{3.62}$$

Then the coefficient $C_{nm}$ is

$$C_{nm} = \int \psi_n^*(x)\psi_m^*(x') f(x,x')dxdx'. \tag{3.63}$$

We can evaluate this integral using the generating function for the Hermite polynomials (Magnus and Oberhettinger 1949):

$$\exp\left(2sx - s^2\right) = \sum_n \frac{s^2}{n!} H_n(x). \tag{3.64}$$

In terms of the oscillator wave functions, this generating function takes the form

$$\left(\frac{1}{\pi}\right)^{1/4} \exp\left(2sx - s^2 - x^2/2\right) = \sum_n s^n \left(2^n/n!\right)^{1/2} \psi_n(x). \tag{3.65}$$

Thus

$$\left(\frac{1}{\pi}\right)^{1/2}\int f(x,x')\exp\left(2sx+2rx'-s^2-r^2-(x^2+x'^2)/2\right)dxdx'$$
$$=\sum_m\sum_n s^n r^m\left(2^n 2^m/(n!m!)\right)^{1/2}C_{nm} \qquad (3.66)$$

After the evaluation of the integral,

$$\left(\sqrt{\pi}/\cosh\eta\right)\exp\left(2rs(\tanh\eta)\right)=\sum_m\sum_n s^n r^m\left(2^n 2^m/(n!m!)\right)^{1/2}C_{nm}. \qquad (3.67)$$

We can now expand the left-hand side in power series of $(2rs)$ to get

$$C_{nm}=\left(\sqrt{\pi}/\cosh\eta\right)(\tanh\eta)^n\delta_{nm}. \qquad (3.68)$$

As a consequence,

$$\rho_T(x,x')=\left(\frac{\tanh(1/2kT)}{\pi}\right)^{1/2}\exp\left\{-\left(\frac{1}{4}\right)\left((x+x')^2\tanh(1/2kT)\right.\right.$$
$$\left.\left.+(x-x')^2\coth(1/2kT)\right)\right\}. \qquad (3.69)$$

This density reproduces $Tr(\rho)=1$ and $Tr(\rho^2)=\tanh(1/2kT)$ given in Eq. 2.91 derived rom the series form of $\rho_T(x,x')$. It becomes that of a pure state in the zero-temperature limit.

From the above expression for $\rho_T(x,x')$, we can derive

$$W_T(x,p)=\left(\frac{\tanh(1/2kT)}{\pi}\right)\exp\left\{-\left(x^2+p^2\right)\tanh(1/2kT)\right\}. \qquad (3.70)$$

This expression becomes that of the ground-state harmonic oscillator when $T$ approaches zero. Furthermore, the traces of the density matrix can be calculated as

$$Tr\left(\rho_T\right)=\int W_T(x,p)dxdp=1, \qquad (3.71)$$

$$Tr\left(\rho_T\rho_T\right)=2\pi\int\left(W_T(x,p)\right)^2dxdp=\tanh(1/2kT). \qquad (3.72)$$

## 3.6 Measurable Quantities

In the Schrödinger picture of quantum mechanics, measurable quantities are calculated from transition probabilities and expectation values. Since the Wigner function is constructed from the Schrödinger wave function, it should be possible to calculate these quantities directly from the Wigner function.

In order to study the transition probability, let us first consider two different wave functions $\psi(x)$ and $\phi(x)$, and their corresponding Wigner function $W_\psi(x,p)$

and $W_\phi(x,p)$. Then the transition probability takes the form of Eq. 3.12. Let us rewrite Eq. 3.12.

$$|\,(\phi(x),\psi(x)\,)\,|^2 = 2\pi\int W_\psi(x,p)W_\phi(x,p)dx\,dp. \tag{3.73}$$

This expression is valid when both of the Wigner functions in the integrand are constructed from the pure states of $\phi(x)$ and $\psi(x)$. For non-pure states, the density matrix is the appropriate quantity. Let us rewrite Eq. 3.57

$$Tr\,(\rho_1\rho_2) = 2\pi\int W_1(x,p)W_2(x,p)dx\,dp. \tag{3.74}$$

The Wigner function accommodates both pure and non-pure states, and Eq. 3.58 is a special case of Eq. 3.57.

For example, $W_0(x,p)$ and $W_1(x,p)$ of Eq. 3.51 are pure-state Wigner functions for the ground and first-excited oscillator states respectively. By direct calculation, it is possible to verify

$$2\pi\int (W_0(x,p))^2\,dx\,dp = 2\pi\int (W_1(x,p))^2\,dx\,dp = 1, \tag{3.75}$$

and

$$2\pi\int W_0(x,p)W_1(x,p)dx\,dp = 0. \tag{3.76}$$

Furthermore, using the differential equation of Eq. 3.24, we can show the orthogonality relation:

$$\int W_n(x,p)W_m(x,p)dx\,dp = 0, \tag{3.77}$$

if $n$ and $m$ are different.

$W_T(x,p)$ of Eq. 3.61 is the Wigner function for a thermally excited state, which is a non-pure state. The probability of $W_T$ being in the n$^{\text{th}}$ excited state is

$$P_n = 2\pi\int W_T(x,p)W_n(x,p)dx\,dp = (-1)^n\tanh\left(\frac{1}{2kT}\right)$$
$$\times\int_0^\infty L_n(\rho^2)\exp\left\{-\left(\rho^2/2\right)\left(1+\tanh\left(\frac{1}{2kT}\right)\right)\right\}(2\rho)d\rho. \tag{3.78}$$

This integral can be readily evaluated from the formula (Gradshteyn and Ryzhik 1965):

$$\int_0^\infty e^{-bx}L_n(x)dx = \frac{1}{b}\left(1-\frac{1}{b}\right)^n. \tag{3.79}$$

The result of the calculation is

$$P_n = \left(1-e^{-1/kT}\right)e^{-n/kT}, \tag{3.80}$$

which is expected from the definition of $\rho(x,x')$.

Let us next consider the expectation value of an operator applicable to $\psi(x)$ or the momentum wave function $\chi(p)$. If the operator $Q$ is of the form

$$Q = A(x) + B(p), \tag{3.81}$$

where $A(x)$ and $B(p)$ depend only on $x$ and $p$ respectively, then the expectation value is

$$\begin{aligned} <Q> &= \left(\psi(x), \hat{A}(x)\psi(x)\right) + \left(\chi(p), \hat{B}(p)\chi(p)\right) \\ &= \int \left(A(x) + B(p)\right) W(x,p)dxdp, \end{aligned} \tag{3.82}$$

where $\hat{A}$ and $\hat{B}$ are the q-number operators in the Schrödinger picture. It is clear that we are using the word "expectation value" for both pure and non-pure states. The above formula is valid for both cases.

For example, for the energy operator for the harmonic oscillator:

$$\hat{H} = \frac{1}{2}\left(x^2 - \left(\frac{\partial}{\partial x}\right)^2\right), \tag{3.83}$$

the expectation value is

$$\begin{aligned} <H> &= \frac{1}{2}\left\{\left(\psi(x), x^2\psi(x)\right) + \left(\chi(p), p^2\chi(p)\right)\right\} \\ &= \frac{1}{2}\int \left(x^2 + p^2\right) W(x,p)dxdp. \end{aligned} \tag{3.84}$$

If we use the Wigner function for the $n^{th}$ excited state oscillator, $<H>$ should be $(n + 1/2)$. For the thermally excited state,

$$\begin{aligned} <H> &= \frac{1}{2}\int \left(x^2 + p^2\right) W_T(x,p)dxdp \\ &= \frac{1}{2}\left\{\tanh\left(\frac{1}{2kT}\right)\right\}. \end{aligned} \tag{3.85}$$

This is also what is expected from the density matrix.

If the operator $Q$ depends on both $x$ and $p$, it is easier to start from the phase-space picture. It is straight-forward to show that (Han *et al.* 1988 and 1989b)

$$\int xpW(x,p)dxdp = -\left(\frac{i}{2}\right)\left(\psi(x), \left(x\frac{\partial}{\partial x} + \frac{\partial}{\partial x}x\right)\psi(x)\right), \tag{3.86}$$

$$\begin{aligned} &\int x^2p^2W(x,p)dxdp \\ &= -\left(\frac{1}{4}\right)\left(\psi(x), \left(x^2\left(\frac{\partial}{\partial x}\right)^2 + 2x\left(\frac{\partial}{\partial x}\right)^2 x + \left(\frac{\partial}{\partial x}\right)^2 x^2\right)\psi(x)\right). \end{aligned} \tag{3.87}$$

These results are special cases of a more general relation:

$$\int W(x,p)\,(x^n p^m)\,dx\,dp$$

$$= (-i)^m \left(\frac{1}{2}\right)^n \sum_{r=0}^{n} \binom{n}{r} \left(\psi(x), x^{n-r} \left(\frac{\partial}{\partial x}\right)^m x^r \psi(x)\right). \tag{3.88}$$

This is derivable from the theorem stating

$$\left(\psi(x), \exp\left(-i\sigma x - \tau\frac{\partial}{\partial x}\right)\psi(x)\right) = \int W(x,p)e^{-i(\sigma x+\tau p)}dx\,dp, \tag{3.89}$$

where $\sigma$ and $\tau$ are scalar parameters. We can derive Eq. 3.88 by expanding the exponential functions of both sides and by choosing the coefficients of $\sigma^n\tau^m$.

Let us prove the above theorem (Moyal 1949, Hillery *et al.* 1984). From the Baker-Campbell-Hausdorff formula (Miller 1972)

$$\exp\left(-i\sigma x - \tau\frac{\partial}{\partial x}\right) = \left(\exp(i\sigma\tau/2)\right)\left(\exp(-i\sigma x)\right)\left(\exp\left(-\tau\frac{\partial}{\partial x}\right)\right). \tag{3.90}$$

Thus the left-hand side of Eq. 3.89 becomes

$$e^{i\sigma\tau/2}\left(\psi(x), e^{-i\sigma x}\psi(x-\tau)\right). \tag{3.91}$$

By changing the variable $y$ to $y/2$, we can write the Wigner function given in Eq. 3.6 as

$$W(x,p) = \frac{1}{2\pi}\int \psi^*(x+y/2)\psi(x-y/2)e^{ipy}dy. \tag{3.92}$$

Then the right-hand side of Eq. 3.89 takes the form

$$\frac{1}{2\pi}\int \psi^*(x+y/2)\psi(x-y/2)e^{ip(y-\tau)}e^{i\sigma x}dx\,dy\,dp. \tag{3.93}$$

The integrations of this expression over $p$ and $y$ lead to

$$\int \psi^*(x+\tau/2)\psi(x-\tau/2)e^{-i\sigma x}dx. \tag{3.94}$$

By changing the variable $x$ to $x-\tau/2$, we arrive at the form

$$e^{i\sigma\tau/2}\int \psi^*(x)\psi(x-\tau)e^{-i\sigma x}dx, \tag{3.95}$$

which is identical to the expression of Eq. 3.91. The proof is now complete.

It is straight-forward to generalize Eq. 3.89 to

$$\int A(\sigma,\tau)\left\{\int W(x,p)e^{-i(\sigma x+\tau p)}dx\,dp\right\}d\sigma\,d\tau$$

$$= \int A(\sigma,\tau)\left[\psi(x), \exp\left(-i\sigma x - \tau\frac{\partial}{\partial x}\right)\psi(x)\right]d\sigma\,d\tau. \tag{3.96}$$

Thus the c-number operator

$$Q(x,p) = \int A(\sigma,\tau)e^{-i(\sigma x+\tau p)}d\sigma d\tau \tag{3.97}$$

corresponds to the $q$ number operator

$$\hat{Q}(x,p) = \int A(\sigma,\tau)\exp\left(-i\sigma x - \tau\frac{\partial}{\partial x}\right)d\sigma d\tau \tag{3.98}$$

in the Schrödinger picture.

## 3.7 Early and Recent Applications

In 1932, Wigner was interested in the thermodynamic behavior of macroscopic objects (Wigner 1932a). At low temperature, quantum effects can become important, and this manifests itself in the equation of state of the helium gas. It was therefore natural to develop a substitute for the classical expression for the density in phase space which forms the basis for the calculation of the thermodynamic behavior in the temperature region in which classical mechanics can be assumed to be valid, and which easily provides a good approximation in the temperature region not too far away from the validity of classical physics. This means a probability function of the position and momentum variables defined in terms of the wave function or the density matrix.

The first application of the Wigner function was concerned with the equation of state of the He gas (Wigner 1932b and 1938). At very low temperatures, the experimental results deviated considerably from that given by the classical distribution function. The correction introduced by the Wigner function was in the right direction but accounted only for about 2/3 of the deviations from the experimental values. It is possible that the reason for this was that the potential energy function was not known well enough. It would therefore be worthwhile to repeat that calculation.

Since the Wigner function started as a theoretical tool in statistical mechanics, its implications in thermodynamics and statistical physics have been extensively studied in the literature (Ross and Kirkwood 1954, Kadanoff and Baym 1962, Fujita 1966, Imre *et al.* 1967, Davies and Davies 1975, Hakim 1978, Royer 1985, Alonso 1985, Vasak *et al.* 1987, Ekert and Knight 1989, Kim and Li 1989).

Condensed matter physics is the natural test ground for new theories in statistical mechanics. The Wigner function may become a new theoretical tool in condensed matter physics (Sumi 1983 and 1984, Jose 1984, O'Connell and Wang 1985, Barocchi *et al.* 1985, Dickman and O'Connell 1985, O'Connell 1987).

It is possible to formulate nonrelativistic potential scattering problems in terms of the Wigner function (Carruthers and Zachariasen 1983). Thus, it is possible to calculate the cross sections from the Wigner function. For this reason, the phase-space distribution is an effective new theoretical tool in nuclear physics (Nix 1969, Balazs and Zippel 1973, Remler 1975 and 1977, Remler and Sathe 1975 and 1978,

Thies 1979, Kuratsuji 1981, Durand *et al.* 1985), and also in atomic physics (Lee and Scully 1980, 1982 and 1983, Dahl 1982, Dahl and Springborg 1982, Eu 1983, Ree 1983, Gracia-Bondia 1984, Cohen 1984, Takatsuka and Nakamura 1985, Sundberg *et al.* 1985).

While it is fundamentally difficult to make quantum mechanics relativistic, it is not uncommon to develop a relativistic scattering theory by making a Lorentz generalization of cross sections derivable from nonrelativistic potential scattering. The Wigner function is useful also in high-energy physics (Carruthers and Zachariasen 1976, Whitenton *et al.* 1983, Carruthers and Shi 1983, Lesche 1984, Franca and Thomas 1985, Petroni *et al.* 1985, Winter 1985, Zuk 1985, Zachariasen 1987).

In optical science, the separation of signals is a very important problem. This depends on how light waves are localized. The Wigner function is a very useful theoretical tool also in this branch of modern physics (Klauder and Sudarshan 1968, Louisell 1973, Kumar and Carroll 1984, Easton *et al.* 1984, Procida and Lee 1984, Subotic and Saleh 1984, Conner and Li 1985, McDonald and Kaufman 1985, Serima *et al.* 1986, Kim 1986, Agarwal 1987, Szu 1987, Kim and Wigner 1987b).

While being the quantity derivable from quantum mechanics, the Wigner function shares some important properties with classical mechanics. For this reason, the distribution function plays the pivotal role in semi-classical Physics (Heller 1976 and 1977, Berry 1977, Davis and Heller 1979, Balazs 1980 and 1981, Korsch and Berry 1981, Maslov and Fedoriuk 1981, Lee and Scully 1983, Almeida 1983, Sirugue *et al.* 1984, Feingold and Peres 1985, Shlomo 1985, Chang and Shi 1985, Molzahn and Osborn 1986, Osborn and Molzahn 1986, Littlejohn 1986).

The most ambitious program in quantum phase-space physics is to see whether the Wigner function can reveal the basic aspects of quantum mechanics which are not seen in the Schrödinger or Heisenberg picture. Efforts have been made along this direction (Groenewold 1946, Moyal 1949, Bartlett and Moyal 1949, Takabayasi 1954, Cohen 1966, Shlomo and Prakash 1981, Bohm and Hiley 1981, O'Connell and Wigner 1981a and 1981b, O'Connell and Rajagopal 1982, Emch 1982, Shlomo 1983, Bertrand *et al.* 1983, Werner 1984, Janssen 1985, Lesche and Seligman 1986, Narcowich and O'Connell 1986, Wigner 1987, Moshinsky 1987). It now appears that the phase space can interpret the uncertainty relations more accurately than the Schrödinger or Heisenberg picture. This is what we intend to discuss further in this book.

For this purpose, we shall study in detail the canonical transformation properties of the Wigner function applicable to quantum optics and to Lorentz-covariant phase space. Quantum optics is a relatively new subject, and the squeezed state of lights is of current interest. We shall see that the Wigner function is the simplest scientific language for coherent and squeezed states. C The construction of a covariant phase space has been a challenging problem since the formulation of the present form of nonrelativistic quantum mechanics. It is an urgent problem now since there are many experimental data to be explained in high-energy physics. It will be very difficult to construct a complete relativistic dynamics (Dirac 1949, Van Dam and Wigner 1965). However, it is not unreasonable to study the relativistic kinematics of the uncertainty relations. This is precisely the purpose of the covariant formulation of phase space which we propose to study.

# Chapter 4

# LINEAR CANONICAL TRANSFORMATIONS IN QUANTUM MECHANICS

The Wigner phase-space distribution function is strictly defined within the framework of quantum mechanics. On the other hand, since both $x$ and $p$ are c-numbers, it is defined in phase space just as in the case of the Hamiltonian formulation of classical mechanics. The purpose of this Chapter is to study the linear canonical transformation properties of the Wigner function in phase space. We shall see that the rules of canonical transformations in classical mechanics are directly applicable to the Wigner function. We shall note also that canonical transformations in the phase-space picture correspond to unitary transformations in the Schrödinger picture of quantum mechanics.

The simplest canonical transformations are those in phase space consisting of one pair of canonical variables. It will be noted that the symmetry group of linear canonical transformations for one pair of canonical variables is the inhomogeneous symplectic group commonly denoted by ISp(2). This group has the two-dimensional Euclidean group and the two-dimensional homogeneous symplectic group as subgroups. These groups are useful in studying the representations of the Poincaré group (Wigner 1939, Kim and Noz 1986).

Included in this Chapter is a discussion of the group of homogeneous linear canonical transformations in the four-dimensional phase space consisting of two pairs of canonical variables. It will be noted that this Group is locally isomorphic to thc (2 + 1)-dimensional de Sitter group.

## 4.1 Canonical Transformations in Two-Dimensional Phase Space

We noted in Chapter 1 that canonical transformations are those transformations which preserve the Poisson brackets. For one-pair of canonical variables, the condition for the transformation from $(x,p)$ to $(x',p')$ to be canonical can be stated as

$$\frac{\partial x'}{\partial x}\frac{\partial p'}{\partial p} - \frac{\partial p'}{\partial x}\frac{\partial x'}{\partial p} = 1. \tag{4.1}$$

Thus canonical transformations in two-dimensional phase space are area-preserving transformations. As was noted in Chapter 1, the group of homogeneous linear canonical transformations consists of rotations and squeezes.

Translations in phase space are inhomogeneous linear transformations which take the form

$$x' = x + a, \qquad p' = p + b. \tag{4.2}$$

These are inhomogeneous linear transformations and can be represented as

$$\begin{pmatrix} x' \\ p' \\ 1 \end{pmatrix} = \begin{pmatrix} 1 & 0 & a \\ 0 & 1 & b \\ 0 & 0 & 1 \end{pmatrix}\begin{pmatrix} x \\ p \\ 1 \end{pmatrix}. \tag{4.3}$$

If we define the translation matrix $T(a,b)$ as

$$T(a,b) = \begin{pmatrix} 1 & 0 & a \\ 0 & 1 & b \\ 0 & 0 & 1 \end{pmatrix}, \tag{4.4}$$

two translation matrices $T(a,b)$ and $T(a',b')$ commute with each other, and translations form a commutative or Abelian group. We are now interested in augmenting this translation group to the homogeneous linear canonical transformation discussed in Chapter 1.

In this three-by-three representation, the matrix performing the rotation around the origin by $\theta/2$ takes the form

$$R(\theta) = \begin{pmatrix} \cos(\theta/2) & -\sin(\theta/2) & 0 \\ \sin(\theta/2) & \cos(\theta/2) & 0 \\ 0 & 0 & 1 \end{pmatrix}. \tag{4.5}$$

Rotations form a one-parameter group. This rotation matrix can be multiplied by the translation matrix of Eq. 4.4 or vice versa, but the two matrices do not commute with each other, as can be seen from

$$T(a,b)R(\theta) = \begin{pmatrix} \cos(\theta/2) & -\sin(\theta/2) & a \\ \sin(\theta/2) & \cos(\theta/2) & b \\ 0 & 0 & 1 \end{pmatrix},$$

$$R(\theta)T(a,b) = \begin{pmatrix} \cos(\theta/2) & -\sin(\theta/2) & a' \\ \sin(\theta/2) & \cos(\theta/2) & b' \\ 0 & 0 & 1 \end{pmatrix}, \tag{4.6}$$

with

$$\begin{pmatrix} a' \\ b' \end{pmatrix} = \begin{pmatrix} \cos(\theta/2) & -\sin(\theta/2) \\ \sin(\theta/2) & \cos(\theta/2) \end{pmatrix}\begin{pmatrix} a \\ b \end{pmatrix}. \tag{4.7}$$

The group of transformations in two-dimensional space consisting of rotations and translations is called the two-dimensional Euclidean group, and occupies an important place in the theory of massless particles (Wigner 1939, Kim and Noz 1986). As we shall see in Chapter 5, this group is the fundamental language for the coherent state representation.

As was noted in Chapter 1, it is also possible to squeeze the system along an arbitrary direction $\phi/2$. If we write this matrix as $S(\phi, \eta)$, then $S(0, \eta)$ takes the form

$$S(0,\eta) = \begin{pmatrix} e^{\eta/2} & 0 & 0 \\ 0 & e^{-\eta/2} & 0 \\ 0 & 0 & 1 \end{pmatrix}. \tag{4.8}$$

The elongation along the $x$ axis is necessarily the contraction along the $p$ axis to preserve the area in phase space. For the squeeze along the $\phi/2$ direction, the first two-by-two portion of the above matrix is replaced by the two-by-two matrix of Eq. 1.39 with $\lambda$ replaced by $\eta$.

Since a canonical transformation followed by another one is also a canonical transformation, the most general form of the transformation matrix is a product of the above three forms of matrices. We can simplify this mathematics by using the *generators* of the transformation matrices. The translation matrix $T(a, b)$ given in Eq. 4.4 can be written as

$$T(u,v) = e^{-i(aN_1 + bN_2)}, \tag{4.9}$$

where

$$N_1 = \begin{pmatrix} 0 & 0 & i \\ 0 & 0 & 0 \\ 0 & 0 & 0 \end{pmatrix}, \quad N_2 = \begin{pmatrix} 0 & 0 & 0 \\ 0 & 0 & i \\ 0 & 0 & 0 \end{pmatrix}.$$

These generators commute with each other:

$$[N_1, N_2] = 0. \tag{4.10}$$

The rotation matrix is generated by

$$L = \begin{pmatrix} 0 & -i/2 & 0 \\ i/2 & 0 & 0 \\ 0 & 0 & 0 \end{pmatrix}, \tag{4.11}$$

and

$$R(\theta) = e^{-i\theta L}. \tag{4.12}$$

This generator satisfies the following commutation relations with $N_1$ and $N_2$:

$$[N_1, L] = (i/2)N_2, \qquad [N_2, L] = (-i/2)N_1. \tag{4.13}$$

Indeed, $L$, $N_1$, and $N_2$ satisfy closed commutation relations. They generate the two-dimensional Euclidean group consisting of rotations and translations in a two-dimensional space (Wigner 1939, Han *et al.* 1982).

The squeeze matrix of Eq. 4.8 can be written as

$$S(0,\eta) = e^{-i\eta K_1}, \tag{4.14}$$

with

$$K_1 = \begin{pmatrix} i/2 & 0 & 0 \\ 0 & -i/2 & 0 \\ 0 & 0 & 0 \end{pmatrix}.$$

In addition, if we introduce the matrix $K_2$ defined as

$$K_2 = \begin{pmatrix} 0 & i/2 & 0 \\ i/2 & 0 & 0 \\ 0 & 0 & 0 \end{pmatrix}, \tag{4.15}$$

which generates a squeeze along the direction which makes a 45° angle with the $x$ axis, then the matrices $L$, $K_1$ and $K_2$ satisfy the following commutation relations.

$$[K_1, K_2] = -iL, \quad [K_1, L] = -iK_2, \quad [K_2, L] = iK_1. \tag{4.16}$$

This set of commutation relations is identical to that for the generators of the (2 + 1)-dimensional Lorentz group (Kim and Noz 1983), as will be seen in Section 4.5. The group generated by the above three operators is known also as the symplectic group Sp(2) (Guilemin and Sternberg 1984, Perelomov 1986), and its connection with the Lorentz group has been extensively discussed in the literature (Kim and Noz 1983, Han *et al.* 1988 and 1989a).

If we take into account the translation operators, the commutation relations become

$$[K_1, N_1] = \left(\frac{i}{2}\right) N_1, \quad [K_1, N_2] = \left(-\frac{i}{2}\right) N_2$$

$$[K_2, N_1] = \left(\frac{i}{2}\right) N_2, \quad [K_2, N_2] = \left(\frac{i}{2}\right) N_1. \tag{4.17}$$

These commutators together with those of Eq. 4.16 form the set of closed commutation relations (or Lie algebra) of the group of canonical transformations. This group is the inhomogeneous symplectic group in the two dimensions or ISp(2) (Han *et al.* 1988).

## 4.2 Linear Canonical Transformations in Quantum Mechanics

Since the Wigner function is real and defined over the phase space of $x$ and $p$, we can perform area-preserving canonical transformations as in the case of classical mechanics. Let us consider first linear canonical transformations applicable to a function of $x$ and $p$.

The generators of translations are

$$N_1 = -i\frac{\partial}{\partial x}, \quad N_2 = -i\frac{\partial}{\partial p}, \tag{4.18}$$

while rotations around the origin are generated by

$$L = \frac{i}{2}\left(p\frac{\partial}{\partial x} - x\frac{\partial}{\partial p}\right). \tag{4.19}$$

The squeezes along the $x$ axis and along the direction which makes an angle of 45° with it are generated by

$$K_1 = -\left(\frac{i}{2}\right)\left(x\frac{\partial}{\partial x} - p\frac{\partial}{\partial p}\right), \quad K_2 = -\left(\frac{i}{2}\right)\left(x\frac{\partial}{\partial p} + p\frac{\partial}{\partial x}\right), \tag{4.20}$$

respectively. These operators satisfy the commutation relations for the generators of the group of linear canonical transformations given in Eq. 4.16 and Eq. 4.17. We can therefore continue using the matrix formalism of classical mechanics in the phase-space picture of quantum mechanics.

Next, let us study how these canonical transformations are translated into the language of the Schrödinger picture. In this picture, the wave function is a function of $x$ or $p$, but not of both. Since the Wigner function is derived from the Schrödinger wave function, we are interested in knowing what operations in the Schrödinger picture lead to the generators of canonical transformations of the Wigner function defined in phase space. For this purpose, we write the transformation on $W(x,p)$ as

$$e^{-i\varepsilon G(x,p)}W(x,p), \tag{4.21}$$

where $G$ is the generator of transformations applicable to the Wigner function, and $\varepsilon$ is the transformation parameter. We then write the corresponding transformation in the Schrödinger picture as

$$e^{-i\varepsilon\hat{G}(x)}\psi(x). \tag{4.22}$$

The operator $\hat{G}(x)$ depends only on $x$. For small $\varepsilon$,

$$G(x,p)W(x,p) = \frac{1}{\pi}\int e^{2ipy}\left\{\psi^*(x+y)\left(\hat{G}(x-y)\psi(x-y)\right) - \left(\hat{G}(x+y)\psi(x+y)\right)^*\psi(x-y)\right\}dy. \tag{4.23}$$

Then the operation of $\hat{G}(x) = i\left(x\frac{\partial}{\partial x} + \frac{\partial}{\partial x}x\right)$ on $\psi(x)$ leads the right-hand-side of the above expression to

$$\frac{i}{\pi}\int e^{2ipy}\left\{\psi^*(x+y)\left((x-y)\frac{\partial}{\partial x}\psi(x-y) + \frac{\partial}{\partial x}(x-y)\psi(x-y)\right) + \left((x+y)\frac{\partial}{\partial x}\psi(x+y) + \frac{\partial}{\partial x}(x+y)\psi(x+y)\right)^*\psi(x-y)\right\}dy$$

$$= \frac{i}{\pi} \int e^{2ipy} \left\{ 2x \frac{\partial}{\partial x} (\psi^*(x+y)\psi(x-y)) \right.$$

$$\left. + 2y \frac{\partial}{\partial y} (\psi^*(x+y)\psi(x-y)) \right\} dy. \tag{4.24}$$

In the above expression, $y\frac{\partial}{\partial y}$ can be replaced by $-p\frac{\partial}{\partial p}$. Thus the net effect is the application of $2i\left(x\frac{\partial}{\partial x} - p\frac{\partial}{\partial p}\right)$ to the Wigner function. This means that the operation of $K_1$ of Eq. 4.20 in phase space corresponds to

$$\hat{K}_1 = -\left(\frac{i}{4}\right)\left(x\frac{\partial}{\partial x} + \frac{\partial}{\partial x}x\right). \tag{4.25}$$

Under the multiplication of $\psi(x)$ by $x^2$ , the right-hand side of Eq. 4.23 becomes

$$\frac{1}{\pi} \int e^{2ipy} \left((x-y)^2 - (x+y)^2\right) \psi^*(x+y)\psi(x-y)dy$$

$$= \frac{1}{\pi} \int e^{2ipy}(-4xy)\psi^*(x+y)\psi(x-y)dy. \tag{4.26}$$

Thus $x^2$ on the Schrödinger wave function leads to $2ix\frac{\partial}{\partial p}$. Likewise, the operation of $\left(\frac{\partial}{\partial x}\right)^2$ becomes $2ip\frac{\partial}{\partial x}$. This means that $L$ of Eq. 4.19 and $K_2$ of Eq. 4.20 will lead to

$$\hat{L} = \frac{1}{4}\left(\left(\frac{\partial}{\partial x}\right)^2 - x^2\right), \quad \hat{K}_2 = -\left(\frac{1}{4}\right)\left(x^2 + \left(\frac{\partial}{\partial x}\right)^2\right), \tag{4.27}$$

respectively. The generators applicable to $\psi(x)$ satisfy the same set of commutation relations as Eq. 4.16. Here we write them as

$$[\hat{K}_1, \hat{K}_2] = -i\hat{L}, \quad [\hat{K}_1, \hat{L}] = -i\hat{K}_2, \quad [\hat{K}_2, \hat{L}] = i\hat{K}_1. \tag{4.28}$$

Next, it is straightforward to show that the translation generators $N_1 = -i\frac{\partial}{\partial x}$ and $N_2 = -i\frac{\partial}{\partial p}$ become

$$\hat{N}_1 = -i\frac{\partial}{\partial x}, \quad \hat{N}_2 = x, \tag{4.29}$$

respectively in the Schrödinger picture. Here again, we can reproduce the commutation relations of Eq. 4.17:

$$[\hat{K}_1, \hat{N}_1] = (i/2)\hat{N}_1, \quad [\hat{K}_1, \hat{N}_2] = (-i/2)\hat{N}_2,$$

$$[\hat{K}_2, \hat{N}_1] = (i/2)\hat{N}_2, \quad [\hat{K}_2, \hat{N}_2] = (i/2)\hat{N}_1. \tag{4.30}$$

However, the operators $\hat{N}_1$ and $\hat{N}_2$ do not commute with each other:

$$[\hat{N}_1, \hat{N}_2] = -i, \tag{4.31}$$

while $N_1$ and $N_2$ commute with each other in phase space. This causes a factor of modulus unity when the translation along $p$ is commuted with the translation along the $x$ direction, and this cannot be explained in terms of canonical transformations in classical mechanics.

## 4.3 Wave Packet Spreads in Terms of Canonical Transformations

There is yet another set of area-conserving linear canonical transformations:

$$\begin{pmatrix} x' \\ p' \\ 1 \end{pmatrix} = \begin{pmatrix} 1 & \alpha & 0 \\ 0 & 1 & 0 \\ 0 & 0 & 1 \end{pmatrix} \begin{pmatrix} x \\ p \\ 1 \end{pmatrix},$$

$$\begin{pmatrix} x' \\ p' \\ 1 \end{pmatrix} = \begin{pmatrix} 1 & 0 & 0 \\ \beta & 1 & 0 \\ 0 & 0 & 1 \end{pmatrix} \begin{pmatrix} x \\ p \\ 0 \end{pmatrix}. \tag{4.32}$$

These are shears which transform a rectangle into a parallelogram.

The wave packet spread discussed in Section 3.3 can be represented by the linear transformation:

$$\begin{pmatrix} x(t) \\ p(t) \\ 1 \end{pmatrix} = \begin{pmatrix} 1 & t/m & 0 \\ 0 & 1 & 0 \\ 0 & 0 & 1 \end{pmatrix} \begin{pmatrix} x \\ p \\ 0 \end{pmatrix}. \tag{4.33}$$

This is a shear transformation, as is clearly illustrated in Figure 3.1.

It is of interest to see whether these shears can be represented in terms of the rotations and squeezes discussed in Section 4.2. The above shears are generated by

$$F_1 = \begin{pmatrix} 0 & i & 0 \\ 0 & 0 & 0 \\ 0 & 0 & 0 \end{pmatrix}, \quad F_2 = \begin{pmatrix} 0 & 0 & 0 \\ i & 0 & 0 \\ 0 & 0 & 0 \end{pmatrix}. \tag{4.34}$$

It is clear that these shear generators can be obtained from $L$ and $K_2$:

$$F_1 = K_2 - L, \quad F_2 = K_2 + L. \tag{4.35}$$

Thus, the shear is generated from repeated applications of rotations and squeezes.

Another way to establish a connection between the shears and the group generators $K_1$, $K_2$ and $L$ is to consider the transformation of $K_2$:

$$K_1' = 2\left(e^{-\eta}\right)\left(e^{-i\eta K_1}\right) K_2 \left(e^{i\eta K_1}\right), \tag{4.36}$$

with

$$e^{-i\eta K_1} = \begin{pmatrix} e^{\eta/2} & 0 & 0 \\ 0 & e^{-\eta/2} & 0 \\ 0 & 0 & 1 \end{pmatrix}.$$

Since $e^{-i\eta K_1}$ is a canonical transformation, $K_1'$ is a generator of a canonical transformation. After matrix algebra,

$$K_1' = \begin{pmatrix} 0 & i & 0 \\ ie^{-2\eta} & 0 & 0 \\ 0 & 0 & 0 \end{pmatrix}. \tag{4.37}$$

In the limit of large $\eta$, $K_1'$ becomes $F_1$. Likewise, $F_2$ can be obtained from the large-$\eta$ limit of

$$K_1'' = 2\left(e^{-\eta}\right)\left(e^{i\eta K_1}\right)K_2\left(e^{-i\eta K_1}\right). \tag{4.38}$$

This limiting process will be discussed further in Chapter 10 in connection with space-time symmetries of relativistic particles.

## 4.4 Harmonic Oscillators

The Wigner function for the one-dimensional harmonic oscillator has been discussed in Section 3.4. It was shown there that there is a canonical coordinate system in which the ground-state Wigner function takes a simple Gaussian form and is concentrated within a circular region: $x^2 + p^2 < 1$. It was shown further that, in this coordinate system, all excited-state Wigner functions are rotationally invariant. In this section, we are interested in canonical transformations starting from one of these rotationally invariant states. Thus, for all practical purposes, we are making canonical transformations of the unit circle centered around the origin in phase space:

$$x^2 + p^2 = 1. \tag{4.39}$$

If the center is at $(q,p) = (a,b)$, the equation is

$$(x-a)^2 + (p-b)^2 = 1. \tag{4.40}$$

Certainly, this circle is a translation of the circle of Eq. 4.39.

With this point in mind, we can drop the translation from our consideration, and write the generators of the homogeneous symplectic group in the two-by-two matrix form:

$$K_1 = \begin{pmatrix} i/2 & 0 \\ 0 & -i/2 \end{pmatrix}, \quad K_2 = \begin{pmatrix} 0 & i/2 \\ i/2 & 0 \end{pmatrix},$$
$$L = \begin{pmatrix} 0 & -i/2 \\ i/2 & 0 \end{pmatrix}. \tag{4.41}$$

These generators satisfy the commutation relations of Eq. 4.16. From these generators, we can construct the squeeze and rotation matrices:

$$R(\theta) = e^{-i\theta L}, \quad S(0,\lambda) = e^{-i\lambda K_1}, \quad S(\pi/2,\eta) = e^{-i\eta K_2}. \tag{4.42}$$

The squeeze along the $\theta/2$ direction is represented by

$$S(\theta,\lambda) = R(\theta)S(0,\lambda)R(-\theta)$$
$$= \begin{pmatrix} \cosh(\lambda/2) + (\cos\theta)\sinh(\lambda/2) & (\sin\theta)\sinh(\lambda/2) \\ (\sin\theta)\sinh(\lambda/2) & \cosh(\lambda/2) - (\cos(\theta)\sinh(\lambda/2) \end{pmatrix}, \tag{4.43}$$

as was shown in Eq. 1.39 for canonical transformation in classical mechanics.

If we perform the rotation $R(\theta)$ on the circle of Eq. 4.39 centered around the origin, it remains invariant. If the same rotation is applied to the circle of Eq. 4.40 which is not centered around the origin, its effect is

$$(x - a')^2 + (p - b')^2 = 1, \tag{4.44}$$

where

$$\begin{pmatrix} a' \\ b' \end{pmatrix} = \begin{pmatrix} \cos(\theta/2) & -\sin(\theta/2) \\ \sin(\theta/2) & \cos(\theta/2) \end{pmatrix} \begin{pmatrix} a \\ b \end{pmatrix}.$$

This is a reflection of the transformation property given in Eq. 4.6.

Under the transformation of $S(\theta, \lambda)$, the circle of Eq. 4.39 becomes a tilted ellipse:

$$e^{-\lambda}\left(x\cos\frac{\theta}{2} + p\sin\frac{\theta}{2}\right)^2 + e^{\lambda}\left(x\sin\frac{\theta}{2} - p\cos\frac{\theta}{2}\right)^2 = 1, \tag{4.45}$$

and the circle of Eq. 4.40 becomes a tilted and displaced ellipse:

$$\begin{aligned} e^{-\lambda}&\left((x - a'')\cos\frac{\theta}{2} + (p - b'')\sin\frac{\theta}{2}\right)^2 \\ &+ e^{\lambda}\left((x - a'')\sin\frac{\theta}{2} - (p - b'')\cos\frac{\theta}{2}\right)^2 = 1, \end{aligned} \tag{4.46}$$

with

$$\begin{pmatrix} a'' \\ b'' \end{pmatrix} = \begin{pmatrix} \cosh\frac{\lambda}{2} + (\cos\theta)\sinh\frac{\lambda}{2} & (\sin\theta)\sinh\frac{\lambda}{2} \\ (\sin\theta)\sinh\frac{\lambda}{2} & \cosh\frac{\lambda}{2} - (\cos\theta)\sinh\frac{\lambda}{2} \end{pmatrix} \begin{pmatrix} a \\ b \end{pmatrix}.$$

These are of course area-preserving transformations. Eq. 4.45 is a special case of Eq. 4.46.

Let us next consider repeated squeezes. If we apply $S(\theta, \lambda)$ after $S(0, \eta)$ on the circle centered around the origin, the net effect is another tilted ellipse which can be obtained from the operation of $S(\phi, \xi)$ on the circle, with (Han *et al.* 1988)

$$\begin{aligned} \cosh\xi &= (\cosh\eta)\cosh\lambda + (\sinh\eta)(\sinh\lambda)\cos\theta, \\ \tan\phi &= \frac{(\sin\theta)[\sinh\lambda + (\tanh\eta)(\cosh\lambda - 1)\cos\theta]}{(\sinh\lambda)\cos\theta + (\tanh\eta)[1 + (\cosh\lambda - 1)(\cos\theta)^2]}. \end{aligned} \tag{4.47}$$

This means that the resulting ellipse is that of Eq. 4.46 where $\theta$ and $\lambda$ are replaced by $\phi$ and $\xi$ respectively. Thus we are tempted to conclude that $S(\theta, \lambda)S(0, \eta) = S(\phi, \xi)$ which leads to $[S(\phi, \xi)]^{-1}S(\theta, \lambda)S(0, \eta) = I$, where $I$ is the two-by-two identity matrix. *This is not correct!* The matrix multiplication of the left-hand side gives

$$[S(\phi, \xi)]^{1}S(\theta, \lambda)S(0, \eta) = R(\Omega), \tag{4.48}$$

where

$$\tan\frac{\Omega}{2} = \frac{(\sin\theta)[\tanh(\lambda/2)][\tanh(\eta/2)]}{1+[\tanh(\lambda/2)][\tanh(\eta/2)](\cos\theta)}.$$

Since the circle centered at the origin is invariant under rotations, the effect of $R(\Omega)$ is the same as the identity transformation.

Indeed, if we apply the above three successive squeezes on the circle of Eq. 4.28 not centered at the origin, the net effect is

$$(x-f)^2 + (p-g)^2 = 1, \tag{4.49}$$

where

$$\begin{pmatrix} f \\ g \end{pmatrix} = \begin{pmatrix} \cos(\Omega/2) & -\sin(\Omega/2) \\ \sin(\Omega/2) & \cos(\Omega/2) \end{pmatrix} \begin{pmatrix} a \\ b \end{pmatrix}.$$

This is clearly a rotation. We shall continue the discussion of this angle in Section 4.5 in connection with the Wigner rotation.

The canonical transformation of the ground-state harmonic oscillator provides the basic mathematical language for coherent and squeezed states of light, which will be discussed extensively in Chapter 5.

## 4.5 (2 + 1)-Dimensional Lorentz Group

Lorentz transformations on the coordinate system of $(x, y, t)$ can be represented by three-by-three matrices. The Lorentz boost along the $x$ direction takes the form

$$S_x(\eta) = \begin{pmatrix} \cosh\eta & 0 & \sinh\eta \\ 0 & 1 & 0 \\ \sinh\eta & 0 & \cosh\eta \end{pmatrix}. \tag{4.50}$$

The boost matrix along the $y$ direction is

$$S_y(\lambda) = \begin{pmatrix} 1 & 0 & 0 \\ 0 & \cosh\lambda & \sinh\lambda \\ 0 & \sinh\lambda & \cosh\lambda \end{pmatrix}. \tag{4.51}$$

The rotation around the z axis is represented by

$$R(\theta) = \begin{pmatrix} \cos\theta & -\sin\theta & 0 \\ \sin\theta & \cos\theta & 0 \\ 0 & 0 & 1 \end{pmatrix}. \tag{4.52}$$

These matrices can be written in terms of their respective generators as

$$S_x(\eta) = e^{-i\eta K_1}, \quad S_y(\lambda) = e^{-i\eta K_2}, \quad R(\theta) = e^{-i\eta L}, \tag{4.53}$$

where

$$K_1 = \begin{pmatrix} 0 & 0 & i \\ 0 & 0 & 0 \\ i & 0 & 0 \end{pmatrix}, \quad K_2 = \begin{pmatrix} 0 & 0 & 0 \\ 0 & 0 & i \\ 0 & i & 0 \end{pmatrix},$$

$$L = \begin{pmatrix} 0 & -i & 0 \\ i & 0 & 0 \\ 0 & 0 & 0 \end{pmatrix}. \tag{4.54}$$

Then these generators satisfy

$$[K_1, K_2] = -iL, \quad [K_1, L] = -iK_2, \quad [K_2, L] = iK_1. \tag{4.55}$$

This set of commutation relations is identical to that given in Eq. 4.16. The (2 + 1)-dimensional Lorentz group is locally isomorphic to Sp(2) or the group of homogeneous linear canonical transformations in two-dimensional phase space. It is thus possible to carry out calculations of the Lorentz group using the two-by-two matrix representation of Sp(2) as is shown in Section 4.4. It is also quite appropriate to use the same notation for boosts and rotation. The rotation matrix $R(\theta)$ of Eq. 4.52 corresponds to the first two-by-two portion $R(\theta)$ given in Eq. 4.5 or the rotation matrix of Eq. 4.42.

The squeeze matrix $S(\theta, \lambda)$ of Eq. 4.43 corresponds to the boost along the direction which makes an angle $\theta$ with the $x$ axis. In this case, the boost matrix is

$$S(\theta, \lambda) = R(\theta)S(0, \lambda)R(-\theta)$$

$$= \begin{pmatrix} 1 + (\cosh\lambda - 1)\cos^2\theta & (\cosh\lambda - 1)(\sin\theta)\cos\theta & (\sinh\lambda)\cos\theta \\ (\cosh\lambda - 1)(\sin\theta)\cos\theta & 1 + (\cosh\lambda - 1)\sin^2\theta & (\sinh\lambda)\sin\theta \\ (\sinh\lambda)\cos\theta & (\sinh\lambda)\sin\theta & \cosh\lambda \end{pmatrix}. \tag{4.56}$$

The algebra of these matrices is the same as that of the two-by-two matrices discussed in Section 4.4. As a consequence, we can carry out coplanar kinematics using two-by-two matrices (Kim and Noz 1983). This is not unlike the case where the algebra of three-dimensional rotations can be carried out in the two-by-two Pauli matrix formalism.

Let us go back to the Lorentz group. In connection with the Thomas precession, Goldstein (1980) discusses in his book repeated Lorentz boosts in different directions. This means that we can use the formula of Eq. 4.48 for two-by-two matrices to calculate

$$[S(\phi, \xi)]^{-1} S(\theta, \lambda) S(0, \eta) = R(\Omega), \tag{4.57}$$

with $\phi$, $\xi$ and $\Omega$ given in Eq. 4.46 and Eq. 4.48. The matrices in the above expressions are now three-by-three Lorentz boost and rotation matrices.

The physical interpretation is also different from that of canonical transformations. If we start with a particle at rest, the right-hand side means a simple rotation.

Table 4.1: Table of generators of the groups which are locally isomorphic to the (2 + 1)-dimensional Lorentz group. The first row consists of the generators of homogeneous linear canonical transformations in phase space. The second line consists of those in the Schrödinger picture. The third row gives the generators of Sp(2), which is the matrix form for the first row. The fourth row gives the generators SU(1,1).

| Lorentz Group | $J_3$ | $K_1$ | $K_2$ |
|---|---|---|---|
| Phase-Space Picture | $\frac{i}{2}\left(p\frac{\partial}{\partial x} - x\frac{\partial}{\partial p}\right)$ | $\frac{i}{2}\left(p\frac{\partial}{\partial p} - x\frac{\partial}{\partial x}\right)$ | $-\frac{i}{2}\left(x\frac{\partial}{\partial p} + p\frac{\partial}{\partial x}\right)$ |
| Schrödinger Picture | $\frac{1}{4}\left(x^2 - \left(\frac{\partial}{\partial x}\right)^2\right)$ | $-\frac{i}{4}\left(x\frac{\partial}{\partial x} + \frac{\partial}{\partial x}x\right)$ | $\frac{1}{4}\left(x^2 + \left(\frac{\partial}{\partial x}\right)^2\right)$ |
| Sp(2), SL(2,R) | $\frac{1}{2}\begin{pmatrix} 0 & -i \\ i & 0 \end{pmatrix}$ | $\frac{1}{2}\begin{pmatrix} i & 0 \\ 0 & -i \end{pmatrix}$ | $\frac{1}{2}\begin{pmatrix} 0 & i \\ i & 0 \end{pmatrix}$ |
| SU(1,1) | $\frac{1}{2}\begin{pmatrix} 1 & 0 \\ 0 & -1 \end{pmatrix}$ | $\frac{1}{2}\begin{pmatrix} 0 & i \\ i & 0 \end{pmatrix}$ | $\frac{1}{2}\begin{pmatrix} 0 & 1 \\ -1 & 0 \end{pmatrix}$ |

The left-hand side means that this particle is boosted three times. It is first boosted along the $x$ axis. Next, it is boosted along the direction which makes an angle $\theta$ with the $x$ axis. It is then boosted in such a way that the particle is at rest. The net effect is a rotation. If the particle at rest is spherically symmetric with zero spin, the rotation matrix $R(\Omega)$ gives no effect. If on the other hand, the particle has a spin, then $R(\Omega)$ changes the direction of the spin. This rotation at the rest frame is known as the Wigner rotation (Ritus 1961, Chakrabarti 1964, Han *et al.* 1987a and 1987b).

Let us go back to the harmonic oscillator of Section 4.4. The spinless particle corresponds to the circle centered around the origin of phase space. This circle is invariant under rotations. The particle with spin corresponds to a circle whose center is not at the origin. This circle is not invariant under rotations.

The group Sp(2) can be represented by two-by-two real matrices. For this reason, it is sometimes called SL(2,$r$) which means the group of real linear matrices with unit determinant. The (2 + 1)-dimensional Lorentz group is known to be locally isomorphic to SU(1,1) generated by

$$K_1 = \frac{i}{2}\sigma_x, \quad K_2 = \frac{i}{2}\sigma_y, \quad L = \frac{i}{2}\sigma_z, \tag{4.58}$$

where $\sigma_x$, $\sigma_y$ , $\sigma_z$ are the Pauli spin matrices. These generators also satisfy the commutation relations given in Eq. 4.16 and Eq. 4.55. While both Sp(2) and SU(1,1) are represented by two-by-two matrices, Sp(2) is real while SU(1,1) is not. They are however unitarily equivalent. Table (4.1) lists the generators of the groups which are locally isomorphic to the (2 + 1)-dimensional Lorentz group.

## 4.6 Canonical Transformations in Four-Dimensional Phase Space

The generators of homogeneous linear canonical transformations in four-dimensional phase space are given in the four-by-four matrix representation in Section 1.5 of Chapter 1. Those matrices can be translated into the differential forms applicable to functions of $x_1$, $x_2$, $p_1$ and $p_2$:

$$
\begin{aligned}
J_1 &= -\left(\frac{i}{2}\right)\left\{\left(p_2\frac{\partial}{\partial x_1} + p_1\frac{\partial}{\partial x_2}\right) - \left(x_2\frac{\partial}{\partial p_1} + x_1\frac{\partial}{\partial p_2}\right)\right\}, \\
J_2 &= -\left(\frac{i}{2}\right)\left\{\left(x_1\frac{\partial}{\partial x_2} - x_2\frac{\partial}{\partial x_1}\right) + \left(p_1\frac{\partial}{\partial p_2} - p_2\frac{\partial}{\partial p_1}\right)\right\}, \\
J_3 &= -\left(\frac{i}{2}\right)\left\{\left(p_1\frac{\partial}{\partial x_1} - p_2\frac{\partial}{\partial x_2}\right) - \left(x_1\frac{\partial}{\partial p_1} - x_2\frac{\partial}{\partial p_2}\right)\right\}, \\
J_0 &= -\left(\frac{i}{2}\right)\left\{\left(p_1\frac{\partial}{\partial x_1} + p_2\frac{\partial}{\partial x_2}\right) - \left(x_1\frac{\partial}{\partial p_1} + x_2\frac{\partial}{\partial p_2}\right)\right\}, \\
K_1 &= -\left(\frac{i}{2}\right)\left\{\left(p_1\frac{\partial}{\partial x_1} - p_2\frac{\partial}{\partial x_2}\right) + \left(x_1\frac{\partial}{\partial p_1} - x_2\frac{\partial}{\partial p_2}\right)\right\}, \\
K_2 &= +\left(\frac{i}{2}\right)\left\{-\left(x_1\frac{\partial}{\partial x_1} + x_2\frac{\partial}{\partial x_2}\right) + \left(p_1\frac{\partial}{\partial p_1} + p_2\frac{\partial}{\partial p_2}\right)\right\}, \\
K_3 &= +\left(\frac{i}{2}\right)\left\{\left(p_2\frac{\partial}{\partial x_1} + p_1\frac{\partial}{\partial x_2}\right) + \left(x_2\frac{\partial}{\partial p_1} + x_1\frac{\partial}{\partial p_2}\right)\right\}, \\
Q_1 &= +\left(\frac{i}{2}\right)\left\{\left(x_1\frac{\partial}{\partial x_1} - x_2\frac{\partial}{\partial_2}\right) - \left(p_1\frac{\partial}{\partial p_1} - p_2\frac{\partial}{\partial p_2}\right)\right\}, \\
Q_2 &= -\left(\frac{i}{2}\right)\left\{\left(p_1\frac{\partial}{\partial x_1} + p_2\frac{\partial}{\partial x_2}\right) + \left(x_1\frac{\partial}{\partial p_1} + x_2\frac{\partial}{\partial p_2}\right)\right\}, \\
Q_3 &= -\left(\frac{i}{2}\right)\left\{\left(x_2\frac{\partial}{\partial x_1} + x_1\frac{\partial}{\partial x_2}\right) - \left(p_2\frac{\partial}{\partial p_1} + p_1\frac{\partial}{\partial p_2}\right)\right\}.
\end{aligned}
\tag{4.59}
$$

In addition, there are translations. Since, however, they can be augmented with homogeneous linear transformations as in the case of the one-pair of canonical variables, we shall concentrate our efforts on the homogeneous transformations.

We are interested in canonical transformations on the Wigner function of two pairs of canonical variables defined as

$$W(x_1, x_2, p_1, p_2) = \left(\frac{1}{\pi}\right)^2 \int \exp\{2i(p_1 y_1 + p_2 y_2)\} \psi^*(x_1 + y_1, x_2 + y_2)$$
$$\times \psi(x_1 - y_1, x_2 - y_2)\, dy_1 dy_2. \tag{4.60}$$

As in Section 4.4, let us study the transformation properties using the ground-state harmonic oscillator. The Wigner function for the two-oscillator system:

$$\psi(x_1, x_2) = \left(\frac{1}{\pi}\right)^{1/2} \exp\left\{-\left(\frac{1}{2}\right)\left(x_1^2 + x_2^2\right)\right\}. \tag{4.61}$$

Then the Wigner function will be

$$W(x_1, x_2, p_1, p_2) = \left(\frac{1}{\pi}\right)^2 \exp\left\{-\left(x_1^2 + x_2^2 + p_1^2 + p_2^2\right)\right\}. \tag{4.62}$$

We are now interested in performing rotations and squeezes with respect to two pairs of variables. There are three possible ways of choosing two pairs among the four variables for the above Wigner function:

$$\begin{aligned}
W(x_1, x_2, p_1, p_2) &= \left(\frac{1}{\pi}\right) \exp\left\{-\left(x_1^2 + p_1^2\right)\right\} \exp\left\{-\left(x_2^2 + p_2^2\right)\right\} \\
&= \left(\frac{1}{\pi}\right) \exp\left\{-\left(x_1^2 + p_2^2\right)\right\} \exp\left\{-\left(x_2^2 + p_1^2\right)\right\} \\
&= \left(\frac{1}{\pi}\right) \exp\left\{-\left(x_1^2 + x_2^2\right)\right\} \exp\left\{-\left(p_1^2 + p_2^2\right)\right\}.
\end{aligned} \tag{4.63}$$

It is thus convenient to write the generators given in Eq. 4.59 in terms of rotation or squeeze generators. There are four generators of rotations:

$$\begin{aligned}
J_1 &= +\left(\frac{i}{2}\right)\left\{\left(x_1 \frac{\partial}{\partial p_2} - p_2 \frac{\partial}{\partial x_1}\right) + \left(x_2 \frac{\partial}{\partial p_1} - p_1 \frac{\partial}{\partial x_2}\right)\right\}, \\
J_2 &= -\left(\frac{i}{2}\right)\left\{\left(x_1 \frac{\partial}{\partial x_2} - x_2 \frac{\partial}{\partial x_1}\right) + \left(p_1 \frac{\partial}{\partial p_2} - p_2 \frac{\partial}{\partial p_1}\right)\right\}, \\
J_3 &= +\left(\frac{i}{2}\right)\left\{\left(x_1 \frac{\partial}{\partial p_1} - p_1 \frac{\partial}{\partial x_1}\right) - \left(x_2 \frac{\partial}{\partial p_2} - p_2 \frac{\partial}{\partial x_2}\right)\right\}, \\
J_0 &= +\left(\frac{i}{2}\right)\left\{\left(x_1 \frac{\partial}{\partial p_1} - p_1 \frac{\partial}{\partial x_1}\right) + \left(x_2 \frac{\partial}{\partial p_2} - p_2 \frac{\partial}{\partial x_2}\right)\right\},
\end{aligned} \tag{4.64}$$

and there are six squeeze operators:

$$K_1 = -\left(\frac{i}{2}\right)\left\{\left(x_1\frac{\partial}{\partial p_1} - p_1\frac{\partial}{\partial x_1}\right) + \left(x_2\frac{\partial}{\partial p_2} - p_2\frac{\partial}{\partial x_2}\right)\right\},$$

$$K_2 = -\left(\frac{i}{2}\right)\left\{\left(x_1\frac{\partial}{\partial x_1} - p_1\frac{\partial}{\partial p_1}\right) + \left(x_2\frac{\partial}{\partial x_2} - p_2\frac{\partial}{\partial p_2}\right)\right\},$$

$$K_3 = +\left(\frac{i}{2}\right)\left\{\left(x_1\frac{\partial}{\partial p_2} - p_2\frac{\partial}{\partial x_1}\right) + \left(x_2\frac{\partial}{\partial p_1} - p_1\frac{\partial}{\partial x_2}\right)\right\},$$

$$Q_1 = +\left(\frac{i}{2}\right)\left\{\left(x_1\frac{\partial}{\partial x_1} - p_1\frac{\partial}{\partial p_1}\right) + \left(x_2\frac{\partial}{\partial x_2} - p_2\frac{\partial}{\partial p_2}\right)\right\},$$

$$Q_2 = -\left(\frac{i}{2}\right)\left\{\left(x_1\frac{\partial}{\partial p_1} - p_1\frac{\partial}{\partial x_1}\right) + \left(x_2\frac{\partial}{\partial p_2} - p_2\frac{\partial}{\partial x_2}\right)\right\},$$

$$Q_3 = -\left(\frac{i}{2}\right)\left\{\left(x_2\frac{\partial}{\partial x_1} - x_1\frac{\partial}{\partial x_2}\right) + \left(p_2\frac{\partial}{\partial p_1} - p_1\frac{\partial}{\partial p_2}\right)\right\}. \tag{4.65}$$

According to the above expression, we can separate the four-dimensional phase space into a pair of two-dimensional spaces, and perform canonical transformations in each space. If all the transformations are done within a given pair, the symmetry group is the same as the one in the two-dimensional phase space. Let us for example consider the following generators in the $(x_1p_1)$ and $(x_2p_2)$ spaces.

$$L^{(1)} = \left(\frac{i}{2}\right)\left(x_1\frac{\partial}{\partial p_1} - p_1\frac{\partial}{\partial x_1}\right), \quad L^{(2)} = \left(\frac{i}{2}\right)\left(x_2\frac{\partial}{\partial p_2} - p_2\frac{\partial}{\partial x_2}\right),$$

$$K_1^{(1)} = \left(\frac{i}{2}\right)\left(x_1\frac{\partial}{\partial p_1} - p_1\frac{\partial}{\partial x_1}\right), \quad K_1^{(2)} = \left(\frac{i}{2}\right)\left(x_2\frac{\partial}{\partial p_2} - p_2\frac{\partial}{\partial x_2}\right),$$

$$K_2^{(1)} = \left(\frac{i}{2}\right)\left(x_1\frac{\partial}{\partial x_1} - p_1\frac{\partial}{\partial p_1}\right), \quad K_2^{(2)} = \left(\frac{i}{2}\right)\left(x_2\frac{\partial}{\partial x_2} - p_2\frac{\partial}{\partial p_2}\right). \tag{4.66}$$

Then the canonical transformations are carried out separately in each phase space. If we combine them as

$$L = L^{(1)} + L^{(2)}, \quad B_1 = K_1^{(1)} - K_1^{(2)}, \quad B_2 = K_2^{(1)} - K_2^{(2)}, \tag{4.67}$$

then $L$, $B_1$ and $B_2$ are $J_0$, $K_1$ and $Q_1$ respectively. The transformations are carried out separately but simultaneously in both phase spaces, and the above generators satisfy the commutation relations for the single phase space given in Eq. 4.16. In the language of group theory, the group Sp(4) has as a subgroup the direct product of two Sp(2) groups.

The group Sp(4) is much richer than the above mentioned subgroup. As we shall see in Section 4.8, it is locally isomorphic to the (3 + 2)- dimensional Lorentz group. There is also a group locally isomorphic to Sp(4) in the Schrödinger picture of quantum mechanics, which will be discussed in Section 4.7.

## 4.7 The Schrödinger Picture of Two-Mode Canonical Transformations

It was noted in Section 4.2 that linear canonical transformations in phase space can be translated into unitary transformations on the Schrödinger wave function. We are therefore interested in establishing the connection between unitary transformations on $\psi(x_1, x_2)$ and canonical transformations on $W(x_1, x_2, p_1, p_2)$.

For this purpose, we can work with the generator $\hat{G}(x, x)$ in the Schrödinger picture and $G(x_1, x_2, p_1, p_2)$ applicable to the Wigner function. Then, the relation given for the single pair of canonical variables in Eq. 4.23 is valid also for these two pairs of canonical variables. We can see from Eq. 4.23 that the application of $\hat{G} = x_i x_j$ on $\psi(x_1, x_2)$ leads to

$$G = i\left(x_i \frac{\partial}{\partial p_j} + x_j \frac{\partial}{\partial p_i}\right) \tag{4.68}$$

on the Wigner function $W(x_1, x_2, p_1, p_2)$. This is valid also for $i = j$. The operator

$$\hat{G} = \left(-i\frac{\partial}{\partial x_i}\right)\left(-i\frac{\partial}{\partial x_j}\right) \tag{4.69}$$

applicable to $\psi(x_1, x_2)$ corresponds to

$$G = -i\left(p_i \frac{\partial}{\partial x_j} + p_j \frac{\partial}{\partial x_i}\right) \tag{4.70}$$

in phase space. The operation of

$$\hat{G} = -\left(\frac{i}{2}\right)\left(x_i \frac{\partial}{\partial x_j} + \frac{\partial}{\partial x_j} x_i\right) \tag{4.71}$$

in the Schrödinger picture leads to

$$G = -i\left(x_i \frac{\partial}{\partial x_j} - p_j \frac{\partial}{\partial p_i}\right) \tag{4.72}$$

applicable to the Wigner function.

It is now possible to translate the operators in the phase-space picture given in Eq. 4.64 and Eq. 4.65 into those applicable to the Schrödinger wave function. They take the form (Dirac 1963)

$$\hat{J}_1 = +\frac{1}{2}\left(x_1 x_2 + \hat{p}_2 \hat{p}_1\right), \qquad \hat{J}_2 = \frac{1}{2}\left(x_1 \hat{p}_2 - x_2 \hat{p}_1\right),$$

$$\hat{J}_3 = +\frac{1}{4}\left\{\left(x_1 x_1 + \hat{p}_1 \hat{p}_1\right) - \left(x_2 x_2 + \hat{p}_2 \hat{p}_2\right)\right\},$$

$$\hat{J}_0 = +\frac{1}{4}\{(x_1x_1 + \hat{p}_1\hat{p}_1) + (x_2x_2 + \hat{p}_2\hat{p}_2)\},$$

$$\hat{K}_1 = -\frac{1}{4}\{(x_1x_1 - \hat{p}_1\hat{p}_1) - (x_2x_2 - \hat{p}_2\hat{p}_2)\},$$

$$\hat{K}_2 = +\frac{1}{4}(x_1\hat{p}_1 + \hat{p}_1x_1 + x_2\hat{p}_2 + \hat{p}_2x_2),$$

$$\hat{K}_3 = +\frac{1}{2}(x_1x_2 - \hat{p}_1\hat{p}_2),$$

$$\hat{Q}_1 = -\frac{1}{4}(x_1\hat{p}_1 + \hat{p}_1x_1 - x_2\hat{p}_2 - \hat{p}_2x_2),$$

$$\hat{Q}_2 = -\frac{1}{4}\{(x_1x_1 - \hat{p}_1\hat{p}_1) + (x_2x_2 - \hat{p}_2\hat{p}_2)\},$$

$$\hat{Q}_3 = +\frac{1}{2}(x_1\hat{p}_2 + x_2\hat{p}_1), \tag{4.73}$$

where $\hat{p}_1 = -i\frac{\partial}{\partial x_1}$, and $\hat{p}_2 = -i\frac{\partial}{\partial x_2}$.

We shall continue the discussion of these generators in Chapter 6 in connection with two-mode squeezed states. It is interesting to note that the group Sp(4) is locally isomorphic to the (3 + 2)-dimensional Lorentz group which is commonly known as the (3 + 2)-dimensional de Sitter group. We shall study this isomorphism in Section 4.8.

## 4.8 (3 + 2)-Dimensional de Sitter Group

In the usual (3 + 1)-dimensional Lorentz group, transformations on the vector space $(x, y, z, t)$ leave $x^2 + y^2 + z^2 - t^2$ invariant. In the (3 + 2)-dimensional de Sitter group, transformations on the vector space $(x, y, z, s, t)$ leave the quantity

$$x^2 + y^2 + z^2 - s^2 - t^2 \tag{4.74}$$

invariant. In this space, there are two time-like variables.

There are three generators of rotations in the space of $x$, $y$, and $z$. They are

$$J_1 = \begin{pmatrix} 0 & 0 & 0 & 0 & 0 \\ 0 & 0 & -i & 0 & 0 \\ 0 & i & 0 & 0 & 0 \\ 0 & 0 & 0 & 0 & 0 \\ 0 & 0 & 0 & 0 & 0 \end{pmatrix}, \quad J_2 = \begin{pmatrix} 0 & 0 & i & 0 & 0 \\ 0 & 0 & 0 & 0 & 0 \\ -i & 0 & 0 & 0 & 0 \\ 0 & 0 & 0 & 0 & 0 \\ 0 & 0 & 0 & 0 & 0 \end{pmatrix},$$

$$J_3 = \begin{pmatrix} 0 & -i & 0 & 0 & 0 \\ i & 0 & 0 & 0 & 0 \\ 0 & 0 & 0 & 0 & 0 \\ 0 & 0 & 0 & 0 & 0 \\ 0 & 0 & 0 & 0 & 0 \end{pmatrix}. \tag{4.75}$$

These generators satisfy the commutation relations for the three- dimensional rotation group:

$$[J_i, J_j] = i\varepsilon_{ijk}J_k. \tag{4.76}$$

In addition, it is possible to perform rotations in the plane of $t$ and $s$, generated by

$$J_0 = \begin{pmatrix} 0 & 0 & 0 & 0 & 0 \\ 0 & 0 & 0 & 0 & 0 \\ 0 & 0 & 0 & 0 & 0 \\ 0 & 0 & 0 & 0 & -i \\ 0 & 0 & 0 & i & 0 \end{pmatrix}. \tag{4.77}$$

This matrix commute with the generators of the three-dimensional rotation group given in Eq. 4.75:

$$[J_0, J_i] = 0. \tag{4.78}$$

There are three generators of boosts with respect to three space-like directions and the first time variable $t$. They are

$$K_1 = \begin{pmatrix} 0 & 0 & 0 & i & 0 \\ 0 & 0 & 0 & 0 & 0 \\ 0 & 0 & 0 & 0 & 0 \\ i & 0 & 0 & 0 & 0 \\ 0 & 0 & 0 & 0 & 0 \end{pmatrix}, \quad K_2 = \begin{pmatrix} 0 & 0 & 0 & 0 & 0 \\ 0 & 0 & 0 & i & 0 \\ 0 & 0 & 0 & 0 & 0 \\ 0 & i & 0 & 0 & 0 \\ 0 & 0 & 0 & 0 & 0 \end{pmatrix},$$

$$K_3 = \begin{pmatrix} 0 & 0 & 0 & 0 & 0 \\ 0 & 0 & 0 & 0 & 0 \\ 0 & 0 & 0 & i & 0 \\ 0 & 0 & i & 0 & 0 \\ 0 & 0 & 0 & 0 & 0 \end{pmatrix}. \tag{4.79}$$

These generators satisfy the commutation relations for the (3 + 1)- dimensional Lorentz group:

$$[K_i, K_j] = -i\varepsilon_{ijk}J_k, \quad [K_i, J_j] = i\varepsilon_{ijk}K_k. \tag{4.80}$$

In addition, there are three boosts with respect to the second time variable $s$. They are generated by

$$Q_1 = \begin{pmatrix} 0 & 0 & 0 & 0 & i \\ 0 & 0 & 0 & 0 & 0 \\ 0 & 0 & 0 & 0 & 0 \\ 0 & 0 & 0 & 0 & 0 \\ i & 0 & 0 & 0 & 0 \end{pmatrix}, \quad Q_2 = \begin{pmatrix} 0 & 0 & 0 & 0 & 0 \\ 0 & 0 & 0 & 0 & i \\ 0 & 0 & 0 & 0 & 0 \\ 0 & 0 & 0 & 0 & 0 \\ 0 & i & 0 & 0 & 0 \end{pmatrix},$$

$$Q_3 = \begin{pmatrix} 0 & 0 & 0 & 0 & 0 \\ 0 & 0 & 0 & 0 & 0 \\ 0 & 0 & 0 & 0 & i \\ 0 & 0 & 0 & 0 & 0 \\ 0 & 0 & i & 0 & 0 \end{pmatrix}. \tag{4.81}$$

These generators also satisfy the commutation relations for the (3 + 1)- dimensional Lorentz group:

$$[Q_i, Q_j] = -i\epsilon_{ijk}J_k, \quad [Q_i, J_j] = i\epsilon_{ijk}Q_k. \tag{4.82}$$

These Q matrices satisfy the following three sets of commutation relations, with $K_i$ and $J_0$.

$$[K_i, J_0] = iQ_i, \quad [Q_i, J_0] = iK_i, \quad [K_i, Q_j] = i\delta_{ij}J_0, \tag{4.83}$$

which are like those for the (2 + 1)-dimensional Lorentz group. Indeed, they generate such a group. However, they are not usual transformations with two space-like coordinates and one time-like coordinate. Instead, they represent transformations with two time-like coordinates and one space-like direction. This is characteristic of the O(3,2) group which has two time-like directions.

# Chapter 5

# COHERENT AND SQUEEZED STATES

The harmonic oscillator serves many different purposes in modern physics. It describes a bound state with equal-spaced energy levels governed by the position-momentum uncertainty relation, known as the first quantization. It is also the basic language for second quantization which is the theoretical framework for creation and annihilation of particles. That Hilbert space which consists of the harmonic oscillator states whose $n^{th}$ excited state corresponds to the state of n particles is called the Fock space. Quantum field theory is based on this Fock space.

However, since we deal with a small number of particles in quantum field theory, there is no need for studying the uncertainty relation in Fock space beyond the equal spacing rule for the oscillator energy levels. The story is quite different for modern optics. In quantum optics, states consisting of one or two photons are important, as in the case of quantum electrodynamics. However, equally important are both coherent and incoherent mixtures of multi-photon states.

It is known that the uncertainty product is minimum for the ground-state harmonic oscillator or for the zero-photon state in Fock-space language. The coherent state is a superposition of the multi-photon state which preserves the minimality in uncertainty, and which has a Poisson distribution in photon numbers. Since it has the minimum uncertainty, the wave function in Fock space takes a Gaussian form. The Gaussian form can change its width without altering the uncertainty product (Stoler 1970). This change in width may change the Poisson distribution to a different distribution. The minimum-uncertainty states which do not have a Poisson distribution in photon number are called squeezed states.

In this Chapter, we shall discuss first in Section 5.1 the uncertainty relation in Fock space which governs the creation and annihilation of particles. The algebra of these operators leads to the conventional formalism of coherent and squeezed states which does not rely on the Wigner function. Much of this conventional formalism relies on the mathematical theorem known as the Baker-Campbell-Hausdorff formula and accompanying lemmas. This theorem is studied in detail in Section 5.2. In Section 5.3 - 5.5, we present the conventional formalism for coherent and squeezed states. Section 5.6 deals with the decoherence process (Hartle 1990) in which the failure to observe one mode in a two-mode states leads to the situation like an incoherent thermal excitation.

## 5.1 Phase-Number Uncertainty Relation

The concept of free photons is well established in terms of their creation and annihilation operators. The creation and annihilation can be defined in terms of the mathematics describing the energy levels of the harmonic oscillator. The oscillator in this case is not a mechanical oscillator, but is defined in the space commonly known as the Fock space. The creation and annihilation of photons are therefore manifestations of the uncertainty relations in Fock space. Let the "spatial" coordinate in this Fock space be $q$. Then we can write the annihilation and creation operators as

$$\hat{a} = \left(1/\sqrt{2}\right)\left(q + \frac{\partial}{\partial q}\right), \quad \hat{a}^{\dagger} = \left(1/\sqrt{2}\right)\left(q - \frac{\partial}{\partial q}\right), \tag{5.1}$$

respectively.

The uncertainty principle in the Fock space is represented by

$$[\hat{a}, \hat{a}^{\dagger}] = 1, \quad [\hat{a}, \hat{a}] = [\hat{a}^{\dagger}, \hat{a}^{\dagger}] = 0. \tag{5.2}$$

If $|n>$ is the state of $n$ photons,

$$\hat{a}|n> = \sqrt{n}|n-1>, \quad \hat{a}^{\dagger}|n> = \sqrt{n+1}|n+1>. \tag{5.3}$$

From these relations, it is possible to define the number operator

$$\hat{N} = \hat{a}^{\dagger}\hat{a}, \tag{5.4}$$

with $\hat{N}|n> = n|n>$.

Since the photon number is the quantity we observe in the real world, let us write $\hat{a}$ and $\hat{a}^{\dagger}$ in terms of the polar representation:

$$\hat{a} = e^{i\hat{\phi}}\left(\hat{N}\right)^{1/2}, \quad \hat{a}^{\dagger} = \left(\hat{N}\right)e^{-i\hat{\phi}}, \tag{5.5}$$

where both $\left(\hat{N}\right)^{1/2}$ and $\hat{\phi}$ are assumed to be Hermitian operators. In terms of these polar variables, the commutation relation $[\hat{a}, \hat{a}^{\dagger}] = 1$ can be transformed into (Dirac 1927, Heitler 1954)

$$[\hat{\phi}, \hat{N}] = -i, \tag{5.6}$$

with the uncertainty relation

$$(\Delta\phi)(\Delta N) \geq 1. \tag{5.7}$$

On the other hand, while the above uncertainty relation is consistent with real world, the commutation relation of Eq. 5.6 is not consistent with with the condition that both $\phi$ and $N$ be Hermitian (Louisell 1963, Susskind and Gologower 1964).

The uncertainty relation without commutation relation is not new in physics (Dirac 1927, Hussar *et al.* 1985). In addition to Heisenberg's position-momentum uncertainty relations, there is an uncertainty relation between the time and energy variables. However, since the time variable is a c-number, the commutator between the time variable and the Hamiltonian vanishes:

$$[t, H] = 0, \tag{5.8}$$

even though the relation

$$(\Delta t)(\Delta E) \geq 1 \tag{5.9}$$

is firmly established and is universally observed, as is discussed further in Chapter 9. The relation given in Eq. 5.9 is called the c-number time-energy uncertainty relation (Dirac 1927). The question then is whether the phase-intensity relation of Eq. 5.7 is a manifestation of the c-number time-energy uncertainty relation.

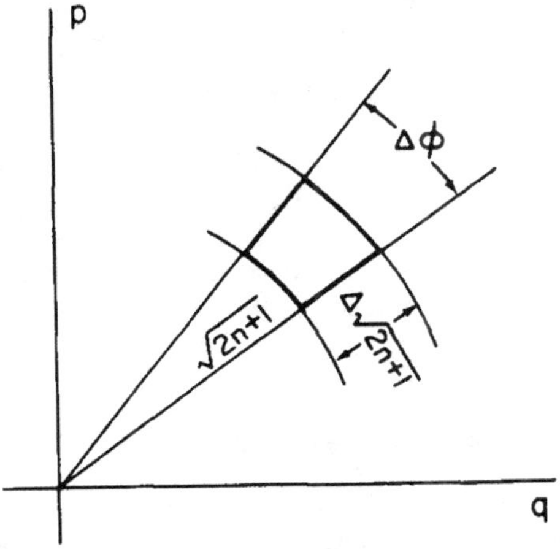


Figure 5.1: Polar coordinate system in phase space. Since both $q$ and $p$ are c numbers, it is possible to make the canonical transformation given in Eq. 5.14. Since this canonical transformation preserves the area element, the minimum uncertainty in $n$ and $\phi$ means the minimum uncertainty in $q$ and $p$.

In order to address this question, let us consider a plane light wave traveling along the $z$ direction, which is commonly described as

$$Ae^{ik(z-t)}, \tag{5.10}$$

where $A$ is the component a vector perpendicular to the direction of propagation. If we do not ask the question of localization, this describes the light wave corresponding to $|A|^2$ photons with momentum $k$ or energy $\omega$. Since this light is monochromatic, the uncertainty in energy is caused by the photon number $|A|^2$ or $N$:

$$\Delta E = \omega(\Delta N). \tag{5.11}$$

On the other hand, the uncertainty in phase is that of $k(z-t)$. If the photon energy is constant, then

$$\Delta\phi = \omega\left(\Delta(z-t)\right). \tag{5.12}$$

If the measurement is made only on the time variable, the uncertainty in $(z-t)$is the uncertainty in $t$: $\Delta t = \Delta(z-t)$. If we combine Eqs. 5.7, 5.9 and 5.11, then the phase-intensity relation is the same as the time-energy uncertainty relation (Goldin 1982):

$$(\Delta\phi)(\Delta N) = (\Delta t)(\Delta E). \tag{5.13}$$

One way to circumvent the difficulty with writing down the commutation relation of Eq. 5.6 is to use the Wigner function. The phase-number uncertainty relation $(\Delta\phi)(\Delta N) \cong 1$ can be canonically transformed into the phase space of $q$ and $p$. Let us consider the transformation from the Cartesian coordinate system into a polar coordinate system with (Dirl *et al.* 1988)

$$N = \frac{1}{2}\left(q^2 + p^2 - 1\right), \quad \phi = \tan^{-1}(p/q). \tag{5.14}$$

This is a canonical transformation from the Cartesian coordinate of $q$ and $p$ with

$$\frac{\partial N}{\partial q}\frac{\partial \phi}{\partial p} - \frac{\partial \phi}{\partial q}\frac{\partial N}{\partial p} = 1. \tag{5.15}$$

The c-number uncertainty relation in $p$ and $q$ in the phase-space picture of quantum mechanics is canonically equivalent to the uncertainty in $\phi$ and $N$. The uncertainty relation $(\Delta q)(\Delta p) = 1$ is translated into $(\Delta\phi)(\Delta N) = 1$, as is illustrated in Figure 5.1.

## 5.2 Baker-Campbell-Hausdorff Relation

In all branches of physics, we often encounter exponentiation of operators and product of exponentiations. We then quote the mathematical theorem commonly known as the Baker-Campbell-Hausdorff relation. In this section, we derive this relation which will be used in Sections 5.3 and 5.4. The discussion given in this section is based on Miller's book on symmetry groups and their applications (Miller 1972).

If $A$ is an operator, its exponentiation is defined as

$$e^A = \sum_n \frac{1}{n!}A^n, \tag{5.16}$$

if the series converges. Thus, for two operators $A$ and $B$, the product of their exponentiations $e^A e^B$ is defined according to the above definition. If $A$ and $B$ commute with each other,

$$e^A e^B = e^{(A+B)}. \tag{5.17}$$

If they do not commute, the above relation is not necessarily valid. It is governed by the Baker-Campbell-Hausdorff formula. For simplicity, we shall hereafter call this formula the BCH formula.

Let us start with the product

$$e^{A}Be^{-A}, \tag{5.18}$$

which we encounter very often in physics. In order to calculate this quantity, we consider

$$B(t) = e^{tA}Be^{-tA}. \tag{5.19}$$

Then $B(0) = B$, and $B(1)$ is the quantity we would like to calculate. $B(t)$ can be expanded as

$$B(t) = \sum_{n=0}^{\infty} \frac{1}{n!} C_n t^n. \tag{5.20}$$

Then

$$e^{A}Be^{-A} = \sum_{n=0}^{\infty} \frac{1}{n!} C_n. \tag{5.21}$$

The remaining task is to calculate $C_n$:

$$C_n = \frac{d^n}{dt^n} B(t) \mid_{t=0} . \tag{5.22}$$

The derivative of $B(t)$ with respect to $t$ is

$$\dot{B}(t) = Ae^{tA}Be^{-tA} - e^{tA}Be^{-tA}A = [A, B(t)]. \tag{5.23}$$

Then, by repeating this process, we can show that

$$C_n = [A, [A, [A, ....[A, B], ..] \quad \text{(n times)}. \tag{5.24}$$

For simplicity, let us write the above quantity as $(Ad^n(A)B)$. Then

$$e^{A}Be^{-A} = (\exp [Ad(A)])\, B. \tag{5.25}$$

As we shall see in the rest of this Chapter, this formula plays the central role in the conventional formalism of coherent and squeezed states.

We also encounter in physics the form

$$e^{C(t)} \frac{d}{dt} e^{-C(t)}, \tag{5.26}$$

where $C(t)$ is an operator or matrix analytic in $t$. In order to compute this quantity, let us set

$$L(s,t) = e^{sC(t)} \frac{d}{dt} e^{-sC(t)}. \tag{5.27}$$

Then $L(0,t) = 0$, and $L(1,t)$ is the quantity to be calculated. $L(s,t)$ can be expanded as

$$L(s,t) = \sum_{n=0}^{\infty} \frac{1}{n!} D_n(t) s^n. \tag{5.28}$$

Now from Eq. 5.27,

$$\frac{\partial}{\partial s}L(s,t) = C(t)L(s,t) - L(s,t)C(t) - \dot{C}(t), \tag{5.29}$$

and

$$\left(\frac{\partial}{\partial s}\right)^n L(s,t) = (Ad^n(C)L(s,t)) - \left(Ad^{n-1}(C)\dot{C}(t)\right). \tag{5.30}$$

Therefore, $D_0(t) = 0$, and

$$D_n(t) = -\left(Ad^{n-1}(C)\dot{C}(t)\right), \tag{5.31}$$

for $n$ greater than zero. This leads to

$$e^{C(t)}\frac{d}{dt}e^{-C(t)} = \sum_{n=0}^{\infty}\frac{1}{(n+1)!}\left(Ad^n(C)\dot{C}(t)\right). \tag{5.32}$$

This series expansion can be written as

$$e^{C(t)}\frac{d}{dt}e^{-C(t)} = -f\left(Ad(C)\right)\dot{C}(t), \tag{5.33}$$

where

$$f(z) = \frac{e^z - 1}{z} = 1 + \frac{z}{2!} + \frac{z^2}{3!} + \cdots, \tag{5.34}$$

If we define another function $g(z)$ as

$$g(z) = \frac{\ln(z)}{z-1} = 1 + \frac{1}{2}(1-z)^2 + \frac{1}{3}(1-z)^3 + \cdots, \tag{5.35}$$

then

$$\{f\left(\ln(z)\right)\}^{-1} = g(z), \tag{5.36}$$

which will be useful in deriving the BCH formula.

We are now ready to prove the theorem of Baker, Campbell and Hausdorff, which states that, for $A$, $B$, and $C = \ln\left(e^A e^B\right)$,

$$C = B + \int_0^1 g\left\{\exp\left(t[Ad(A)]\right)\exp[Ad(B)]\right\}(A)dt. \tag{5.37}$$

According to this theorem,

$$C = A + B + \frac{1}{2}[A,B] + \frac{1}{12}[A,[A,B]] - \frac{1}{12}[B,[B,A]] + \cdots. \tag{5.38}$$

If $[A,B]$ commutes with both $A$ and $B$, then

$$C = A + B + \frac{1}{2}[A,B], \tag{5.39}$$

which leads to

$$e^A e^B = e^{(A+B)} e^{(1/2)[A,B]}. \tag{5.40}$$

This is the form of the BCH formula seen most often in physics, particularly in the theory of coherent states of light.

Let us prove Eq. 5.37. Let $C(t) = \ln\left(e^{tA}e^B\right)$. Then for an operator $H$,

$$\begin{aligned}(\exp[Ad(C(t))])\,H &= e^{tA}\left(e^B H e^{-B}\right)e^{-tA} \\ &= (\exp[Ad(tA)])\,(\exp[Ad(B)])\,(H).\end{aligned} \tag{5.41}$$

Thus

$$Ad[C(t)] = \ln\left(\exp[Ad(tA)]\exp[Ad(B)]\right). \tag{5.42}$$

By definition

$$e^{C(t)}\frac{d}{dt}e^{-C(t)} = -e^{tA}e^B e^{-B} e^{-tA} A = -A. \tag{5.43}$$

Thus, according to Eq. 5.33,

$$f\left(Ad(C(t))\right)\dot{C} = A, \tag{5.44}$$

which, according to Eq. 5.36 and Eq. 5.42, becomes

$$\dot{C}(t) = g\left(\exp[Ad(tA)]\exp[Ad(B)]\right)(A). \tag{5.45}$$

Therefore

$$C(t) = B + \int_0^t g\left\{\exp\left(t'[Ad(A)]\right)\exp[Ad(B)]\right\}(A)dt', \tag{5.46}$$

which takes into account $C(0) = B$. When $t = 1$, the above formula leads to the theorem of Baker, Hausdorff and Campbell given in Eq. 5.37.

## 5.3 Coherent States of Light

The coherent state of light is defined as (Glauber 1963, Sudarshan 1963, Klauder and Skagerstam 1985)

$$|\alpha\rangle = e^{-\alpha\alpha^*/2}\sum_n \frac{\alpha^n}{\sqrt{n!}}|n\rangle, \tag{5.47}$$

where $\alpha$ is a complex number, which may be written in terms of two real numbers $\alpha_1$ and $\alpha_2$ as

$$\alpha = \alpha_1 + i\alpha_2. \tag{5.48}$$

This and other forms of coherent states have been discussed extensively in the literature (Klauder and Sudarshan 1968, Perina 1971, Sargent *et al.* 1974, Goldin 1982, Zhang *et al.* 1990). The expression for the coherent state given in Eq. 5.47 is normalized. The probability of being in the n-photon state is

$$P_n = (\alpha\alpha^*)^n \exp\left(-\alpha\alpha^*\right)/n!. \tag{5.49}$$

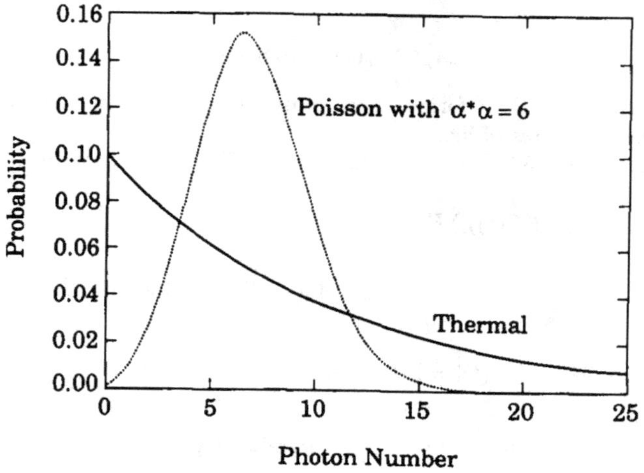


Figure 5.2: Poisson distribution and thermal distribution. It was shown experimentally that the photon distribution in laser is Poissonian (Arrechi 1965). Photons without coherence exhibit an exponential distribution.

This means that the number of photons in the coherent state has a Poisson distribution. This is the distribution expected from an ideal laser (Arecchi 1965). This Poisson distribution is illustrated in Figure 5.2.

The expectation value of the number operator is

$$<\alpha|\hat{N}|\alpha>= \exp(-\alpha\alpha^*)\sum_n \frac{n}{n!}(\alpha\alpha^*)^n = \exp(-\alpha\alpha^*)\frac{1}{n!}(\alpha\alpha^*)^{n+1}. \tag{5.50}$$

Thus

$$<\hat{N}>=<\alpha|\hat{N}|\alpha>= \alpha\alpha^*. \tag{5.51}$$

The expectation value of $\hat{N}^2$ is often useful in quantum optics. The eigenvalue of this operator on $|n>$ is of course $n^2$, which can be written as $n^2 = n(n-1)+n$. This leads to

$$<\hat{N}^2>= (\alpha\alpha^*)^2 + \alpha\alpha^*. \tag{5.52}$$

From this, we can calculate the variance of the number operator defined as

$$<(\Delta\hat{N})^2>=<\left(\hat{N}- <\hat{N}>\right)^2>= \alpha\alpha^*. \tag{5.53}$$

While the n-photon state can be represented by a harmonic oscillator wave function of $q$:

$$<q|n>= \left(1/(\sqrt{\pi}2^n n!)\right)^{1/2} H_n(q)\exp(-q^2/2), \tag{5.54}$$

in momentum space, the wave function becomes

$$<p|n>= (-i)^n \left(1/(\sqrt{\pi}2^n n!)\right)^{1/2} H_n(p)\exp(-p^2/2). \tag{5.55}$$

The coherent state $|\alpha>$ is not an eigenstate of the number operator, and takes the form

$$<q|\alpha> = \left(\frac{1}{\pi}\right)^{1/4} e^{-\alpha\alpha^*/2} \exp\left(-q^2/2\right) \sum_n \frac{\alpha^n}{n!}\left(\frac{1}{2}\right)^{n/2} H_n(q),$$

$$<p|\alpha> = \left(\frac{1}{\pi}\right)^{1/4} e^{-\alpha\alpha^*/2} \exp\left(-p^2/2\right) \sum_n (-)^n \frac{\alpha^n}{n!}\left(\frac{1}{2}\right)^{n/2} H_n(p). \qquad (5.56)$$

This appears to be a complicated function of $q$, but we note that this is an eigenstate of the annihilation operator:

$$\left(q + \frac{\partial}{\partial q}\right) <q|\alpha> = \sqrt{2}\alpha <q|\alpha>,$$

$$\left(p + \frac{\partial}{\partial p}\right) <p|\alpha> = \sqrt{2}\alpha <p|\alpha>. \qquad (5.57)$$

The solution of this differential equation is straight-forward, and takes the form (Goldin 1982):

$$<q|\alpha> = \left(\frac{1}{\pi}\right)^{1/4} e^{-(\alpha_2)^2} \exp\left(-(q - \sqrt{2}\alpha)^2/2\right),$$

$$<p|\alpha> = \left(\frac{1}{\pi}\right)^{1/4} e^{-(\alpha_2)^2} \exp\left(-(p - \sqrt{2}\alpha)^2/2\right), \qquad (5.58)$$

where $\alpha_1$ and $\alpha_2$ are the real and imaginary parts of $\alpha$ as given in Eq. 5.48. This becomes the ground-state harmonic oscillator wave function centered around the origin when $\alpha = 0$.

When $\alpha$ is a real number, it is a displaced ground-state oscillator wave function whose maximum is at $q = \sqrt{2}\alpha$. When $\alpha$ is complex, it is a ground state wave function whose origin is displaced to a complex value. However, the probability distribution becomes

$$|<q|\alpha>|^2 = \left(\frac{1}{\pi}\right)^{1/2} \exp\left\{-\left(q - \sqrt{2}\alpha_1\right)^2\right\},$$

$$|<p|\alpha>|^2 = \left(\frac{1}{\pi}\right)^{1/2} \exp\left\{-\left(p - \sqrt{2}\alpha_2\right)^2\right\}. \qquad (5.59)$$

Indeed, the probability distribution is a Gaussian function for all possible values of $\alpha$. This leads to

$$<(\Delta q)^2>=<(\Delta q^2)>= 1/2, \quad <(\Delta q)^2><(\Delta q)^2>= 1/4. \qquad (5.60)$$

The coherent state is a minimum-uncertainty state.

We are next interested in possible orthogonality relations for the coherent states. The eigenvalue $\alpha$ has two real components. Thus the sum over this continuous eigenvalue is necessarily an integral over the two-dimensional space of $\alpha_1$ and $\alpha_2$. With this point in mind, let us consider the integration

$$\frac{1}{\pi}\int |\alpha\rangle\langle\alpha| d\alpha_1 d\alpha_2. \tag{5.61}$$

The integral over the two-dimensional space can be converted into a polar coordinate with $r = (\alpha_1^2 + \alpha_2^2)^{1/2}$, and $\phi = \tan^{-1}(\alpha_2/\alpha_1)$. Then

$$\alpha\alpha^* = r^2, \qquad \alpha^n = r^n e^{in\phi}, \qquad \text{and} \qquad d\alpha_1 d\alpha_2 = r dr d\phi. \tag{5.62}$$

As a result,

$$\frac{1}{\pi}\int |\alpha\rangle\langle\alpha| d\alpha_1 d\alpha_2 = \sum_n \left\{\frac{2}{n!}\int_0^\infty r^{(2n+1)} e^{-r^2} dr\right\} |n\rangle\langle n|.$$
$$= \sum_n |n\rangle\langle n|. \tag{5.63}$$

Then the integral of Eq. 5.61 is an identity operator.

Next, we are interested in $\langle\beta|\alpha\rangle$. From Eq. 5.47,

$$\langle\beta|\alpha\rangle = \exp\left(-(\alpha\alpha^* + \beta\beta^*)/2\right) \sum_n \frac{1}{n!}(\alpha\beta^*)^n$$
$$= \exp\left(-(\alpha\alpha^* + \beta\beta^* - 2\alpha\beta^*)/2\right), \tag{5.64}$$

which leads to

$$|\langle\beta|\alpha\rangle|^2 = \exp\left(-|\alpha - \beta|^2\right). \tag{5.65}$$

This means that the coherent state is not a complete orthonormal state. It is overcomplete.

## 5.4 Symmetry Groups of Coherent States

Since $|n\rangle = (1/\sqrt{n!})\left(\hat{a}^\dagger\right)^n |0\rangle$, the coherent state of Eq. 5.47 can be written as

$$|\alpha\rangle = e^{-\alpha\alpha^*/2} \sum_n \frac{\alpha^n}{n!}\left(\hat{a}^\dagger\right)^n |0\rangle, \tag{5.66}$$

which then can be written as

$$|\alpha\rangle = e^{-\alpha\alpha^*/2} e^{\alpha\hat{a}^\dagger} |0\rangle. \tag{5.67}$$

Since $e^{-\alpha^*\hat{a}}$ gives no effect on $|0\rangle$, we can insert this factor before $|0\rangle$, and the result is

$$|\alpha\rangle = e^{-\alpha\alpha^*/2} e^{\alpha\hat{a}^\dagger} e^{-\alpha^*\hat{a}} |0\rangle. \tag{5.68}$$

In order to simplify this expression, we can use a special case of the BCH formula given in Eq. 5.40 for two operators resulting in a c-number commutator. Then the coherent state $|\alpha>$ can be written in terms of the Hermitian operator $T(\alpha)$:

$$|\alpha>= \hat{T}(\alpha)|0>, \tag{5.69}$$

where

$$\hat{T}(\alpha) = \exp\left(\alpha\hat{a}^{\dagger} - \alpha^{*}\hat{a}\right),$$

which is unitary. The group generated by $\hat{a}$, $\hat{a}^{\dagger}$ and 1 is called the Heisenberg group.

Let us see the effect of the number operator $\hat{N} = \hat{a}^{\dagger}\hat{a}$. The application of the unitary operator

$$\hat{M}(\theta) = e^{-i\theta\hat{N}} \tag{5.70}$$

to the coherent state results in

$$\begin{aligned}\hat{M}(\theta)|\alpha> &= e^{-\alpha\alpha^{*}/2}\sum_{n}\frac{\alpha^{n}}{n!}e^{-in\theta}\left(\hat{a}^{\dagger}\right)^{n}|0> \\ &= |\alpha e^{-i\theta}> .\end{aligned} \tag{5.71}$$

Thus the operation of $\hat{M}(\theta)$ decreases the phase angle of $\alpha$ by $\theta$, and the operator $\hat{N}$ generates the rotation of the phase angle of $\alpha$. The commutation relations of $\hat{N}$ with $\hat{a}$ and $\hat{a}^{\dagger}$ are

$$[\hat{a},\hat{N}] = \hat{a}, \quad [\hat{a}^{\dagger},\hat{N}] = -\hat{a}^{\dagger}. \tag{5.72}$$

The group generated by $\hat{N}$, $\hat{a}$, $\hat{a}^{\dagger}$, and 1 is called the harmonic oscillator group. The coherent state is a representation of the harmonic oscillator group.

Since the creation and annihilation operators are not Hermitian, we are led to consider

$$\hat{q} = \left(\hat{a} + \hat{a}^{\dagger}\right)/\sqrt{2}, \quad \hat{p} = -i\left(\hat{a} - \hat{a}^{\dagger}\right)/\sqrt{2}, \tag{5.73}$$

which are Hermitian operators. They satisfy the familiar Heisenberg commutation relation:

$$[\hat{q},\hat{p}] = i. \tag{5.74}$$

In terms of these Hermitian operators, the number operator becomes

$$\hat{N} = \frac{1}{2}\left(\hat{p}^{2} + \hat{q}^{2} - 1\right), \tag{5.75}$$

which is also Hermitian. With $\hat{q}$ and $\hat{p}$, its commutation relations are

$$[\hat{N},\hat{q}] = -i\hat{p}, \quad [\hat{N},\hat{p}] = i\hat{q}. \tag{5.76}$$

The operator $\hat{T}(\alpha)$ can then be written as

$$\hat{T}(\alpha) = \exp\left\{i\left(\sqrt{2}\alpha_{2}\right)\hat{q} - i\left(\sqrt{2}\alpha_{1}\right)\hat{p}\right\}. \tag{5.77}$$

where $\alpha_1$ and $\alpha_2$ are real independent parameters. Using the BCH formula, we can transform this into

$$\hat{T}(\alpha) = \exp(i\alpha_1\alpha_2)\exp\left\{i\left(\sqrt{2}\alpha_2\right)\hat{q}\right\}\exp\left\{-i\left(\sqrt{2}\alpha_1\right)\frac{\partial}{\partial q}\right\}. \tag{5.78}$$

If this is applied to the vacuum state wave function $<q|0>$,

$$<q|\hat{T}(\alpha)|0> = \left(\frac{1}{\pi}\right)^{1/4}\exp\left\{i\alpha_1\alpha_2 + i\left(\sqrt{2}\alpha_2\right)q\right\}\exp\left\{-\left(q-\sqrt{2}\alpha_1\right)^2/2\right\}, \tag{5.79}$$

which becomes the expression of Eq. 5.58 for $<q|\alpha>$. It is thus clear that $\hat{p}$ generates translations of wave function along the real $q$ axis, and $-\hat{q}$ generates the addition of the imaginary number to $\alpha$. Thus $\hat{p}$ and $\hat{q}$ appear to generate the translation group in the two-dimensional plane of $\alpha_1$ and $\alpha_2$. However, since they do not commute with each other, they cannot be regarded as the translation generators.

In order to clarify the problem, let us consider two repeated applications of $\hat{T}$. Again from the Baker-Hausdorff-Campbell relation

$$\hat{T}(\beta)\hat{T}(\alpha) = e^{i(Im(\beta\alpha^*))}\hat{T}(\alpha+\beta). \tag{5.80}$$

The complex parameter $\alpha$ is additive in the sense that the right-hand side is $\hat{T}(\alpha+\beta)$ multiplied by a factor of unit modulus. The unitarity is preserved. As a consequence,

$$\hat{T}(\beta)\hat{T}(\alpha)\hat{T}(-\beta) = e^{2i(Im(\beta\alpha^*))}\hat{T}(\alpha). \tag{5.81}$$

Let us see whether this relation can be derived from $\left(\hat{T}(\beta)\hat{a}\hat{T}(-\beta)\right)$ and $\left(\hat{T}(\beta)\hat{a}^\dagger\hat{T}(-\beta)\right)$. $\left(\hat{T}(\beta)\hat{a}\hat{T}(-\beta)\right)$ can be written as

$$\hat{T}(\beta)\hat{a}\hat{T}(-\beta) = e^{(\beta\hat{a}^\dagger-\beta^*\hat{a})}\hat{a}e^{-(\beta\hat{a}^\dagger-\beta^*\hat{a})}. \tag{5.82}$$

Let us go back to Eq. 5.25 Section 5.2. Then, since $[\hat{a}^\dagger,\hat{a}] = -1$, and $[\hat{a}^\dagger,[\hat{a}^\dagger,\hat{a}]] = 0$,

$$\hat{T}(\beta)\hat{a}\hat{T}(-\beta) = \hat{a} - \beta. \tag{5.83}$$

Likewise

$$\hat{T}(\beta)\hat{a}^\dagger\hat{T}(-\beta) = \hat{a}^\dagger - \beta^*. \tag{5.84}$$

Thus

$$\hat{T}(\beta)\hat{T}(\alpha)\hat{T}(-\beta) = \left\{e^{-(\alpha\beta^*-\alpha^*\beta)}\right\}e^{(\alpha\hat{a}^\dagger-\alpha^*\hat{a})}, \tag{5.85}$$

which is the same as Eq. 5.80.

The story is similar for $\hat{M}(\theta)\hat{a}\hat{M}(-\theta)$ and $\hat{M}(\theta)\hat{a}^\dagger\hat{M}(-\theta)$. Again let us go back to Eq. 5.25, and note

$$[\hat{N},\hat{a}] = -\hat{a}, \quad [\hat{N},[\hat{N},\hat{a}]] = (-1)^2\hat{a},$$

$$[\hat{N},[\hat{N},\cdots[\hat{N},[\hat{N},\hat{a}]\cdots] \quad (\text{n times}) = (-1)^n\hat{a}. \tag{5.86}$$

Therefore,

$$\hat{M}(\theta)\hat{a}\hat{M}(-\theta) = e^{-i\theta\hat{N}}\hat{a}e^{i\theta\hat{N}} = e^{i\theta}\hat{a}. \tag{5.87}$$

Likewise

$$\hat{M}(\theta)\hat{a}^{\dagger}\hat{M}(-\theta) = e^{-i\theta}\hat{a}^{\dagger}. \tag{5.88}$$

The symmetry of the coherent state is dictated by the transformations $T(\alpha)$ and $M(\theta)$. $M(\theta)$ is clearly a rotation operator in the complex plane of $\alpha$. $T(\beta)$ appears as a translation operator in Eq. 5.83 and Eq. 5.84. However, Eq. 5.85 shows that it does not represent a commutative group. Thus $T(\alpha)$ represents a multiplier representation of the translation group. This point will be examined further in Chapter 6 where the phase-space picture is discussed.

## 5.5 Squeezed States

In Section 5.1, we noticed that the operator $\hat{N}$ is quadratic in the creation and annihilation operators. If we define $J_3$ as

$$\hat{J}_3 = \frac{1}{4}\left(\hat{a}^{\dagger}\hat{a} + \hat{a}\hat{a}^{\dagger}\right), \tag{5.89}$$

then it is related to $\hat{N}$ by

$$\hat{J}_3 = \frac{1}{2}\left(\hat{N} + \frac{1}{2}\right). \tag{5.90}$$

In addition, we can consider the following two additional quadratic forms:

$$\hat{B}_1 = \frac{1}{4}\left(\hat{a}^{\dagger}\hat{a}^{\dagger} + \hat{a}\hat{a}\right), \quad \hat{B}_2 = \frac{i}{4}\left(\hat{a}^{\dagger}\hat{a}^{\dagger} - \hat{a}\hat{a}\right). \tag{5.91}$$

Then they satisfy the commutation relations:

$$[\hat{B}_1, \hat{B}_2] = -i\hat{J}_3, \quad [\hat{J}_3, \hat{B}_2] = i\hat{B}_1, \quad [\hat{J}_3, \hat{B}_1] = -i\hat{B}_2. \tag{5.92}$$

This set of commutation relations is identical to that of the (2 + 1)- dimensional Lorentz group discussed in Section 4.5

The transformations generated by these operators lead to two-photon coherent states (Yuen 1976) or squeezed states (Caves 1980, Caves 1981, Walls 1983, Bondourant and Shapiro 1984, Reid and Walls 1985, Yurke 1985, Slusher *et al.* 1985, Caves and Schumaker 1985, Collet and Walls 1985, Klauder *et al.* 1986, Yurke *et al.* 1986, Schumaker 1986, Ho *et al.* 1986 and 1987, Wu *et al.* 1986, Ou *et al.* 1987, Schleich *et al.* 1988, Teich and Saleh 1989 and 1990). It is now possible to design experiments to produce squeezed states of light (Yuen and Shapiro 1979, Slusher *et al.* 1985, Teich and Saleh 1985, Shelby *et al.* 1986, Wu *et al.* 1986, Maeda *et al.* 1987, Machida and Yamamoto 1988). This indeed provides a strong motivation to develop a formalism for this new branch of physics devoted to the study of uncertainty relations.

The coherent state a minimum uncertainty state because its wave function in the $q$ space is Gaussian. The Gaussian form remains invariant when the value of $\alpha$ is changed. The Gaussian form still gives a minimal uncertainty even if its width is changed (Stoler 1970, Yuen 1976). The squeezed state of light is defined as

$$|\zeta,\alpha> = \hat{S}(\zeta)|\alpha>, \tag{5.93}$$

where

$$\hat{S}(\zeta) = \exp\left\{\frac{1}{4}\left(\zeta\hat{a}^{\dagger}\hat{a}^{\dagger} - \zeta^{*}\hat{a}\hat{a}\right)\right\}.$$

The squeeze parameter $\zeta$ is a complex number, and can be written in terms of two real numbers $\lambda$ and $\phi$ as

$$\zeta = \lambda e^{i\phi}. \tag{5.94}$$

The squeeze operator of Eq. 5.93 is unitary. When $\zeta$ is real and equal to $\lambda$, it takes the form

$$\hat{S}(\lambda) = \exp\left\{\frac{\lambda}{2}\left(q\frac{\partial}{\partial q} + \frac{1}{2}\right)\right\}. \tag{5.95}$$

If this is operated on the coherent state of Eq. 5.58, the result is

$$<q|\hat{S}(\lambda)|\alpha> = \left(\frac{1}{\pi}\right)^{1/4} e^{\lambda/2}\exp\left\{-\left(\frac{1}{2}\right)e^{\lambda}\left(q - e^{-\lambda/2}Re(\sqrt{2}\alpha)\right)^{2}\right\}. \tag{5.96}$$

This wave function still is in a Gaussian form and preserves the minimum uncertainty. However, the effect of $S(\zeta)$ is to reduce or squeeze the width of the Gaussian function. The word "squeeze" can be more precisely defined in the phase-space picture of quantum mechanics, as is seen in Chapter 4.

Since $|\alpha>$ is obtained from the vacuum $|0>$ through $T(\alpha)|0>$, we can consider the squeezed state of the form

$$|\alpha,\zeta> = \hat{T}(\alpha)\hat{S}(\zeta)|0>. \tag{5.97}$$

It is possible to obtain a closed form of the Schrödinger wave function (Rai and Mehta 1988). It is more convenient to write the squeeze operator as (Fisher *et al.* 1984, Truax 1985)

$$\hat{S}(\theta) = \exp\left\{\frac{1}{2}\left(e^{i\phi}\tanh\frac{\lambda}{2}\right)\hat{a}^{\dagger}\hat{a}^{\dagger}\right\}\exp\left\{-\left(\ln\left[\cosh\frac{\lambda}{2}\right]\right)\left(\hat{a}^{\dagger}\hat{a} + \frac{1}{2}\right)\right\}$$
$$\times\exp\left\{-\frac{1}{2}\left(e^{-i\phi}\tanh\frac{\lambda}{2}\right)\hat{a}\hat{a}\right\}, \tag{5.98}$$

where $\zeta = \lambda e^{i\phi}$. If this operator is applied to the vacuum state, then

$$\hat{S}(\zeta)|0> = \left(\cosh\frac{\lambda}{2}\right)^{-1/2}\exp\left\{\left(e^{i\phi}\tanh\frac{\lambda}{2}\right)\hat{a}^{\dagger}\hat{a}^{\dagger}\right\}|0>. \tag{5.99}$$

The Taylor series of the above expression is (Schumaker and Caves 1985)

$$\hat{S}(\zeta)|0\rangle = \left(\cosh\frac{\lambda}{2}\right)^{1/2} \sum_n \frac{\sqrt{(2n)!}}{n!}\left(e^{i\phi}\tanh\frac{\lambda}{2}\right)^n |2n\rangle . \tag{5.100}$$

The squeezed vacuum can be written as a superposition of an even number of photon states. For this reason, the squeezed state is sometimes called a two-photon coherent state (Yuen 1976).

Let us now consider the most general form of the squeezed state given in Eq. 5.97. Using the form of $\hat{T}(\alpha)$ discussed in Section 5.3, we can write

$$|\alpha,\zeta\rangle = \left(\cosh\frac{\lambda}{2}\right)^{-1/2}$$

$$\times \sum_n \frac{\sqrt{(2n)!}}{n!}\left(e^{i\phi}\tanh\frac{\lambda}{2}\right)^n \exp\left(\alpha\hat{a}^\dagger - \alpha^*\hat{a}\right)|2n\rangle . \tag{5.101}$$

It is possible to expand $\exp\left(\alpha\hat{a}^\dagger - \alpha^*\hat{a}\right)|2n\rangle$ in a power series of $\alpha$ and $\alpha^*$. However, the result will be a double series expansion of the above expression. We are thus led to look for a simpler mathematical language for the squeezed state. As we shall see in Chapter 6, the phase-space picture of quantum mechanics is the natural language for this state.

Let us next turn our attention to some mathematical questions which arise from the above discussion. In Eq. 5.93, the squeezed state $|\zeta,\alpha\rangle$ is defined as

$$|\zeta,\alpha\rangle = \hat{S}(\zeta)\hat{T}(\alpha)|0\rangle . \tag{5.102}$$

This appears different from the state $|\alpha,\zeta\rangle$ given in Eq. 5.97. Certainly $\hat{S}(\zeta)\hat{T}(\alpha)$ is not the same as $\hat{T}(\alpha)\hat{S}(\zeta)$. Then how are they different?

In order to answer this question, let us write the above expression as

$$|\zeta,\alpha\rangle = \left(\hat{S}(\zeta)\hat{T}(\alpha)\hat{S}(-\zeta)\right)\hat{S}(\zeta)|0\rangle, \tag{5.103}$$

and ask whether $\left(\hat{S}(\zeta)\hat{T}(\alpha)\hat{S}(-\zeta)\right)$ can be written as $\hat{T}(\alpha')$, where $\alpha'$ is to be determined from $\alpha$ and $\zeta$. For this purpose, we can start with $\hat{S}(\zeta)\hat{a}\hat{S}(-\zeta)$. If we use $G$ for the exponent in Eq. 5.93:

$$G = \frac{1}{4}\left(\zeta\hat{a}^\dagger\hat{a}^\dagger - \zeta^*\hat{a}\hat{a}\right), \tag{5.104}$$

then

$$[G,\hat{a}] = -\left(\frac{\zeta}{2}\right)\hat{a}^\dagger, \quad [G,[G,\hat{a}]] = \left(\frac{\zeta}{2}\right)\left(\frac{\zeta}{2}\right)^*\hat{a}. \tag{5.105}$$

Thus the result of the computation is

$$\hat{S}(\zeta)\hat{a}\hat{S}(-\zeta) = \left(\cosh\frac{\lambda}{2}\right)\hat{a} - e^{i\phi}\left(\sinh\frac{\lambda}{2}\right)\hat{a}^\dagger,$$

$$\hat{S}(\zeta)\hat{a}^\dagger\hat{S}(-\zeta) = \left(\cosh\frac{\lambda}{2}\right)\hat{a}^\dagger - e^{-i\phi}\left(\sinh\frac{\lambda}{2}\right)\hat{a}. \tag{5.106}$$

Then

$$\hat{S}(\zeta)\hat{T}(\alpha)\hat{S}(-\zeta) = \hat{T}(\alpha'), \tag{5.107}$$

where

$$\begin{aligned}\alpha' &= \left(\cosh\frac{\lambda}{2}\right)\alpha + e^{i\phi}\left(\sinh\frac{\lambda}{2}\right)\alpha^* \\ &= \left\{\left[\left(\cosh\frac{\lambda}{2}\right) + \left(\sinh\frac{\lambda}{2}\right)\cos\phi\right]\alpha_1 + \left[\left(\sinh\frac{\lambda}{2}\right)\sin\phi\right]\alpha_2\right\} \\ &\quad + i\left\{\left[\left(\sinh\frac{\lambda}{2}\right)\sin\phi\right]\alpha_1 + \left[\left(\cosh\frac{\lambda}{2}\right) - \left(\sinh\frac{\lambda}{2}\right)\cos\phi\right]\alpha_2\right\}. \end{aligned} \tag{5.108}$$

$\alpha_1$ and $\alpha_2$ are the real and imaginary parts of $\alpha$ respectively. Thus

$$\hat{S}(\zeta)\hat{T}(\alpha) = \hat{T}(\alpha')\hat{S}(\zeta). \tag{5.109}$$

This means that $\hat{S}(\zeta)$ and $\hat{T}(\alpha)$ can be transposed if $\hat{T}(\alpha)$ is replaced by $\hat{T}(\alpha')$. The group theoretical implication of this property will be discussed in Chapter 6.

The above analysis enables us to calculate the expectation value of the number operator for the squeezed vacuum. Let us start with $<\zeta|N|\zeta>$. This is the vacuum expectation value of $\left[\hat{S}(-\zeta)\hat{a}^\dagger\hat{a}\hat{S}(\zeta)\right]$:

$$<\zeta|N|\zeta>=<0|\left[\hat{S}(-\zeta)\hat{a}^\dagger\hat{a}\hat{S}(\zeta)\right]|0>. \tag{5.110}$$

This results in

$$<\zeta|N|\zeta>=\left(\sinh\frac{\lambda}{2}\right)^2. \tag{5.111}$$

In this manner, other quantities of interest can be calculated (Henry and Glotzer 1988). However, the calculation can be done more efficiently in terms of the Wigner function, as we shall see in Chapter 6.

Let us finally prove the relation of Eq. 5.98 (Fisher *et al.* 1984, Truax 1985). If we introduce $B_+$ and $B_-$ as $B_\pm = B_1 \pm iB_2$, then

$$B_+ = \left(\frac{1}{2}\right)\hat{a}^\dagger\hat{a}^\dagger, \quad B_- = -\left(\frac{1}{2}\right)\hat{a}\hat{a}. \tag{5.112}$$

They satisfy the commutation relations:

$$[B_+, B_-] = 2J_3, \quad [J_3, B_\pm] = \pm B_\pm. \tag{5.113}$$

The simplest solution of the above commutation relations is

$$B_+ = \begin{pmatrix} 0 & 1 \\ 0 & 0 \end{pmatrix}, \quad B_- = \begin{pmatrix} 0 & 0 \\ 1 & 0 \end{pmatrix}, \quad J_3 = \begin{pmatrix} 1/2 & 1 \\ 0 & -1/2 \end{pmatrix}. \tag{5.114}$$

Then $S(\zeta)$ can be represented as

$$S(\zeta) = \exp\left\{\left(\frac{1}{2}\right)\begin{pmatrix} 0 & \zeta \\ \zeta^* & 0 \end{pmatrix}\right\}$$

$$= \begin{pmatrix} 1 & 0 \\ 0 & 1 \end{pmatrix}\sum_n \frac{1}{(2n)!}\left(\frac{\lambda}{2}\right)^{2n} + \left(\frac{1}{2}\right)\begin{pmatrix} 0 & \zeta \\ \zeta^* & 0 \end{pmatrix}\sum_n \frac{1}{(2n+1)!}\left(\frac{\lambda}{2}\right)^{2n}$$

$$= \begin{pmatrix} \cosh(\lambda/2) & e^{i\phi}\sinh(\lambda/2) \\ e^{-i\phi}\sinh(\lambda/2) & \cosh(\lambda/2) \end{pmatrix}. \tag{5.115}$$

The above two-by-two matrix can be decomposed into

$$S(\zeta) = \begin{pmatrix} 1 & e^{i\phi}\tanh(\lambda/2) \\ 0 & 1 \end{pmatrix}\begin{pmatrix} 1/\cosh(\lambda/2) & 0 \\ 0 & \cosh(\lambda/2) \end{pmatrix}$$

$$\times \begin{pmatrix} 1 & 0 \\ e^{i\phi}\tanh(\lambda/2) & 1 \end{pmatrix}. \tag{5.116}$$

This can be written as

$$S(\zeta) = \exp\left\{\left(e^{i\phi}\tanh\frac{\lambda}{2}\right)B_+\right\}\exp\left\{-2\left(\ln\left(\cosh\frac{\lambda}{2}\right)\right)J_3\right\}$$

$$\times \exp\left\{\left(e^{-i\phi}\tanh\frac{\lambda}{2}\right)B_-\right\}. \tag{5.117}$$

If we write the operators $B_+$, $B_-$, $J_3$ in terms of $\hat{a}$ and $\hat{a}^\dagger$, in the above expression, we arrive at Eq. 5.98.

## 5.6 Two-Mode Squeezed States

In Section 5.4, we studied photon states generated by two repeated operations of the annihilation or creation, and by the number operator which consists the creation preceded by the annihilation operator. We assumed that these operators are applicable to only one kind of photon. However, in laboratory experiments, it is more common that these two photons are different.

Indeed, it is possible to design interferometers (Yurke *et al.* 1986, Campos *et al.* 1989) using the following two-photon operators.

$$\hat{J}_1 = \frac{1}{2}\left(\hat{a}_1^\dagger\hat{a}_2 + \hat{a}_2^\dagger\hat{a}_1\right), \quad \hat{J}_2 = \frac{1}{2i}\left(\hat{a}_1^\dagger\hat{a}_2 - \hat{a}_2^\dagger\hat{a}_1\right),$$

$$\hat{J}_3 = \frac{1}{2}\left(\hat{a}_1^\dagger\hat{a}_1 - \hat{a}_2^\dagger\hat{a}_2\right). \tag{5.118}$$

These operators satisfy the commutation relations:

$$[\hat{J}_i, \hat{J}_j] = i\epsilon_{ijk}\hat{J}_k, \tag{5.119}$$

which are identical to those for the three-dimensional rotation group or the group SU(2). The interferometer based on this algebra is commonly called the SU(2) interferometer.

It is also possible to design interferometers (Yurke *et al.* 1986, Singer *et al.* 1989) using the following set of two-photon operators.

$$\hat{K}_3 = \frac{1}{2}\left(\hat{a}_1^\dagger \hat{a}_2^\dagger + \hat{a}_1\hat{a}_2\right), \quad \hat{Q}_3 = \left(\frac{i}{2}\right)\left(\hat{a}_1^\dagger \hat{a}_2^\dagger - \hat{a}_1\hat{a}_2\right),$$
$$\hat{J}_0 = \frac{1}{2}\left(\hat{a}_1^\dagger \hat{a}_1 + \hat{a}_2\hat{a}_2^\dagger\right). \tag{5.120}$$

These operators satisfy the commutation relations:

$$[\hat{K}_3, \hat{Q}_3] = i\hat{J}_0, \quad [\hat{J}_0, \hat{K}_3] = -i\hat{Q}_3, \quad [\hat{J}_0, \hat{Q}_3] = -i\hat{K}_3. \tag{5.121}$$

This set of commutation relations is identical to that of the (2 + 1)- dimensional Lorentz group (Bargmann and Wigner 1948), which is locally isomorphic to the group SU(1,1) (Kim and Noz 1983). The interferometer based on the above set of generators is called the SU(1,1) interferometer.

It is not inconceivable to build an interferometer combining both the SU(2) and SU(1,1) generators (Han *et al.* 1990b). Then, would the generators of the SU(2) interferometer and those for SU(1,1) form a closed set of commutation relations. In order to answer this question, let us take commutation relations of $\hat{J}_0$ with $\hat{J}_i$ .

$$[\hat{J}_0, \hat{J}_1] = 0, \quad [\hat{J}_0, \hat{J}_2] = 0, \quad [\hat{J}_0, \hat{J}_3] = 0. \tag{5.122}$$

It commutes with $\hat{J}_i$.

Then would $\hat{K}_3$ commutes with $\hat{J}_i$? The answer is NO. Indeed,

$$[\hat{J}_1, \hat{K}_3] = -i\hat{K}_2, \quad [\hat{J}_2, \hat{K}_3] = i\hat{K}_1, \quad [\hat{J}_3, \hat{K}_3] = 0, \tag{5.123}$$

where

$$\hat{K}_1 = -\left(\frac{1}{4}\right)\left(\hat{a}_1^\dagger \hat{a}_1^\dagger + \hat{a}_1\hat{a}_1 - \hat{a}_2^\dagger\hat{a}_2^\dagger - \hat{a}_2\hat{a}_2\right),$$
$$\hat{K}_2 = +\left(\frac{i}{4}\right)\left(\hat{a}_1^\dagger \hat{a}_1^\dagger - \hat{a}_1\hat{a}_1 + \hat{a}_2^\dagger\hat{a}_2^\dagger - \hat{a}_2\hat{a}_2\right). \tag{5.124}$$

As for $\hat{Q}_3$, the commutation relations are

$$[\hat{J}_1, \hat{Q}_3] = -i\hat{Q}_2, \quad [\hat{J}_2, \hat{Q}_3] = i\hat{Q}_1, \quad [\hat{J}_3, \hat{Q}_3] = 0, \tag{5.125}$$

where

$$\hat{Q}_1 = +\left(\frac{i}{4}\right)\left(\hat{a}_1^\dagger \hat{a}_1^\dagger - \hat{a}_1\hat{a}_1 - \hat{a}_2^\dagger\hat{a}_2^\dagger - \hat{a}_2\hat{a}_2\right),$$
$$\hat{Q}_2 = -\left(\frac{1}{4}\right)\left(\hat{a}_1^\dagger \hat{a}_1^\dagger + \hat{a}_1\hat{a}_1 + \hat{a}_2^\dagger\hat{a}_2^\dagger + \hat{a}_2\hat{a}_2\right). \tag{5.126}$$

Thus, the attempt to combine the commutation relations of Eq. 5.119 and Eq. 5.121 produces the new generators $\hat{K}_1$, $\hat{K}_2$, $\hat{Q}_1$ and $\hat{Q}_2$.

If we combine the generators for the SU(2) and SU(1,1) interferometers and the new generators given in Eq. 5.124 and Eq. 5.126, there are ten generators. Would they form a closed set of commutation relations? In order to answer this question, let us first take the commutation relations among the new generators given in Eq. 5.124 and Eq. 5.126:

$$[\hat{K}_1, \hat{Q}_1] = [\hat{K}_2, \hat{Q}_2] = +i\hat{J}_0,$$

$$[\hat{K}_1, \hat{Q}_2] = [\hat{K}_2, \hat{Q}_1] = 0,$$

$$[\hat{K}_1, \hat{K}_2] = [\hat{Q}_1, \hat{Q}_2] = -i\hat{J}_3. \tag{5.127}$$

Let us next see the commutation relations of the new operators with those for the SU(2) and SU(1,1) interferometers. They are

$$[\hat{K}_1, \hat{J}_1] = 0, \quad [\hat{K}_1, \hat{J}_2] = +i\hat{K}_3, \quad [\hat{K}_1, \hat{J}_3] = -i\hat{K}_2,$$

$$[\hat{K}_2, \hat{J}_1] = -i\hat{K}_3, \quad [\hat{K}_2, \hat{J}_2] = 0, \quad [\hat{K}_2, \hat{J}_3] = +i\hat{K}_1,$$

$$[\hat{Q}_1, \hat{J}_1] = 0, \quad [\hat{Q}_1, \hat{J}_2] = +i\hat{Q}_3, \quad [\hat{Q}_1, \hat{J}_3] = -i\hat{Q}_2,$$

$$[\hat{Q}_2, \hat{J}_1] = -i\hat{Q}_3, \quad [\hat{Q}_2, \hat{J}_2] = 0, \quad [\hat{Q}_2, \hat{J}_3] = +i\hat{Q}_1. \tag{5.128}$$

With respect to $\hat{J}_0$, the commutation relations are

$$[\hat{K}_1, \hat{J}_0] = +i\hat{Q}_1, \quad [\hat{K}_2, \hat{J}_0] = +i\hat{Q}_2,$$

$$[\hat{Q}_1, \hat{J}_0] = -i\hat{K}_1, \quad [\hat{Q}_2, \hat{J}_0] = -i\hat{K}_2. \tag{5.129}$$

As for $\hat{K}_3$ and $\hat{Q}_3$,

$$[\hat{Q}_1, \hat{K}_3] = [\hat{Q}_2, \hat{K}_3] = [\hat{K}_1, \hat{Q}_3] = [\hat{K}_2, \hat{Q}_3] = 0,$$

$$[\hat{K}_1, \hat{K}_3] = [\hat{Q}_1, \hat{Q}_3] = +i\hat{J}_2,$$

$$[\hat{K}_2, \hat{K}_3] = [\hat{Q}_2, \hat{Q}_3] = -i\hat{J}_1. \tag{5.130}$$

Since there are ten generators, there are one hundred commutation relations. Since they are antisymmetric, there are forty five non-trivial commutators. We have discussed all of them.

These ten generators form a closed set of commutation relations. It is indeed remarkable that this set of commutation relations is identical to that for the (3 +

2)-dimensional de Sitter group discussed in Chapter 4, which in turn shares the same set of commutation relations as the group $Sp(4)$ discussed in Chapter 1. We shall continue this discussion in terms of the Wigner function in Chapter 6.

## 5.7 Density Matrix through Two-Mode Squeezed States

As was noted in Chapter 2, the density matrix is the principal language for a quantum system involving an ensemble average (Von Neumann, 1927 and 1955). Once the average is made, the system is no longer in a coherent state. This decoherence (Hartle 1990) is caused by the incompleteness in measurement. We are now interested in whether the decoherence effect of the ensemble average can be derived from the two-mode squeezed state. This effect can best be derived in terms of the density matrix (Feynman 1973).

Let us go back to Section 2.5 of Chapter 2, and start with the orthonormal expansion of the wave function:

$$\psi(q_1) = \sum_n C_n \psi_n(q_1), \tag{5.131}$$

where $x$ in Eq. 2.71 is replaced by $q_1$. The density matrix in this case is

$$\rho(q_1, q_1') = \sum_n |C_n|^2 \psi_n(q_1)\psi_n^*(q_1'). \tag{5.132}$$

This is different from $\psi(q_1)\psi^*(q_1')$ which contains the terms $C_m^* C_n \psi_n(q_1)\psi_m^*(q_1')$ for different values of $m$ and $n$.

One way of obtaining the density matrix of Eq. 5.132 is to start with a wave function of two variables (Fetter and Walecka 1971, Umezawa *et al.* 1982, Mann and Revzen 1989):

$$\psi(q_1, q_2) = \sum_n C_n \psi_n(q_2)\psi_n(q_1). \tag{5.133}$$

This wave function is also normalized in the space of $q_1$ and $q_2$, and is in a pure state. The density matrix $\rho(q_1, q_2; q_1', q_2')$ is

$$\rho(q_1, q_2; q_1', q_2') = \psi(q_1, q_2)\psi^*(q_1', q_2'), \tag{5.134}$$

and its trace is 1 in the sense that

$$\int \rho(q_1, q_2; q_1, q_2) dq_1 dq_2 = 1. \tag{5.135}$$

On the other hand, we are only interested is the Hilbert space of $\psi_n(q_1)$. The wave functions $\psi_n(q_2)$ are introduced purely for mathematical convenience. We are thus led to consider $\rho(q_1, q_1')$ defined as

$$\begin{aligned}\rho(q_1, q_1') &= \int \rho(q_1, q_2; q_1', q_2) dq_2 \\ &= \int \psi(q_1, q_2)\psi^*(q_1', q_2) dq_2.\end{aligned} \tag{5.136}$$

This integral leads to the expression for $\rho(q_1, q_1')$ of Eq. 5.132.

The basic advantage of using the two-coordinate system is that the transformation of one state of $\psi(q_1, q_2)$ to another is achieved through a unitary transformation. For example, let us consider the thermally excited oscillator state with the density matrix (Pathria 1972, Feynman 1973):

$$\rho_T(q_1, q_1') = \left(1 - e^{-\omega/kT}\right) \sum_n e^{-n\omega/kT} \psi_n(q_1)\psi_n^*(q_1'), \tag{5.137}$$

which was discussed in Section 2.5. In order to obtain this expression, we can start with the ground-state harmonic oscillator wave function

$$\psi_0(q_1, q_2) = \psi_0(q_1)\psi_0(q_2) = \left(\frac{1}{\pi}\right)^{1/2} \exp\left\{-(m\omega/2)\left(q_1^2 + q_2^2\right)\right\}. \tag{5.138}$$

It is now possible to make a coordinate transformation:

$$\begin{pmatrix} q_1' \\ q_2' \end{pmatrix} = \begin{pmatrix} \frac{1}{\left(1-e^{-\omega/2kT}\right)^{1/2}} & \frac{e^{-\omega/2kT}}{\left(1-e^{-\omega/2kT}\right)^{1/2}} \\ \frac{e^{-\omega/2kT}}{\left(1-e^{-\omega/2kT}\right)^{1/2}} & \frac{1}{\left(1-e^{-\omega/2kT}\right)^{1/2}} \end{pmatrix} \begin{pmatrix} q_1 \\ q_2 \end{pmatrix}, \tag{5.139}$$

which results in

$$\psi_T(q_1, q_2) = \left(\frac{1}{\pi}\right)^{1/2} \exp\left\{-\left(\frac{m\omega}{4}\right)\left[\left(\tanh\frac{\omega}{2kT}\right)(q_1 + q_2)^2 + \left(\coth\frac{\omega}{2kT}\right)(q_1 - q_2)^2\right]\right\}. \tag{5.140}$$

Indeed, the transformation from $\psi_0(q_1, q_2)$ of Eq. 5.138 to $\psi_T(q_1, q_2)$ of Eq. 5.140 is unitary. The above expression can be expanded as (Kim and Noz 1986)

$$\psi_T(q_1, q_2) = (1 - \exp(-\omega/T))^{1/2} \sum_n (\exp(-\omega/2T))^n \psi_n(q_1)\psi_n(q_2). \tag{5.141}$$

The transformation of Eq. 5.138 to the above series through the coordinate transformation of Eq. 5.139 is often called the Bogoliubov transformation, and its mathematics is that of the Lorentz boost. The transformation is generated by $\hat{Q}_3$ which takes the form

$$\hat{Q}_3 = -\left(\frac{i}{2}\right)\left(q_1\frac{\partial}{\partial q_2} + q_2\frac{\partial}{\partial q_1}\right). \tag{5.142}$$

The squeeze operator is

$$\exp\left\{-i\left[\ln\left(\tanh\frac{\omega}{2kT}\right)\right]\hat{Q}_3\right\}. \tag{5.143}$$

This performs a unitary transformation of Eq. 5.138 to Eq. 5.140. $\hat{Q}_3$ is the generator of the Bogoliubov transformation.

In this section, the parameter $q_2$ was introduced strictly for mathematical convenience and has no physical meaning. In Chapter 9, we shall discuss a example in which this auxiliary variable carries a physical meaning.

# Chapter 6

# PHASE-SPACE PICTURE OF COHERENT AND SQUEEZED STATES

Coherent and squeezed states constitute the backbone of the theoretical framework of modern optics. As was shown in Chapter 5, this framework starts from the creation and annihilation operators, which are harmonic-oscillator step-up and step-down operators in Fock space respectively. If a unitary transformation from the vacuum state is generated by these operators, it produces coherent states. If the transformation is generated by their quadratic forms, it leads to a squeezed state. It is possible to form the quadratic form of the creation and annihilation operators for two different photons. In this case, the transformation leads to a two-mode squeezed state. Thus, the theory of coherent and squeezed states in the Schrödinger picture of quantum mechanics is essentially an algebra of creation and annihilation operators applicable to harmonic-oscillator states in Fock space.

In this Chapter, we present the Wigner phase-space picture for the same physical problem. As was emphasized in Chapter 2, the major advantage of this phase-space picture is that both the position and momentum variables are c-numbers. The Wigner function takes a particularly simple form for harmonic oscillators. Furthermore, every coherent or squeezed state can be constructed from the vacuum state through canonical transformations in phase space.

Therefore, the best way to present the phase-space picture of coherent and squeezed states is to translate Chapter 5 into the language of Chapter 4 developed for canonical transformations in quantum phase space. Harmonic oscillators were discussed in considerable detail in Chapter 4. The difference is that the oscillators are now in Fock space. In order to emphasize this difference, we used in Chapter 5 the $q$ coordinate instead of $x$. The phase space therefore consists of $q$ and $p$.

In order to facilitate the theoretical development in Chapter 5, we discussed in Section 5.1 the Baker-Campbell-Hausdorff relation. The mathematical theorem useful in this Chapter is the theorem on invariant subgroups. This theorem will further simplify the theoretical treatment of coherent and squeezed states. According to this theorem, coherent states can be understood in terms of translations in two-dimensional space, and squeezed states as translated squeezed vacuum. We shall therefore start this Chapter with the discussion of invariant subgroups.

## 6.1 Invariant Subgroups

The concept of invariant subgroups will play a very important role in canonical transformations of coherent and squeezed states. We shall therefore study the theory of invariant subgroups in this section.

Let us start with a set of elements $g$ forming a group $G$, satisfying the axioms of group multiplication. Namely, if $g_1$ and $g_2$ are two elements of $G$, then the result of the multiplication $g_1 g_2$ is also in the group; the group has an identity element and each element has its own inverse in the group. In addition, multiplication is associative.

If a subset $H$ consisting of elements of $G$ satisfies the group axioms, then $H$ is called a subgroup of $G$ (Wigner 1959, Gilmore 1974, Kim and Noz 1986). The set $gH$ with $g$ in $G$ is called the *left coset* of $H$. The set $Hg$ is called the *right coset* of $H$. This means that, if $h_1, h_2, \ldots, h_m$ are the element of the subgroup $H$, then the left coset consists of $gh_1, gh_2, \ldots, gh_m$, while the right coset consists of $h_1 g, h_2 g, \ldots, h_m g$. The cosets are not necessarily groups. The right coset is not necessarily the same as the left coset.

On the other hand, it is possible that the right coset be the same as the left coset. Let $G$ have n elements: $g_1, g_2, \ldots, g_n$, with $n > m$, and let the first $m$ of them be the elements of the subgroup $H$. Then it is clear that the cosets $g_i H$ and $H g_i$ are $H$, if $i$ is smaller than $m$. If the subgroup $H$ satisfies the condition

$$g_i H = H g_i \tag{6.1}$$

for all remaining elements of $g$, then $H$ is called an invariant subgroup of $G$. The above condition does not necessarily mean that $g_i h_j = h_j g_i$. It means that the set of $g_i h_1$, $g_i h_2$, $\ldots$, $g_i h_m$ is the same as the set of $h_1 g_i, h_2 g_i, \ldots, h_m g_i$ for every $g_i$.

For practical purposes, the condition for a subgroup to be invariant can be translated into

$$H = g_i H g_i^{-1}, \tag{6.2}$$

for all $g_i$. This means that $g_i h_j g_i^{-1}$ is an element of $H$, but is not necessarily $h_j$ itself.

For example, let us consider the three-dimensional rotation group. In this case, it is possible to make rotations around three orthogonal axes. Rotations around the $z$ axis form a subgroup. Is this an invariant subgroup? For an element of this subgroup, we choose

$$h = \begin{pmatrix} \cos\phi & -\sin\phi & 0 \\ \sin\phi & \cos\phi & 0 \\ 0 & 0 & 1 \end{pmatrix}. \tag{6.3}$$

As for an arbitrary element of $G$, we choose the rotation around the $x$ direction by 90°:

$$g = \begin{pmatrix} 1 & 0 & 0 \\ 0 & 0 & -1 \\ 0 & 1 & 0 \end{pmatrix}, \quad g^{-1} = \begin{pmatrix} 1 & 0 & 0 \\ 0 & 0 & 1 \\ 0 & -1 & 0 \end{pmatrix}. \tag{6.4}$$

Then

$$ghg^{-1} = \begin{pmatrix} \cos\phi & 0 & -\sin\phi \\ 0 & 1 & 0 \\ \sin\phi & 0 & \cos\phi \end{pmatrix}, \tag{6.5}$$

which is clearly a rotation around the $y$ axis. Thus the subgroup consisting of rotations around the $z$ axis is not an invariant subgroup of the three-dimensional rotation group.

Let us consider next the permutation group of three objects. There are six elements in this group - three even permutations and three odd permutations. The three even permutations form a subgroup, while the odd permutations do not. The question is whether the group of even permutations is an invariant subgroup.

In order to answer this question, let us carry out the group multiplications explicitly. We can write the three even permutations as

$$P_1 = \begin{pmatrix} 1 & 2 & 3 \\ 1 & 2 & 3 \end{pmatrix}, \quad P_2 = \begin{pmatrix} 1 & 2 & 3 \\ 2 & 3 & 1 \end{pmatrix}, \quad P_3 = \begin{pmatrix} 1 & 2 & 3 \\ 3 & 1 & 2 \end{pmatrix}, \tag{6.6}$$

where each number in the top row is to be replaced by the number directly below. $P_1$ is the identity element. $P_2$ can be written as

$$P_2 = \begin{pmatrix} 3 & 1 & 2 \\ 1 & 2 & 3 \end{pmatrix}. \tag{6.7}$$

Thus $P_3$ followed by $P_2$ can be written as

$$\begin{aligned} P_2P_3 &= \begin{pmatrix} 1 & 2 & 3 \\ 2 & 3 & 1 \end{pmatrix}\begin{pmatrix} 1 & 2 & 3 \\ 3 & 1 & 2 \end{pmatrix} = \begin{pmatrix} 3 & 1 & 2 \\ 1 & 2 & 3 \end{pmatrix}\begin{pmatrix} 1 & 2 & 3 \\ 3 & 1 & 2 \end{pmatrix} \\ &= \begin{pmatrix} 1 & 2 & 3 \\ 1 & 2 & 3 \end{pmatrix} = P_1. \end{aligned} \tag{6.8}$$

Likewise $P_3P_2 = P_1$. Indeed, $P_1, P_2$, and $P_3$ form a group.

We can consider three additional permutations:

$$P_4 = \begin{pmatrix} 1 & 2 & 3 \\ 2 & 1 & 3 \end{pmatrix}, \quad P_5 = \begin{pmatrix} 1 & 2 & 3 \\ 1 & 3 & 2 \end{pmatrix}, \quad P_6 = \begin{pmatrix} 1 & 2 & 3 \\ 3 & 2 & 1 \end{pmatrix}, \tag{6.9}$$

which are odd permutations. Each of these three permutation is its own inverse, but they do not form a group because the multiplication of two odd permutations results in an even permutation. The multiplication of one even permutation and one odd permutation results in an odd permutation. Therefore, the subgroup of even permutations is an invariant subgroup. More explicitly,

$$P_4P_2P_4^{-1} = P_3, \quad P_5P_2P_5^{-1} = P_3, \quad P_6P_2P_6^{-1} = P_3,$$

$$P_4P_3P_4^{-1} = P_2, \quad P_5P_3P_5^{-1} = P_2, \quad P_6P_3P_6^{-1} = P_2. \tag{6.10}$$

The specific invariant subgroup which is useful in this Chapter is the translation subgroup of the Euclidean group in two-dimensional space consisting of rotations around the origin and translations in two-dimensional space (Wigner 1939, Kim and Noz 1986). Both translations and rotations form their own subgroups. The translations subgroup is an invariant subgroup, while the rotation subgroup is not. We shall see in Section 6.2 that the symmetry in phase space of the coherent state is that of the two-dimensional Euclidean group.

## 6.2 Coherent States

From the expression in Eq. 5.68 for the coherent state in the $q$ or $p$ space, we can write the corresponding Wigner function as

$$W(\alpha; q, p) = \frac{1}{\pi}\exp\left\{-(q-a)^2 - (p-b)^2\right\}, \tag{6.11}$$

where $a = \sqrt{2}Re(\alpha)$ and $b = \sqrt{2}Im(\alpha)$. If $\alpha = 0$, then both $a$ and $b$ vanish, and the above function is concentrated within a circular region around the origin. This Wigner function reproduces the probability distribution functions of Eq. 5.59 through

$$|<q|\alpha>|^2 = \int W(\alpha; q, p)dp, \quad |<p|\alpha>|^2 = \int W(\alpha; q, p)dq. \tag{6.12}$$

The Wigner function of Eq. 6.11 leads to

$$|<\beta|\alpha>|^2 = \int W(\beta; q, p)W(\alpha; q, p)dqdp$$

$$= \exp\left(-|\alpha - \beta|^2\right), \tag{6.13}$$

which is consistent with one of the basic properties of the Wigner function given in Eq. 3.12.

We can obtain the Wigner function of Eq. 6.11 by making canonical transformations of the vacuum-state Wigner function:

$$W(0; q, p) = \frac{1}{\pi}\exp\left\{-\left(q^2 + p^2\right)\right\}. \tag{6.14}$$

More specifically, the Wigner function of Eq. 6.11 is translation of the vacuum state along the $q$ axis by $a$ and along the $p$ axis by $b$.

In Section 5.2, we obtained the coherent state by applying a unitary transformation to the vacuum state. The generators of transformations were $\hat{a}\hat{a}^{\dagger}$, and $\hat{a}^{\dagger}\hat{a}$. It is possible to write these generators in terms of $q$ and $p$ as given in Eq. 5.73, which satisfy Heisenberg's commutation relation of Eq. 5.74. They are equivalent to $\hat{P}_2$ and $\hat{P}_1$ given in Eq. 4.29 respectively. These operators correspond to

$$P_1 = -i\frac{\partial}{\partial q}, \quad P_2 = -i\frac{\partial}{\partial p}, \tag{6.15}$$

in the phase-space picture. $P_1$ and $P_2$ commute with each other, while $\hat{P}_2$ and $\hat{P}_1$ do not. This indeed is one of the advantages of using the phase-space picture of quantum mechanics. As for $\hat{a}^{\dagger}\hat{a}$, it corresponds to

$$L = \frac{i}{2}\left(p\frac{\partial}{\partial q} - q\frac{\partial}{\partial p}\right), \tag{6.16}$$

in phase space. The commutation relations of this operator with $P_1$ and $P_2$ are the same as in the case of the Schrödinger picture.

The Wigner function given in Eq. 6.11 is localized within the circle

$$(q-a)^2 + (p-b)^2 = 1. \tag{6.17}$$

Thus all the instruments developed in Chapter 4 to deal with canonical transformations of a circle are applicable to the Wigner function for the coherent state.

It was noted in Chapter 4 that $P_1, P_2$ and $L$ are the generators of the two-dimensional Euclidean group. Rotations around the origin form the one-parameter subgroup generated by $L$. Translations generated by $P_1$ and $P_2$ form a two-parameter subgroup. The translation operator applicable to the Wigner function is

$$T(\alpha) = \exp\left(-a\frac{\partial}{\partial q} - b\frac{\partial}{\partial p}\right), \tag{6.18}$$

which leads to

$$W(\alpha; q, p) = T(\alpha)W(0; q, p). \tag{6.19}$$

If we multiply $\alpha$ by $e^{i\phi/2}$, the circle corresponding to Eq. 6.17 becomes rotated around the origin, and the resulting equation is

$$(q-a')^2 + (p-b')^2 = 1, \tag{6.20}$$

where

$$\begin{pmatrix} a' \\ b' \end{pmatrix} = \begin{pmatrix} \cos(\phi/2) & -\sin(\phi/2) \\ \sin(\phi/2) & \cos(\phi/2) \end{pmatrix} \begin{pmatrix} a \\ b \end{pmatrix}.$$

The rotation operator applicable to the Wigner function is

$$R(\phi) = \exp\left\{-\frac{\phi}{2}\left(q\frac{\partial}{\partial p} - p\frac{\partial}{\partial q}\right)\right\}, \tag{6.21}$$

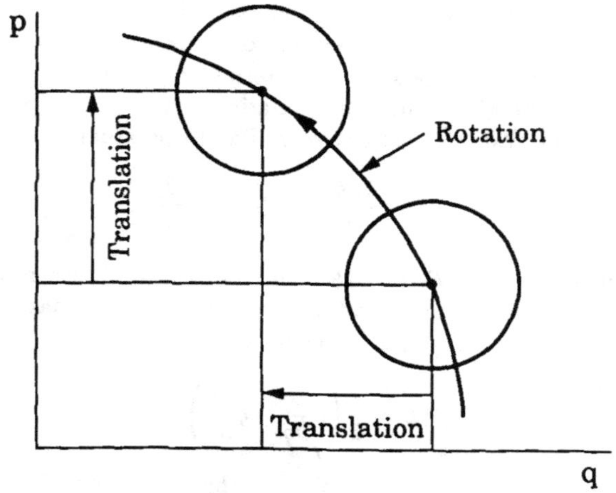


Figure 6.1: Euclidean transformations in phase space. A rotation around the origin results in a translation.

which leads to

$$R(\phi)W(\alpha;q,p) = W(\alpha e^{i\phi/2};q,p). \tag{6.22}$$

As is shown in Figure 6.1, the result of the above rotation is only a translation of the circle centered at the origin to $(a', b')$. Therefore, do we really need the rotation operator? The answer to this question is yes, since the translation subgroup is an invariant subgroup. In order to apply this theorem to the present problem, let us write $T(\alpha)$ and $R(\phi)$ in matrix form:

$$T(\alpha) = \begin{pmatrix} 1 & 0 & a \\ 0 & 1 & b \\ 0 & 0 & 1 \end{pmatrix}, \quad R(\phi) = \begin{pmatrix} \cos\phi & -\sin\phi & 0 \\ \sin\phi & \cos\phi & 0 \\ 0 & 0 & 1 \end{pmatrix}. \tag{6.23}$$

Then it is straight-forward to compute

$$T(\beta)T(\alpha)\,[T(\beta)]^{-1} = T(\alpha), \quad R(\phi)T(\alpha)\,[R(\phi)]^{-1} = T(\alpha'). \tag{6.24}$$

On the other hand, the rotation subgroup is not an invariant subgroup as can be seen from

$$T(\alpha)R(\phi)\,[T(\alpha)]^{-1} = T(-\alpha')T(\alpha)R(\phi). \tag{6.25}$$

In general, the transformation consists of rotations and translations in an arbitrary order. Two repeated rotations can be regarded as a one rotation. Two repeated translations can also be regarded as one translation. Thus, the most general form of transformation takes the form

$$R_n T_n R_{n-1} T_{n-1} \ldots R_1 T_1. \tag{6.26}$$

The invariance property of the translation subgroup allows us to write the most general form of transformation as a rotation followed by a translation, because a translation followed by a rotation can always be written as a translation preceded by a rotation. Let us start with $R(\phi)T(\alpha)$. Then

$$R(\phi)T(\alpha) = \left(R(\phi)T(\alpha)R^{-1}(\phi)\right) R(\phi) = T(\alpha')R(\phi). \tag{6.27}$$

In this way, all the rotation operators can be moved to the right. If the rotation operator is applied to the vacuum state represented by a circle around the origin, it gives no effect. Thus, the net effect is a translation.

## 6.3 Single-Mode Squeezed States

As for the single-mode squeezed state, the squeeze and rotation generators in the Schrödinger picture are given in Eq. 4.25 and Eq. 4.27, where $x$ is to be replaced by $q$. In terms of the creation and annihilation operators, they are in Eq. 5.89 and Eq. 5.92. The rotation generator applicable to the Wigner function is given in Eq. 6.16. The squeeze generators are

$$B_1 = -\left(\frac{i}{2}\right)\left(q\frac{\partial}{\partial q} - p\frac{\partial}{\partial p}\right), \quad B_2 = -\left(\frac{i}{2}\right)\left(q\frac{\partial}{\partial p} + p\frac{\partial}{\partial q}\right). \tag{6.28}$$

Unlike the case of coherent states, the squeeze generators in phase space satisfy the same set of commutation relations as that for the Schrödinger picture. The commutation relations in phase space were discussed in Section 4.1.

For the squeezed state given in Eq. 5.96, the Wigner function takes the form

$$W(\alpha, \lambda; q, p) = \frac{1}{\pi}\exp\left\{-\left[e^{-\lambda}(q - e^{\lambda/2}a)^2 + e^{\lambda}(p - e^{-\lambda/2}b)^2\right]\right\}. \tag{6.29}$$

We can obtain this expression from the definition of the Wigner function or by applying the squeeze operator

$$S(\lambda) = \exp\left\{-\frac{\lambda}{2}\left(q\frac{\partial}{\partial q} - p\frac{\partial}{\partial p}\right)\right\}, \tag{6.30}$$

to the Wigner function for a coherent state given in Eq. 6.11. Unlike the case of coherent states, the distribution for a squeezed state is elliptic. The area in phase space of this ellipse is the same as the circle. The squeezed state is also a minimum uncertainty state.

The squeeze can also be made along an arbitrary direction. Using the matrix representation for the rotation operator and the squeeze generators given in Eq. 4.14 and Eq. 4.15,

$$R(\phi)B_1R(-\phi) = (\cos\phi)B_1 + (\sin\phi)B_2. \tag{6.31}$$

The squeeze therefore requires two parameters, namely $\lambda$ and $\phi$. These parameters are specified by one complex variable $\zeta$ as is given in Eq. 5.98. If the squeeze is made along the direction which makes an angle $\phi/2$ with the $q$ axis, the squeeze operator is

$$S(\zeta) = R(\phi)S(\lambda)R(-\phi)$$

$$= \exp\left\{-\frac{\lambda}{2}(\cos\phi)\left(q\frac{\partial}{\partial q} - p\frac{\partial}{\partial p}\right) - \frac{\lambda}{2}(\sin\phi)\left(q\frac{\partial}{\partial p} + p\frac{\partial}{\partial q}\right)\right\}. \quad (6.32)$$

The effect of this squeeze on the vacuum or coherent state is the same as that one discussed the harmonic oscillator in Section 4.4.

Squeezes do not form a group, but squeezes and rotations form a group. This is called the group of homogeneous linear canonical transformations discussed in detail in Chapter 4. The matrix form of the squeeze operator is

$$S(\zeta) = R(\phi)S(\lambda)R(-\phi)$$

$$= \begin{pmatrix} \cosh\frac{\lambda}{2} + (\cos\phi)\sinh\frac{\lambda}{2} & (\sin\phi)\sinh\frac{\lambda}{2} & 0 \\ (\sin\phi)\sinh\frac{\lambda}{2} & \cosh\frac{\lambda}{2} - (\cos\phi)\sinh\frac{\lambda}{2} & 0 \\ 0 & 0 & 1 \end{pmatrix}, \quad (6.33)$$

applicable to the column matrix of $(q, p, 1)$.

As is seen in Section 4.1, the group of linear canonical transformations in two-dimensional phase space consists of translations, rotations, and squeezes. Translations and rotations have been discussed in detail in Section 6.2. The subgroup of homogeneous linear canonical transformations consists of rotations and squeezes. Translations form their own subgroup. The question then is whether this subgroup is invariant. A simple matrix algebra leads to

$$S(\zeta)T(\alpha)\left[S(\zeta)\right]^{-1} = T(\alpha''), \quad (6.34)$$

with

$$a'' = \left(\cosh\frac{\lambda}{2} + (\cos\phi)\sinh\frac{\lambda}{2}\right)a + \left((\sin\phi)\sinh\frac{\lambda}{2}\right)b,$$

$$b'' = \left((\sin\phi)\sinh\frac{\lambda}{2}\right)a + \left(\cosh\frac{\lambda}{2} - (\cos\phi)\sinh\frac{\lambda}{2}\right)b.$$

Here again, the translation does not commute with squeezes. However, once commuted, it still remains as an element of the translation subgroup even though its parameters are changed. This is a manifestation of the fact that the translation subgroup is an invariant subgroup. Eq. 6.34 can be written as

$$T(\alpha'')S(\zeta) = S(\zeta)T(\alpha). \quad (6.35)$$

The most general form of canonical transformations consists of a product of squeezes, rotations and translations. In view of Eq. 6.32 and Eq. 6.35, the most general form may be written as a translation followed by rotation and squeeze. Let us write this as $TSR$. If this is operated on the vacuum state, $R$ does not give any effect. Thus, the most general form applicable to the vacuum is $TS$, which means a translation of a squeezed vacuum.

## 6.4 Squeezed Vacuum

In view of the conclusion of Section 6.3, the study of squeezed states starts from the squeezed vacuum. Let us study the squeezed vacuum in detail in this section. The Wigner function for the squeezed vacuum takes the form

$$W(0,\lambda;q,p) = \frac{1}{\pi}\exp\left\{-e^{-\lambda}\left(q\cos\frac{\phi}{2}+p\sin\frac{\phi}{2}\right)^2 \right.$$
$$\left. -e^{\lambda}\left(q\sin\frac{\phi}{2}-p\cos\frac{\phi}{2}\right)^2\right\}. \tag{6.36}$$

It is a function of $\lambda$ and $\phi$. The expectation value of $F(q,p)$ is

$$<F>_0 = \frac{1}{\pi}\int F(q,p)\exp\left\{-e^{-\lambda}\left(q\cos\frac{\phi}{2}+p\sin\frac{\phi}{2}\right)^2 \right.$$
$$\left. -e^{\lambda}\left(q\sin\frac{\phi}{2}-p\cos\frac{\phi}{2}\right)^2\right\} dqdp. \tag{6.37}$$

We use the subscript "0" to indicate that the expectation value is taken for the squeezed vacuum. Since the volume element $dqdp$ is invariant under canonical transformations, the integral is invariant under canonical transformation. It is possible to rotate the coordinate to align the major axis of the vacuum state with the $q$ or $p$ axis. Then

$$<F>_0 = \frac{1}{\pi}\int F(q',p')\exp\left\{-\left(q^2+p^2\right)\right\} dqdp, \tag{6.38}$$

where

$$q' = \left(e^{\lambda/2}\cos\frac{\phi}{2}\right)q \ - \left(e^{-\lambda/2}\sin\frac{\phi}{2}\right)p,$$
$$p' = \left(e^{-\lambda/2}\sin\frac{\phi}{2}\right)q + \left(e^{\lambda/2}\cos\frac{\phi}{2}\right)p.$$

This means that $<q>_0$ and $<p>_0$ are zero, but

$$<q^2> = \frac{\cosh\lambda + (\sinh\lambda)\cos\phi}{2},$$
$$<p^2> = \frac{\cosh\lambda - (\sinh\lambda)\cos\phi}{2}. \tag{6.39}$$

As a consequence

$$<q^2><p^2> = \frac{1+(\sin\phi)^2(\sinh\lambda)^2}{4}. \tag{6.40}$$

This is the same as the result obtained in the Schrödinger picture which is mentioned frequently in the literature (Fisher *et al.* 1984), Henry and Glotzer 1988, Schumaker and Caves 1985). In order for the squeezed state to be a minimal uncertainty state, the above quantity can not be greater than 1/4. How can we resolve this puzzle? The answer to this question is in Figure 6.2. The phase-space picture

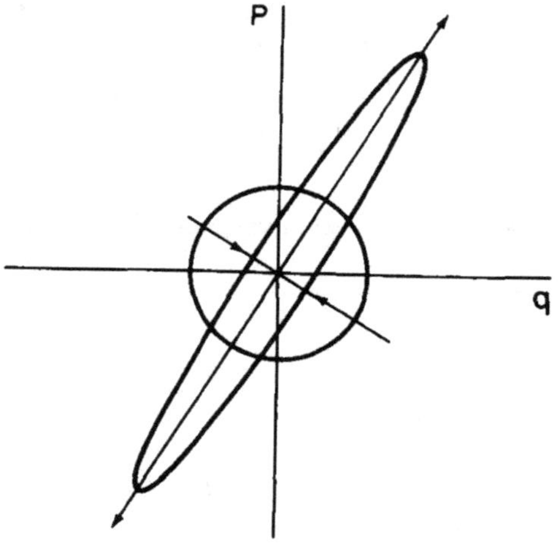


Figure 6.2: Vacuum and squeezed vacuum states. The circle in this figure corresponds to the vacuum state. The tilted ellipse is for the squeezed vacuum. The squeezed vacuum is not a zero-photon state, but is a minimum uncertainty state in the sense that the area of the ellipse is the same as that of the circle. This interpretation is similar to the case of wave-packet spreads discussed in Section 3.3.

allows canonical transformations. The rotation of $W(0,\lambda;q,p)$ to

$$W(0,\lambda;q,p) = \frac{1}{\pi}\exp\left\{-\left(e^{-\lambda}q^2 + e^{\lambda}p^2\right)\right\} \tag{6.41}$$

is a canonical transformation (Han *et al.* 1989b). This expression of course gives the minimal uncertainty product. Indeed, it is possible to quantify the uncertainty in terms of the area in phase space in which the Wigner function is concentrated. This situation is very similar to the case of wave packet spread in which the amount of uncertainty is constant in time, while the uncertainty product in the Schrödinger picture increases as time progresses or regresses. The wave packet spread was discussed in detail in Section 3.3 and Section 4.3.

Among the many operators in quantum mechanics, the photon number and the (photon number)$^2$ operators are the two most important operators in quantum optics. In the Schrödinger picture, the number operator takes the form

$$\hat{N} = \frac{1}{2}\left\{q^2 - \left(\frac{\partial}{\partial q}\right)^2 - 1\right\}. \tag{6.42}$$

Thus, in the phase-space picture, the expectation value of this operator is

$$<N>= \frac{1}{2}\int \left(q^2+p^2-1\right) W(q,p)dqdp, \tag{6.43}$$

which is a straight-forward application of Eq. 3.82. The number operator in phase space is

$$N = \frac{1}{2}\left(q^2+p^2-1\right). \tag{6.44}$$

The expression for $N^2$ is more complicated. However, a straight-forward application of Eq. 3.82 and Eq. 3.87 leads to

$$<N^2>= \frac{1}{4}\int \left\{\left(q^2+p^2-1\right)^2-1\right\} W(q,p)dqdp. \tag{6.45}$$

Thus

$$N^2 = \frac{1}{4}\left\{\left(q^2+p^2-1\right)^2-1\right\} \tag{6.46}$$

Let us next consider $<N>_0$. According to the expression of Eq. 6.44,

$$<N>_0 = \frac{1}{2\pi}\int \left(q^2+p^2-1\right) \exp\left\{-e^{-\lambda}\left(q\cos\frac{\phi}{2}+p\sin\frac{\phi}{2}\right)^2 \right.$$
$$\left. -e^{\lambda}\left(q\sin\frac{\phi}{2}-p\cos\frac{\phi}{2}\right)^2\right\} dqdp, \tag{6.47}$$

Since $(q^2+p^2-1)$ is invariant under rotations, the above integral can be reduced to

$$<N>_0= \frac{1}{2\pi}\int \left(q^2+p^2-1\right) \exp\left\{-\left(e^{-\lambda}q^2+e^{\lambda}p^2\right)\right\} dqdp, \tag{6.48}$$

which can be written as the vacuum expectation value of $\frac{1}{2}\left(e^{\lambda}q^2+e^{-\lambda}p^2-1\right)$:

$$<N>_0= \frac{1}{2\pi}\int \left(e^{\lambda}q^2+e^{-\lambda}p^2-1\right) \exp\left\{-\left(q^2+p^2\right)\right\} dqdp. \tag{6.49}$$

The result is

$$<N>_0= \frac{1}{2}(\cosh\lambda-1), \tag{6.50}$$

which vanishes when $\lambda = 0$, indicating that the number of photons in an unsqueezed vacuum is zero.

Let us next calculate $<N^2>_0$. Since $N^2$ is invariant under rotations,

$$<N^2>_0= \frac{1}{4\pi}\int \left\{\left(e^{\lambda}q^2+e^{-\lambda}p^2-1\right)^2-1\right\} \exp\left\{-\left(q^2+p^2\right)\right\} dqdp. \tag{6.51}$$

The evaluation of this integral leads to

$$<N^2>_0= \frac{1}{4}\left\{3(\cosh\lambda)^2-2(\cosh\lambda)-1\right\}). \tag{6.52}$$

This expression is also independent of the direction of squeeze, and vanishes when $\lambda = 0$. From this, we can calculate $<(\Delta N)^2>_0 = <N^2>_0 - <N>_0^2$. From Eq. 6.50 and Eq. 6.52,

$$<(\Delta N)^2>_0 = \frac{1}{2}(\sinh\lambda)^2. \tag{6.53}$$

It is possible to derive this result within the framework of the Schrödinger picture discussed in Chapter 5. The advantage of using the phase-space picture is that we can take advantage of symmetry properties in phase space. Let us go back to Figure 6.2, for example. In integrating over the entire phase space, we can use the coordinate system in which the major or minor axis of the ellipse coincides with the coordinate axis. The integration is invariant also under translations in phase space. We shall study this aspect in Section 6.5.

## 6.5 Expectation Values in terms of Vacuum Expectation Values

We have discussed in Section 6.4 the squeezed vacuum and the squeezed vacuum expectation values with the Wigner function given in Eq. 6.36 with $\alpha = 0$. For other squeezed states, the expectation value for the number operator is

$$<N> = \frac{1}{2}\int \left(q^2 + p^2 - 1\right) W(\zeta, \alpha; q, p)dqdp, \tag{6.54}$$

and the expectation value of $N^2$ is

$$<N^2> = \frac{1}{4}\int \left\{\left(q^2 + p^2 - 1\right)^2 - 1\right\} W(\zeta, \alpha; q, p)dqdp. \tag{6.55}$$

These integrals are invariant under canonical transformations.

In this section, we are interested in writing the above expressions in terms of squeezed vacuum expectation values. We obtain the Wigner function $W(\zeta, \alpha; q, p)$ by making canonical transformations on the vacuum state. If we apply the squeeze operator to the vacuum before translation, the resulting state is a translated squeezed vacuum.

$$\begin{aligned} W(\alpha, \zeta; q, p) &= T(\alpha)W(0, \zeta; q, p) \\ &= \left(\frac{1}{\pi}\right)\exp\left\{-e^{-\lambda}\left[(q-a)\cos\frac{\phi}{2} + (p-b)\sin\frac{\phi}{2}\right]^2 \right. \\ &\quad \left. - e^{\lambda}\left[(q-a)\sin\frac{\phi}{2} - (p-b)\cos\frac{\phi}{2}\right]^2\right\}. \end{aligned} \tag{6.56}$$

On the other hand, if the translation is applied before squeeze, then

$$W(\zeta, \alpha; q, p) = S(\zeta)W(0, \alpha; q, p) = S(\zeta)T(\alpha)W(0, 0; q, p). \tag{6.57}$$

Since the translations form an invariant subgroup (Han *et al.* 1989b),

$$S(\zeta)T(\alpha) = T(\alpha'')S(\zeta), \tag{6.58}$$

and

$$W(\zeta,\alpha;q,p) = \left(\frac{1}{\pi}\right)\exp\left\{-e^{-\lambda}\left[(q-a'')\cos\frac{\phi}{2} + (p-b'')\sin\frac{\phi}{2}\right]^2 \right.$$
$$\left. - e^{\lambda}\left[(q-a'')\sin\frac{\phi}{2} - (p-b'')\cos\frac{\phi}{2}\right]^2\right\}. \tag{6.59}$$

where $a''$ and $b''$ are given in Eq. 6.34. The above expression is for a translated squeezed vacuum. Indeed, in phase space, every squeezed state is a translated squeezed state. Without loss of generality, we can use the expression of Eq. 6.56 for every squeezed state.

We can now formulate the following theorem. Consider a function $F(q,p)$ and its expectation value in phase space defined as

$$<F> = \int F(q,p)W(\alpha,\zeta;q,p)dqdp. \tag{6.60}$$

Then it can be written as a vacuum expectation value of a canonically transformed $F(q,p)$.

We can prove this theorem by writing the above integration as:

$$<F> = \int F(q,p)\left(T(\alpha)S(\zeta)W(0,0;q,p)\right)dqdp. \tag{6.61}$$

Since the area element $dqdp$ is canonically invariant,

$$<F> = \int \left([S(\zeta)]^{-1}[T(\alpha)]^{-1}F(q,p)\right)W(0,0;q,p)dqdp. \tag{6.62}$$

The above form is a vacuum expectation value of $([S(\zeta)]^{-1}[T(\alpha)]^{-1}F(q,p))$.

It is also possible to write the above expression as

$$<F> = \int \left([T(\alpha)]^{-1}F(q,p)\right)W(0,\zeta;q,p)dqdp, \tag{6.63}$$

which corresponds to the squeezed vacuum expectation value of the translated function $([T(\alpha)]^{-1}F(q,p))$. This form will be useful in establishing parallel axis theorems for a polynomial form of $F(q,p)$ such as the photon number function.

It is possible to carry out the above analysis in the Schrödinger picture, using the techniques discussed in Chapter 5. The basic advantages of using the phase space representation are listed in Chapter 5. In addition, this representation allows us to use the translation subgroup to deal with the $\alpha$ variable.

Let us now calculate the expectation value of the number operator for a squeezed state. The most general form is

$$<N> = \frac{1}{2\pi}\int \left(q^2 + p^2 - 1\right) \exp\left\{-e^{-\lambda}\left[(q-a)\cos\frac{\phi}{2} + (p-b)\sin\frac{\phi}{2}\right]^2 \right.$$
$$\left. - e^{\lambda}\left[(q-a)\sin\frac{\phi}{2} - (p-b)\cos\frac{\phi}{2}\right]^2\right\} dqdp. \tag{6.64}$$

The center of the elliptical distribution is not at the origin. This is not unlike the moment-of-inertia integral in classical mechanics (Goldstein 1980), whose axis is displaced from $(q,p) = (0,0)$ to $(q,p) = (a,b)$. After the translation,

$$<N> = \frac{1}{2\pi}\int \left[(q+a)^2 + (p+b)^2 - 1\right] \exp\left\{-e^{-\lambda}\left(q\cos\frac{\phi}{2} + p\sin\frac{\phi}{2}\right)^2 \right.$$
$$\left. - e^{\lambda}\left(q\sin\frac{\phi}{2} - p\cos\frac{\phi}{2}\right)^2\right\} dqdp. \tag{6.65}$$

The evaluation of the above integral leads to the parallel axis theorem:

$$<N>=<N>_0 + \alpha^*\alpha, \tag{6.66}$$

where $\alpha^*\alpha = \frac{1}{2}(a^2 + b^2)$, which is the photon number for the coherent state centered at $\alpha$.

Let us next compute $<N^2>$. Again the rotational invariance of $(q^2 + p^2)$ leads the most general expression to

$$<N^2> = \frac{1}{4\pi}\int \left\{\left[(q+a)^2 + (p+b)^2 - 1\right]^2 - 1\right\} \exp\left\{-e^{-\lambda}\left(q\cos\frac{\phi}{2} + p\sin\frac{\phi}{2}\right)^2 \right.$$
$$\left. - e^{\lambda}\left(q\sin\frac{\phi}{2} - p\cos\frac{\phi}{2}\right)^2\right\} dqdp. \tag{6.67}$$

The evaluation of this integral leads to the parallel axis theorem for $<N^2>$:

$$<N^2> = <N^2>_0 + \frac{1}{4}\left\{\left(a^2 + b^2\right)^2 - 2\left(a^2 + b^2\right) + 4\left(a^2 + b^2\right)\cosh\lambda \right.$$
$$\left. + 2(\sinh\lambda)\left[(a^2 - b^2)\cos\phi + 2(ab)\sin\phi\right]\right\}. \tag{6.68}$$

The direction of squeeze is $\phi/2$. We can now consider the phase of $\alpha$, which in this case is $\tan^{-1}(p/q)$. Then we can consider the angle between these two directions. Let us call this angle $\delta$, which is equal to $[\phi/2 - \tan^{-1}(p/q)]$. As a result, $<(\Delta N)^2>$ becomes

$$<(\Delta N)^2>=<(\Delta N)^2>_0 + \alpha^*\alpha\left[\cosh\lambda + (\sinh\lambda)\cos(2\delta)\right]. \tag{6.69}$$

In this Section, we established the theorem that every squeezed state is a translated squeezed vacuum. Accordingly, it is possible to express $<N>$ and $<(\Delta N)^2>$ in terms of $<N>_0$ and $<(\Delta N)^2>_0$ respectively. In the case of $<N>$, the difference $(<N> - <N>_0)$ is simply $\alpha\alpha^*$ which measures the distance between the center of the ellipse and the origin in phase space. On the other hand, for $<(\Delta N)^2>$, the difference depends on $\alpha^*\alpha$, the eccentricity of the ellipse, and the direction of squeeze with respect to the phase angle of $\alpha$.

## 6.6 Overlapping Distribution Functions

Let us now consider the transition probability $|(\phi(x), \psi(x)|^2$ when $\phi(x)$ and $\psi(x)$ are the wave functions for two different squeezed states. For convenience, we write this quantity as $| <\alpha_2, \zeta_2|\alpha_1, \zeta_1> |^2$. In terms of the Wigner functions,

$$| <\alpha_2, \zeta_2|\alpha_1, \zeta_1> |^2 = 2\pi \int W(\alpha_2, \zeta_2; q, p) W(\alpha_1, \zeta_1; q, p) dq dp. \tag{6.70}$$

Because of the canonical invariance, the above integral can be brought to the form

$$| <0, -\eta|\alpha, \zeta> |^2 = 2\pi \int W(0, -\eta; q, p) W(\alpha, \zeta; q, p) dq dp, \tag{6.71}$$

where

$$W(0, -\eta; q, p) = \left(\frac{1}{\pi}\right) \exp\left\{-\left(e^{\eta} q^2 + e^{-\eta} p^2\right)\right\}, \tag{6.72}$$

as is shown in Figure 6.3.

We can now squeeze this distribution along the $q$ direction until it becomes the vacuum state described by a circle centered at the origin in Figure 6.3. Figure 6.3 indicates how the other ellipse, which is for $W(\alpha, \zeta; q, p)$ becomes deformed to another ellipse. As is shown in Figure 6.4, we can rotate the system until the principal axis of the ellipse becomes parallel to the $q$ axis:

$$| <0|\alpha', \zeta'> |^2 = \frac{2}{\pi} \int e^{-(q^2+p^2)} \exp\left\{-e^{-\lambda'}(q - a')^2 - e^{\lambda'}(p - b')^2\right\} dq dp. \tag{6.73}$$

This process is essentially a simultaneous diagonalization of two quadratic forms (Aravind 1989, Han *et al.* 1989b). This leads to a product of two integrals:

$$| <0|\alpha', \zeta'> |^2 = \frac{2}{\pi} \left\{\int \exp\left(-e^{-\lambda'}(q - a')^2 - q^2\right) dq\right\} \times \int \exp\left(-e^{\lambda'}(p - b')^2 - p^2\right) dp. \tag{6.74}$$

The result of this integral is

$$| <0|\alpha, \zeta> |^2 = \left(1/\cosh\frac{\lambda'}{2}\right) \exp\left\{-\left[a'/(1 + e^{\lambda'}) + b'^2/(1 + e^{-\lambda'})\right]\right\}. \tag{6.75}$$

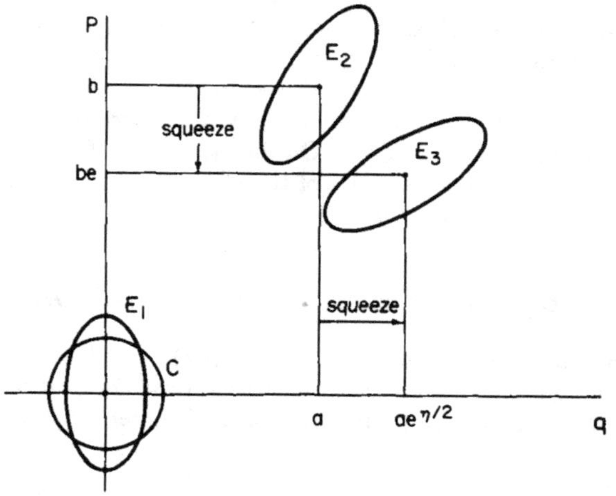


Figure 6.3: Two squeezed states in phase space. Without loss of generality, we can choose the principal axes of the first ellipse ($E_1$) as the coordinate axes. We can squeeze this ellipse until it becomes a circle ($C$) centered at the origin. The other ellipse ($E_2$) becomes another ellipse ($E_3$), with no apparent orientation to the coordinate axes.

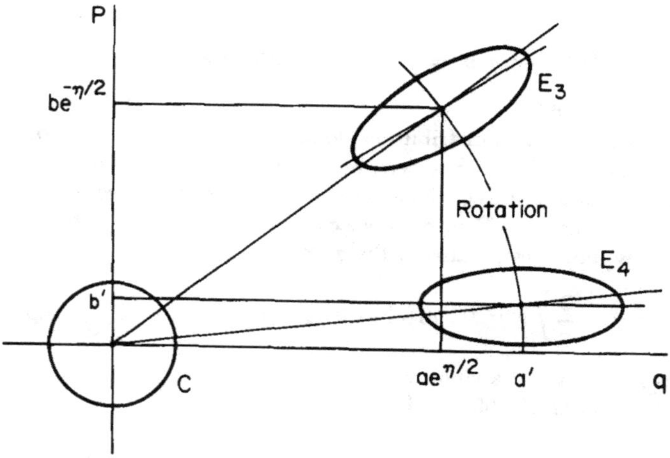


Figure 6.4: Rotation of the vacuum and squeezed states. We can rotate the ellipse ($E_3$) around the origin until its major axis becomes oriented to one of the coordinate axes, as the ellipse ($E_4$) indicates. The center of this ellipse is not necessarily on the coordinate axis.

The above procedure can be extended to the evaluation of the probability of an arbitrary squeezed state being in the n-photon state. The only modification is to replace the vacuum state by the n-photon state:

$$| <n|\alpha', \zeta'> |^2 = 2\int W_n(q,p)\exp\left\{-e^{-\lambda'}(q-a')^2 - e^{\lambda'}(p-b')^2\right\} dqdp, \qquad (6.76)$$

where $W_n(q,p)$ is the Wigner function for the n-photon state, whose form is

$$W_n(q,p) = [(-1)^n/\pi]\left\{L_n\left[2(q^2+p^2)\right]\right\} e^{-(q^2+p^2)}, \qquad (6.77)$$

where $L_n\,[2(q^2+p^2)]$ is the Laguerre polynomial. This expression is invariant under rotations around the origin. Using the decoupling formula for the Laguerre polynomial (Schleich *et al.* 1988), we can write the Wigner function as

$$W_n(q,p) = \frac{1}{\pi}\left(\frac{1}{4}\right)^n e^{-(q^2+p^2)} \sum_{k=0}^{n}\left(\frac{1}{k!(n-k)!}\right) H_{2k}(\sqrt{2}q)H_{2(n-k)}(\sqrt{2}p), \qquad (6.78)$$

where $H_{2k}(\sqrt{2}q)$ and $H_{2(n-k)}(\sqrt{2}p)$ are the Hermite polynomials. We can then go back to the expression for the transition probability in Eq. 6.76, and write the overlap integral as

$$| <n|\alpha', \zeta'> |^2 = \left(\frac{1}{4}\right)^n \sum_{k=0}^{n}\left(\frac{1}{k!(n-k)!}\right) I_k(a', \lambda')I_{n-k}(b', -\lambda'), \qquad (6.79)$$

with

$$I_k(a,\lambda) = \left(\frac{2}{\pi}\right)^{1/2} \int H_{2k}(q)\exp\left[-x^2/2 - e^{-\lambda}(q/\sqrt{2}-a)^2\right] dq.$$

This integral is readily available in the literature (Gradsteyn and Ryzhik 1965, Schleich *et al.* 1988), and the result is

$$I_k(a,\lambda) = \left[1/(e^{-\lambda}+1)\right]\left(-\tanh\frac{\lambda}{2}\right)^{k/2} H_k\left(\sqrt{2}a/(1-e^{\lambda})^{1/2}\right). \qquad (6.80)$$

This function is analytic in $\lambda$ at $\lambda = 0$, and is therefore a well defined function for both positive and negative values of $\lambda$.

## 6.7 Thomas Effect

It was shown in Sections 4.4 and 4.5 that the algebraic property of the group of homogeneous linear canonical transformations is the same as that of the (2 + 1)-dimensional Lorentz group. Two successive Lorentz boosts in different directions, starting from a particle at rest, do not end up with the same result. The difference is a rotation of the particle (Han *et al.* 1988). If the particle has no spin, then it gives no effect. If the particle has an intrinsic spin, it will result in a rotation of the spin. This effect is known in the literature as the Thomas precession (Thomas 1927, Goldstein 1980). As was pointed out in Section 4.5, this is also known as the Wigner rotation (Ritus 1961, Chakrabarti 1964, Han *et al.* 1986 and 1987b).

It is now possible to translate the study of the Thomas rotation in terms of the phase-space formulation of coherent and squeezed states. Let us consider here a special case where the second Lorentz boost is perpendicular to the first boost (Chiao and Jordan 1988), in such a way that the resulting momentum will make an angle of 45° with the direction of the first boost. For this purpose, let us boost the four-momentum of a particle at rest $(0, 0, m)$ along the $x$ direction first. We use here the metric $x^\mu = (x, y, t)$. The $z$ component has been omitted for convenience. The transformation matrix is

$$S_x(\eta) = \begin{pmatrix} \cosh\eta & 0 & \sinh\eta \\ 0 & 1 & 0 \\ \sinh\eta & 0 & \cosh\eta \end{pmatrix}. \tag{6.81}$$

The resulting four-momentum is $m(\sinh\eta, 0, \cosh\eta)$. We then boost this four-momentum along the $y$ direction with the matrix:

$$S_y(\lambda) = \begin{pmatrix} 1 & 0 & 0 \\ 0 & \cosh\lambda & \sinh\lambda \\ 0 & \sinh\lambda & \cosh\lambda \end{pmatrix}, \tag{6.82}$$

Then the four-momentum becomes $m\,(\sinh\eta, (\sinh\lambda)\cosh\eta, (\cosh\lambda)\cosh\eta)$. In order that the new momentum make an angle of 45°, its $x$ and $y$ components must be the same, and therefore

$$\sinh\lambda = \tanh\eta. \tag{6.83}$$

The transformations $S_y(\lambda)S_x(\eta)$ and $S_x(\lambda)S_y(\eta)$, with $\sinh\lambda = \tanh\eta$, on a particle at rest will give the same momentum along the direction which makes an angle of 45° with the $x$ and $y$ axes. However, are the transformation matrices identical to each other? The answer to this question is NO.

$$[S_x(\lambda)S_y(\eta)]^{-1}\, S_y(\lambda)S_x(\eta) = R(\Omega), \tag{6.84}$$

where

$$R(\Omega) = \begin{pmatrix} \cos\Omega & -\sin\Omega & 0 \\ \sin\Omega & \cos\Omega & 0 \\ 0 & 0 & 1 \end{pmatrix},$$

with

$$\sin\Omega = (\tanh\eta)^2. \tag{6.85}$$

We can perform this experiment in an optics laboratory using squeezed states of light. Instead of using $S_x(\eta)$ and $S_y(\lambda)$ for special relativity, we can use the following expression applicable to a coherent state of light with a given parameter $\alpha$.

$$S_x(\eta) = \begin{pmatrix} e^{\eta/2} & 0 \\ 0 & e^{-\eta/2} \end{pmatrix}, \quad S_y(\eta) = \begin{pmatrix} \cosh(\lambda/2) & \sinh(\lambda/2) \\ \sinh(\lambda/2) & \cosh(\lambda/2) \end{pmatrix}. \tag{6.86}$$

Then we can carry out the same calculation as the one leading to Eq. 6.84, with

$$R(\Omega) = \begin{pmatrix} \cos(\Omega/2) & -\sin(\Omega/2) \\ \sin(\Omega/2) & \cos(\Omega/2) \end{pmatrix}. \tag{6.87}$$

Each of these two-by-two matrices represents a concrete measurable operation in modern optics laboratories. $S_y(\lambda)S_x(\eta)$ and $S_x(\lambda)S_y(\eta)$ represent two repeated squeezes. The difference between these two operations is the rotation $R(\theta)$ of Eq. 6.87 which represents a phase change in $\alpha$. This angle can be measured (Han *et al.* 1988, Chiao and Jordan 1988, Chiao 1989, Aravind 1990).

It is interesting to note that quantum optics with squeezed states can serve as an analog computer for special relativity . This analogy is illustrated in Figure 6.5.

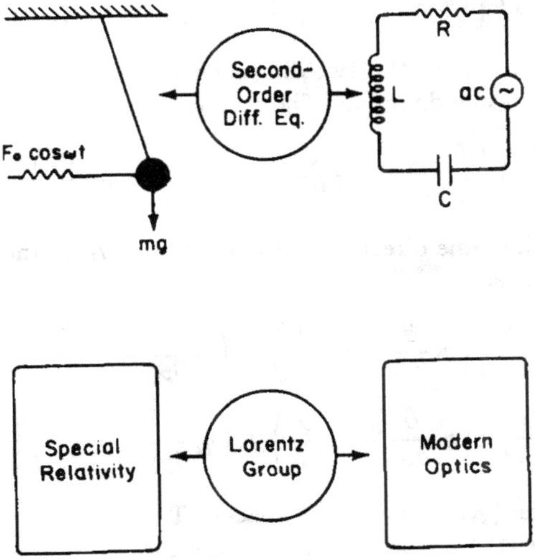


Figure 6.5: Special relativity and modern optics. The analogy between the forced harmonic oscillator and the driven LCR circuit. This analogy will enable us to study special relativity in terms of modern optics and vice versa.

## 6.8 Two-Mode Squeezed States

We have seen in Section 6.7 that the phase-space symmetry of squeezed states of light has an intimate connection with the (2 + 1)-dimensional Lorentz group. It was noted in Section 5.6 that the basic symmetry of two-mode squeezed states is that of O(3,2). While most of the squeezed states produced in laboratories are two-mode states, the symmetry of interest is still that of the (2 + 1)-dimensional

Lorentz group (Yurke *et al.* 1986). We are therefore naturally interested in the (2 + 1)-like symmetry of the two-mode squeezed states.

Let us start with the Wigner function for the two-mode vacuum state:

$$W(q_1, q_2, p_1, p_2) = \left(\frac{1}{\pi}\right)^2 \exp\left\{-\left(q_1^2 + q_2^2 + p_1^2 + p_2^2\right)\right\}. \tag{6.88}$$

The symmetry property of this Wigner function was discussed in Section 4.6. The only difference is that the $x_1$ and $x_2$ variables are now replaced by $q_1$ and $q_2$ respectively. It was noted there that the four-dimensional phase space can be divided into two two-dimensional spaces.

There are three possibilities. The operator $J_1$,

$$J_1 = +\left(\frac{i}{2}\right)\left\{\left(q_1\frac{\partial}{\partial p_2} - p_2\frac{\partial}{\partial q_1}\right) + \left(q_2\frac{\partial}{\partial p_1} - p_1\frac{\partial}{\partial q_2}\right)\right\}, \tag{6.89}$$

generates separated rotations in the two-dimensional spaces of $(q_1p_2)$ and $(q_2p_1)$. The rotations are in the same direction. The operator $J_2$, which takes the form

$$J_2 = -\left(\frac{i}{2}\right)\left\{\left(q_1\frac{\partial}{\partial q_2} - q_2\frac{\partial}{\partial q_1}\right) + \left(p_1\frac{\partial}{\partial p_2} - p_2\frac{\partial}{\partial p_1}\right)\right\}, \tag{6.90}$$

generates rotations in the same direction in the spaces of $(q_1q_2)$ and $(p_1p_2)$. $J_0$ and $J_3$, which can be written as

$$J_3 = +\left(\frac{i}{2}\right)\left\{\left(q_1\frac{\partial}{\partial p_1} - p_1\frac{\partial}{\partial q_1}\right) - \left(q_2\frac{\partial}{\partial p_2} - p_2\frac{\partial}{\partial q_2}\right)\right\},$$

$$J_0 = +\left(\frac{i}{2}\right)\left\{\left(q_1\frac{\partial}{\partial p_1} - p_1\frac{\partial}{\partial q_1}\right) + \left(q_2\frac{\partial}{\partial p_2} - p_2\frac{\partial}{\partial q_2}\right)\right\}, \tag{6.91}$$

generate rotations in the $(q_1p_1)$ and $(q_2p_2)$ spaces. The rotations are in the same direction for $J_0$ and in the opposite directions in the case of $J_3$.

From the expressions given in Eq. 4.64, we rewrite the operators $K_2$ and $K_3$ as

$$K_2 = -\left(\frac{i}{2}\right)\left\{\left(q_1\frac{\partial}{\partial q_1} - p_2\frac{\partial}{\partial p_2}\right) + \left(q_2\frac{\partial}{\partial q_2} - p_1\frac{\partial}{\partial p_1}\right)\right\},$$

$$K_3 = +\left(\frac{i}{2}\right)\left\{\left(q_1\frac{\partial}{\partial p_2} + p_2\frac{\partial}{\partial q_1}\right) + \left(q_2\frac{\partial}{\partial p_1} + p_1\frac{\partial}{\partial q_2}\right)\right\}. \tag{6.92}$$

It is clear that $J_1$, $K_2$, and $K_3$ will satisfy the set of commutation relations for (2 + 1)-like transformations in the $(q_1p_2)$ and $(q_2p_1)$ spaces separately. Likewise, the subgroups generated by $J_2, K_3, K_1$ and by $J_3, K_1, K_2$ perform (2 + 1)-like transformations in two separate two-dimensional phase spaces.

The story is quite different for the three sets of commutation relations

$$[Q_i, Q_j] = -i\varepsilon_{ijk}J_k, \quad [J_i, Q_j] = i\varepsilon_{ijk}Q_k, \tag{6.93}$$

for $J_1, J_2$, and $J_3$. Transformations generated by these operators are also like those of the (2 + 1)-dimensional Lorentz group. However, these transformations are not separable into two two-dimensional phase spaces. On the other hand, according to Eq. 4.83, each $Q_i$ can be transformed into $K_i$ by a rotation generated by $J_0$, while $J_i$ remains invariant under the same rotation according to Eq. 4.78.

Let us next consider the (2 + 1)-like subgroup generated by $J_0, K_1$, and $Q_1$, where

$$K_1 = -\left(\frac{i}{2}\right)\left\{\left(q_1\frac{\partial}{\partial p_1} + p_1\frac{\partial}{\partial q_1}\right) - \left(q_2\frac{\partial}{\partial p_2} + p_2\frac{\partial}{\partial q_2}\right)\right\},$$

$$Q_1 = +\left(\frac{i}{2}\right)\left\{\left(q_1\frac{\partial}{\partial q_1} - p_1\frac{\partial}{\partial p_1}\right) - \left(q_2\frac{\partial}{\partial q_2} - p_2\frac{\partial}{\partial p_2}\right)\right\}. \tag{6.94}$$

It is quite clear that this subgroup also performs transformations in two separate two-dimensional phase spaces. As for $J_0, K_2, Q_2$, we write $Q_2$ as

$$Q_2 = -\left(\frac{i}{2}\right)\left\{\left(q_1\frac{\partial}{\partial p_1} + p_1\frac{\partial}{\partial q_1}\right) + \left(q_2\frac{\partial}{\partial p_2} + p_2\frac{\partial}{\partial q_2}\right)\right\}, \tag{6.95}$$

while $K_2$ is given in Eq. 6.92.

As for $J_0, K_3, Q_3$, where $Q_3$ can be written as

$$Q_3 = -\left(\frac{i}{2}\right)\left\{\left(q_2\frac{\partial}{\partial q_1} + q_1\frac{\partial}{\partial q_2}\right) - \left(p_2\frac{\partial}{\partial p_1} + p_1\frac{\partial}{\partial p_2}\right)\right\}, \tag{6.96}$$

while $K_3$ is given in Eq. 6.92. $J_0$ is separable in $(q_1p_1)(q_2p_2)$, while $K_3$ and $Q_3$ are separable in $(q_1p_2)(q_2p_1)$ and $(q_1q_2)(p_1p_2)$ respectively. However, both $K_3$ and $Q_3$ can be transformed into $K_1$ and $Q_1$ through a rotation generated by $J_2$, while $J_0$ remains invariant under the same rotation.

There are altogether nine (2 + 1)-like subgroups in the (3 + 2)-like symmetry group of two-mode squeezed states. They are either separable or can be transformed into separable representations. Therefore, experiments such as the Thomas rotation discussed in Section 6.7 can be designed in the two-mode framework.

As for other properties of the two mode state, the overlap integral of Eq. 6.70 can generalized to

$$|(\phi(q_1,q_2),\psi(q_1,q_2))|^2$$
$$= (2\pi)^2\int W_\psi(q_1,q_2;p_1,p_2)W_\phi(q_1,q_2;p_1,p_2)dq_1dq_2dp_1dp_2. \tag{6.97}$$

As for the expectation value, the generalization is straight-forward if the operator is an addition of those for the first and second modes. For example

$$<(N_1+N_2)>=<N_1> + <N_2>, \tag{6.98}$$

where

$$<N_1> = <\left(J_0 + J_3 + \frac{1}{2}\right)> = \frac{1}{2}\int \left(q_1^2 + p_1^2 - 1\right) W(q_1, q_2, p_1, p_2) dq_1 dq_2 dp_1 dp_2,$$

$$<N_2> = <\left(J_0 - J_3 + \frac{1}{2}\right)> = \frac{1}{2}\int \left(q_2^2 + p_2^2 - 1\right) W(q_1, q_2, p_1, p_2) dq_1 dq_2 dp_1 dp_2.$$

Likewise,

$$<N_1^2> = \frac{1}{4}\int \left[\left(q_1^2 + p_1^2 - 1\right)^2 - 1\right] W(q_1, q_2, p_1, p_2) dq_1 dq_2 dp_1 dp_2,$$

$$<N_2^2> = \frac{1}{4}\int \left[\left(q_2^2 + p_2^2 - 1\right)^2 - 1\right] W(q_1, q_2, p_1, p_2) dq_1 dq_2 dp_1 dp_2. \quad (6.99)$$

In the above cases, the integration over one of the modes is trivial. The expectation value is not additive in the case of

$$<N_1 N_2> = \frac{1}{4}\int \left(q_2^2 + p_2^2 - 1\right)\left(q_1^2 + p_1^2 - 1\right) W(q_1, q_2, p_1, p_2) dq_1 dq_2 dp_1 dp_2. \quad (6.100)$$

However, also in this case, the calculation is straight-forward. These quantities are needed in calculating the photon number variations: $<(\Delta N)^2>$, $<(\Delta N_1)^2>$, and $<(\Delta N_2)^2>$.

## 6.9 Contraction of Phase Space

We discussed in Section 6.8 two-photon squeezed states assuming that both of the photons are observable. It is quite possible in laboratory experiments that only one of the photons is observable while the other is not (Yurke and Potasek 1987). In the phase-space picture of quantum mechanics, we achieve this procedure by integrating the two-mode Wigner function over the phase space governing the second photon. We call this reduction in the dimension of phase space "contraction of phase space".

If the phase space can be separated into those of the first and second photons, namely $(q_1 p_1)(q_2 p_2)$, the problem is reduced to that of a single-mode squeezed state after the integration of the Wigner function over the second phase space. If the phase space is not separable, what would be the effect of the unobserved photon?

For example, let us consider the squeeze generated by $Q_3$ of Eq. 6.96 which is not separable into the $(q_1 p_1)$ and $(q_2 p_2)$ spaces. If we apply the squeeze operator $\exp(-i\eta Q_3)$ to the vacuum-state Wigner function, then

$$W_\eta(q_1, p_1; q_2, p_2) = \left(\frac{1}{\pi}\right)^2 \exp\left\{-\left(\frac{1}{2}\right)\left(e^{-\eta}(q_1 + q_2)^2 + e^{\eta}(p_1 - p_2)^2\right)\right\}$$
$$\times \exp\left\{-\left(\frac{1}{2}\right)\left(e^{\eta}(q_1 - q_2)^2 + e^{-\eta}(p_1 + p_2)^2\right)\right\}. \quad (6.101)$$

This expression is not separable into the $(q_1p_1)$ and $(q_2p_2)$ spaces, and the integration over the second phase space affects the distribution in the first phase space. Let us contract the phase space by integrating over $q_2$ and $p_2$:

$$W_\eta(q_1,p_1) = \int W_\eta(q_1,p_1;q_2,p_2)\,dq_2dp_2. \tag{6.102}$$

The integration is straight-forward, and the result is

$$W_\eta(q,p) = \left(\frac{1}{\pi}\right)\left(\frac{1}{\cosh\eta}\right)\exp\left\{-\left(q^2+p^2\right)/\cosh\eta\right\}, \tag{6.103}$$

where $q_1$ and $p_1$ have been for simplicity replaced by $q$ and $p$ respectively.

The above Wigner function is normalized but becomes widespread as $\eta$ increases. This expression can now be compared with the Wigner function for a thermally excited oscillator state discussed in Section 3.5. In that case, the Wigner function is

$$W_T(q,p) = \left\{\left(\frac{1}{\pi}\right)\tanh\left(\frac{1}{2kT}\right)\right\} \times\exp\left\{-\left(q^2+p^2\right)\tanh\left(\frac{1}{2kT}\right)\right\}. \tag{6.104}$$

We can now compare $1/\cosh\eta$ with $\tanh(1/2kT)$. If $T$ approaches zero, $\tanh(1/2kT)$ becomes 1 and consequently $\eta$ is zero.

We can obtain the Wigner function $W_T(q,p)$ of Eq. 6.104 by applying the expansion operator $E(T)$ to the $T=0$ Wigner function:

$$W_T(q,p) = E(T)W_0(q,p), \tag{6.105}$$

with

$$E(T) = \exp\left\{-\left(\ln\left[\tanh(1/2T)\right]\right)\left(q\frac{\partial}{\partial q}+p\frac{\partial}{\partial p}+1\right)\right\}.$$

The operator $E(T)$ is invariant under rotations around the origin in phase space. This is not an area-preserving canonical transformation in phase space. Therefore, it is not a unitary transformation in the Schrödinger picture of quantum mechanics. This is a clear indication that thermally excited states are not pure states, as we discussed in Section 2.5.

The generator for radial expansion in phase space is

$$F = -i\left(q\frac{\partial}{\partial q}+p\frac{\partial}{\partial p}+1\right). \tag{6.106}$$

This generator commutes with the rotation generator $L$, and also with the squeeze generators of Eq. 6.28:

$$[F,L]=0, \quad [F,K_1]=0, \quad [F,K_2]=0. \tag{6.107}$$

However, the expansion generator does not commute with the translation generators, and their commutation relations are

$$[F, P_1] = iP_1, \quad [F, P_2] = iP_2. \tag{6.108}$$

The expansion leads to a scale change both in the $q$ and $p$ directions. In the matrix representation applicable to the column vector of $(q, p, 1)$, the expansion generator takes the form

$$F = \begin{pmatrix} i & 0 & 0 \\ 0 & i & 0 \\ 0 & 0 & 0 \end{pmatrix}. \tag{6.109}$$

# Chapter 7

# LORENTZ TRANSFORMATIONS

The group of Lorentz transformations is called the Lorentz group. The group of Lorentz transformations and space-time translations is called the inhomogeneous Lorentz group or the Poincaré group (Wigner 1939, Bargmann and Wigner 1948). Since the Poincaré group governs the basic space-time symmetry of elementary particles, there are many books and review articles on this subject (Kim and Noz 1986, Noz and Kim 1988). In this book, we are interested in the squeeze property of this group.

In Chapters 4 and 6, the Lorentz groups O(2,1) and O(3,2) played important roles in squeezed states of light through their isomorphisms to Sp(2) and Sp(4) respectively. However, we are now interested in the direct role the Lorentz group plays in Lorentz transformations of relativistic extended particles. Is the Lorentz group capable of squeezing those particles? Can we then observe the effect of squeeze in high-energy laboratories? In order to answer these questions, it is necessary to construct a covariant phase-space picture of relativistic extended particles.

In this Chapter, we shall first discuss Wigner's little groups which govern the internal space-time symmetries of relativistic particles. It is known that the little groups are isomorphic to O(3), E(2), and O(2,1) respectively for massive, massless, and imaginary-mass particles. However, we shall discuss a representation which allows us to make analytic continuations from one to another little group.

As for Lorentz boosts along a given direction, it is known that the light-cone coordinate system is the most effective scientific language. It will be seen in this Chapter that the Wigner function in the light-cone coordinate system undergoes a squeeze when the system is boosted. A localized light wave is discussed in detail as an illustrative example. The squeeze property in the covariant harmonic oscillator model will be discussed in Chapter 8.

## 7.1 Group of Lorentz Transformations

In the Minkowskian world, there are three space dimensions and one time dimension. It is therefore possible to make Lorentz boosts along the three different directions. If we boost along the $z$ direction, the transformation matrix is

$$\begin{pmatrix} 1 & 0 & 0 & 0 \\ 0 & 1 & 0 & 0 \\ 0 & 0 & \cosh\eta & \sinh\eta \\ 0 & 0 & \sinh\eta & \cosh\eta \end{pmatrix}, \tag{7.1}$$

applicable to the column vector of $(x, y, z, t)$. This transformation matrix can be exponentiated as $\exp(-i\eta K_3)$, where $K_3$ is the generator of boosts along the $z$ direction and takes the form

$$K_3 = \begin{pmatrix} 0 & 0 & 0 & 0 \\ 0 & 0 & 0 & 0 \\ 0 & 0 & 0 & i \\ 0 & 0 & i & 0 \end{pmatrix}. \tag{7.2}$$

The boost generators along the $x$ and $y$ directions are

$$K_1 = \begin{pmatrix} 0 & 0 & 0 & i \\ 0 & 0 & 0 & 0 \\ 0 & 0 & 0 & 0 \\ i & 0 & 0 & 0 \end{pmatrix}, \quad K_2 = \begin{pmatrix} 0 & 0 & 0 & 0 \\ 0 & 0 & 0 & i \\ 0 & 0 & 0 & 0 \\ 0 & i & 0 & 0 \end{pmatrix}, \tag{7.3}$$

As we noted in Chapters 4 and 6, the commutator $[K_1, K_2]$ results in the generator of rotations around the $z$ axis. The rotation generators are

$$J_1 = \begin{pmatrix} 0 & 0 & 0 & 0 \\ 0 & 0 & -i & 0 \\ 0 & i & 0 & 0 \\ 0 & 0 & 0 & 0 \end{pmatrix}, \quad J_2 = \begin{pmatrix} 0 & 0 & i & 0 \\ 0 & 0 & 0 & 0 \\ -i & 0 & 0 & 0 \\ 0 & 0 & 0 & 0 \end{pmatrix},$$

$$J_3 = \begin{pmatrix} 0 & -i & 0 & 0 \\ i & 0 & 0 & 0 \\ 0 & 0 & 0 & 0 \\ 0 & 0 & 0 & 0 \end{pmatrix}. \tag{7.4}$$

These boost and rotation generators constitute the generators of the Lorentz group, satisfying the commutation relations:

$$[K_i, K_j] = -ie_{ijk}J_k, \quad [J_i, K_j] = ie_{ijk}K_k,$$

$$[J_i, J_j] = ie_{ijk}J_k. \tag{7.5}$$

There are many different forms of operators satisfying the above commutation relations. For functions of $x$, $y$, $z$, at $t$, the generators take the differential form:

$$J_1 = -i\left(y\frac{\partial}{\partial z} - z\frac{\partial}{\partial y}\right), \quad J_2 = -i\left(z\frac{\partial}{\partial x} - x\frac{\partial}{\partial z}\right), \quad J_3 = -i\left(x\frac{\partial}{\partial y} - y\frac{\partial}{\partial x}\right),$$

$$K_1 = -i\left(x\frac{\partial}{\partial t} + t\frac{\partial}{\partial x}\right), \quad K_2 = -i\left(y\frac{\partial}{\partial t} + t\frac{\partial}{\partial y}\right), \quad K_3 = -i\left(z\frac{\partial}{\partial t} + t\frac{\partial}{\partial z}\right). \quad (7.6)$$

In Chapter 8, we shall discuss in detail a normalizable representation space in which the above operators are Hermitian.

The Lorentz group has several interesting three-parameter subgroups. The three rotation generators form a closed set of commutation relations. They generate the rotation subgroup of the Lorentz group. It was noted in Chapter 6 that $K_1$, $K_2$ and $J_3$ generate a (2 + 1)-dimensional Lorentz group. The Lorentz group clearly has two other such subgroups generated by $K_2$, $K_3$, $J_1$ and $K_3$, $K_1$ and $J_2$ respectively. There are other subgroups which will be discussed in Section 7.2.

The smallest matrix representation for the generators of the Lorentz group consists of $J_i = \frac{1}{2}\sigma_i$ and $K_i = \frac{i}{2}\sigma_i$, where $\sigma_i$ are the Pauli spin matrices. The commutation relations remain invariant under the sign change of $K_i$ to $-K_i$. This is why the Dirac matrices are four-by-four which accommodate both signs of the boost generator (Kim and Noz 1986). The representation of the Lorentz group by the two-by-two Pauli matrices is called SL(2,c) in the literature. It has three SU(1,1) subgroups isomorphic to the (2 + 1)-dimensional Lorentz group. These subgroups are in turn unitarily equivalent to the group Sp(2). For instance, the representation of the SU(1,1) group generated by $K_1 = \frac{i}{2}\sigma_1$, $K_2 = \frac{i}{2}\sigma_2$, and $J_3 = \frac{1}{2}\sigma_3$ is a complex representation. This group is unitarily equivalent to the group generated by $K_1 = \frac{i}{2}\sigma_3$, $K_2 = \frac{i}{2}\sigma_1$, and $J_3 = \frac{1}{2}\sigma_2$ (Kim and Noz 1983), which constitute the generators of the group Sp(2) discussed in Chapters 4 and 6.

## 7.2 Little Groups of the Lorentz Group

The little group of the Lorentz group is the maximal subgroup of the Lorentz group whose transformations leave a given four-momentum vector invariant. The little group therefore governs the internal space-time symmetry of relativistic particles.

In his original paper (Wigner 1939) observed that the four-momentum can be time-like, light-like, or space-like. If the momentum is time-like, there is a Lorentz frame in which the particle is at rest or its four-momentum is proportional to $(0,0,0,1)$. If the particle is light-like, there is not a Lorentz frame in which it is at rest. However, there is a frame in which the four-momentum is proportional to $(0,0,1,1)$ or $(0,0,1,-1)$. If the four-momentum is space-like, it is possible to find the frame in which the momentum is proportional to $(0,0,1,0)$.

Let us consider a massive particle at rest. The little group in this case is the rotation subgroup. Since the particle is at rest, its momentum is not affected by a rotation. However, the rotation will change the direction of the particle spin. Indeed, the concept of spin is associated with the symmetry of the little group.

As for a space-like particle whose four-momentum is proportional to $(0,0,1,0)$, the little group is the $(2+1)$-dimensional Lorentz group generated by $J_3$, $K_1$ and $K_2$. Properties of this group have been discussed in Chapters 4 and 6.

For a light-like particle whose four-momentum is $(0,0,1,1)$, rotations around the $z$ axis certainly leave this four-momentum invariant. In addition, the transformation matrices generated by $N_1 = K_1 - J_2$ and $N_2 = K_2 + J_1$ leave the four-momentum invariant. These generators take the form

$$N_1 = \begin{pmatrix} 0 & 0 & -i & i \\ 0 & 0 & 0 & 0 \\ i & 0 & 0 & 0 \\ i & 0 & 0 & 0 \end{pmatrix}, \quad N_2 = \begin{pmatrix} 0 & 0 & 0 & 0 \\ 0 & 0 & -i & i \\ 0 & i & 0 & 0 \\ 0 & i & 0 & 0 \end{pmatrix}. \tag{7.7}$$

These generators satisfy the commutation relations:

$$[N_1, N_2] = 0, \quad [J_3, N_1] = iN_2, \quad [J_3, N_2] = iN_1. \tag{7.8}$$

This set of commutation relations is exactly the same as that for the two- dimensional Euclidean group discussed in Section 6.2 in connection with the coherent states of light. This observation was first made by Wigner in 1939.

The $N_1$ and $N_2$ matrices have the following properties. $N_1 N_2 = N_2 N_1 = 0$, $N_1^3 = N_2^3 = 0$, and

$$N_1^2 = N_2^2 = \begin{pmatrix} 0 & 0 & 0 & 0 \\ 0 & 0 & 0 & 0 \\ 0 & 0 & 1 & -1 \\ 0 & 0 & 1 & -1 \end{pmatrix}. \tag{7.9}$$

This enables us to calculate the transformation matrix:

$$\begin{aligned} D(\sigma, \rho) &= \exp\left(-i\sigma N_1 - i\rho N_2\right) \\ &= I - i\left(\sigma N_1 + \rho N_2\right) - \frac{1}{2}\left(\sigma^2 + \rho^2\right) N_1^2 \\ &= \begin{pmatrix} 1 & 0 & -\sigma & \rho \\ 0 & 1 & -\sigma & \rho \\ \xi & \eta & 1-(\sigma^2+\rho^2)/2 & (\sigma^2+\rho^2/2 \\ \xi & \eta & -(\sigma^2+\rho^2)/2 & 1+(\sigma^2+\rho^2)/2 \end{pmatrix}. \end{aligned} \tag{7.10}$$

In the regime of the two-by-two matrix representation of SL(2,c), $N_1$ and $N_2$ are

$$N_1 = \begin{pmatrix} 0 & i \\ 0 & 0 \end{pmatrix}, \quad N_2 = \begin{pmatrix} 0 & 1 \\ 0 & 0 \end{pmatrix}, \tag{7.11}$$

or

$$N_1 = \begin{pmatrix} 0 & 0 \\ i & 0 \end{pmatrix}, \quad N_2 = \begin{pmatrix} 0 & 0 \\ -1 & 0 \end{pmatrix}, \tag{7.12}$$

depending on sign of the boost generators. In this case, $N_1 N_2 = N_1^2 = N_2^2 = 0$. As a consequence (Wigner 1939)

$$D(\sigma, \rho) = \exp(-i\sigma N_1 - i\rho N_2) = I - i(\sigma N_1 + \rho N_2)$$

$$= \begin{pmatrix} 1 & \sigma - i\rho \\ 0 & 1 \end{pmatrix} \quad \text{or} \quad \begin{pmatrix} 1 & 0 \\ -\sigma + i\rho & 1 \end{pmatrix}. \tag{7.13}$$

As for the Hermitian generators applicable to the functions of space-time variables,

$$N_1 = -i\left(x\frac{\partial}{\partial t} + t\frac{\partial}{\partial x}\right) + i\left(z\frac{\partial}{\partial x} - x\frac{\partial}{\partial z}\right),$$

$$N_2 = -i\left(y\frac{\partial}{\partial t} + t\frac{\partial}{\partial y}\right) - i\left(y\frac{\partial}{\partial z} - z\frac{\partial}{\partial y}\right). \tag{7.14}$$

This form of the $N_1$ and $N_2$ operators will be discussed in detail in Chapter 10.

## 7.3 Massless Particles

For a massive particle, the little group in its rest frame is the three-dimensional rotation group. The connection between this group and the spin of the particle is well known. For a massless particle, the rotation around its momentum is associated with its helicity. On the other hand, the physical interpretation of the $D(\sigma, \rho)$ matrix is not widely known. It performs a gauge transformation on the plane wave propagating along the $z$ direction. The $D(\sigma, \rho)$ matrix therefore plays a very important role of establishing a bridge between Maxwell's equations and Wigner's little groups. The purpose of this section is to elaborate on this point.

Let us consider a free photon moving along the $z$ direction. Then its four-potential is

$$A^\mu(x) = A^\mu e^{i\omega(z-t)}, \tag{7.15}$$

where

$$A^\mu = (A_1, A_2, A_3, A_0).$$

The momentum four-vector is clearly

$$P^\mu = (0, 0, \omega, \omega). \tag{7.16}$$

For this plane wave, without loss of generality, we can impose the Lorentz condition:

$$\frac{\partial}{\partial x^\mu} A^\mu(x) = P^\mu A_\mu(x) = 0, \tag{7.17}$$

resulting in $A_3 = A_0$. Since the third and fourth components are identical, the $N_1$ and $N_2$ matrices of Eq. 7.7 can be replaced respectively by

$$N_1 = \begin{pmatrix} 0 & 0 & 0 & 0 \\ 0 & 0 & 0 & 0 \\ i & 0 & 0 & 0 \\ i & 0 & 0 & 0 \end{pmatrix}, \quad N_2 = \begin{pmatrix} 0 & 0 & 0 & 0 \\ 0 & 0 & 0 & 0 \\ 0 & i & 0 & 0 \\ 0 & i & 0 & 0 \end{pmatrix}, \tag{7.18}$$

and the $D(\sigma,\rho)$ matrix by

$$D(\sigma,\rho)=\begin{pmatrix}1&0&0&0\\0&1&0&0\\\sigma&\rho&1&0\\\sigma&\rho&0&1\end{pmatrix}. \tag{7.19}$$

If this matrix is applied to the column vector of $(A_1, A_2, A_3, A_3)$, the result is the addition of $(0,0,\sigma A_1+\rho A_2, \sigma A_1+\rho A_2)$ to the original vector. This quantity is proportional to the four-momentum. Thus the operation of the $D(\sigma,\rho)$ is a gauge transformation.

Let us next consider the effect of Lorentz boosts along the $z$ direction. The boost is generated by $K_3$. It commutes with the helicity generator $J_3$:

$$[K_3, J_3]=0. \tag{7.20}$$

Its commutation relations with $N_1$ and $N_2$ are

$$[K_3, N_1] = -iN_1,$$

$$[K_3, N_2] = -iN_2. \tag{7.21}$$

Since the operators $N_1$, $N_2$, $J_3$, and $K_3$ form a closed Lie algebra, we shall call the group generated by these four operators the "extended little group" for massless particles (Kim and Wigner 1987b).

Since the Lorentz condition of Eq. 7.17 is invariant under the Lorentz boost along the $z$ direction, we can continue to use the simplified expression for $N_1$ and $N_2$ given in Eq. 7.18, and also for $D(\sigma,\rho)$ of Eq. 7.19 . Furthermore, under the Lorentz boost $B_z(\lambda)=\exp(-i\lambda K_3)$,

$$B_z(\lambda)D(\sigma,\rho)B_z(-\lambda)=D\left(e^{\lambda}\sigma, e^{\lambda}\rho\right). \tag{7.22}$$

This means that the Lorentz boost along the $z$ direction changes only the scale of the gauge transformation generated by $N_1$ and $N_2$.

Let us go back to the expression of the plane wave in Eq. 7.15. Since $N_1$ and $N_2$ generate gauge transformations on the four-potential $A^{\mu}$ which do not lead to observable consequences, $A^{\mu}$ can be written as

$$A^{\mu}=(A_1, A_2, 0, 0). \tag{7.23}$$

$J_3$ generates rotations around the $z$ axis. In this case, the rotation leads to a linear combination of the $x$ and $y$ components. This operation is consistent with the fact that the photon has two independent components, which is thoroughly familiar to us.

Under the boost along the $z$ direction, the frequency of the light wave becomes changed. Thus, the above reasoning for the monochromatic wave remains valid for the case of the superposition of several waves with different frequencies propagating in the same direction:

$$A^{\mu}(x) = \sum_i A_i e^{i(k_i z - \omega_i t)}, \tag{7.24}$$

and the norm:

$$N = \sum_i |A_i|^2 \tag{7.25}$$

remains invariant under transformations of the extended little group.

## 7.4 Decomposition of Lorentz Transformations

Since a massive particle can have its angular momentum parallel or anti-parallel to its momentum, it is possible to define the helicity for both massive and massless particles. It was noted by Wigner in 1957 that there are Lorentz transformations that preserve helicity and those that do not. Wigner (1957) noted also that the difference between these two different sets of transformations may play an important role in understanding the internal space-time symmetries of elementary particles, particularly the symmetry of massless particles as a limiting case of the space-time symmetry of massive particles.

Let us start with a particle at rest with mass $m$ with it spin along the $z$ direction, and then boost it along the $z$ direction with velocity parameter $\alpha$. The resulting four-momentum $p$ is

$$P^{\mu} = m\left(0, 0, \alpha/(1-\alpha^2)^{1/2}, 1/(1-\alpha^2)^{1/2}\right). \tag{7.26}$$

The matrix which boosts the rest-state four-momentum to the above form is

$$A_z(\alpha) = \begin{pmatrix} 1 & 0 & 0 & 0 \\ 0 & 1 & 0 & 0 \\ 0 & 0 & 1/(1-\alpha^2)^{1/2} & \alpha/(1-\alpha^2)^{1/2} \\ 0 & 0 & \alpha/(1-\alpha^2)^{1/2} & 1/(1-\alpha^2)^{1/2} \end{pmatrix}. \tag{7.27}$$

After this boost, the particle is in the positive helicity state.

Without changing the helicity, we can boost this particle with non-zero momentum along the $z$ direction, and also rotate the system (Chou and Zastavenco 1958, Jacob and Wick 1959). We cannot however boost the system along an arbitrary direction without changing the helicity (Wigner 1957). As is indicated in Figure 7.1, let us boost the four-momentum of Eq. 7.26 along the $\phi$ direction in the $xz$ plane with boost parameter $\eta$:

$$P' = B_{\phi}(\eta)P. \tag{7.28}$$

This is not a helicity-preserving transformation.

This boost is not the only transformation which changes the four-momentum $P$ to $P'$. As is shown in Figure 7.1, we can boost $P$ along the $z$ axis first so that its

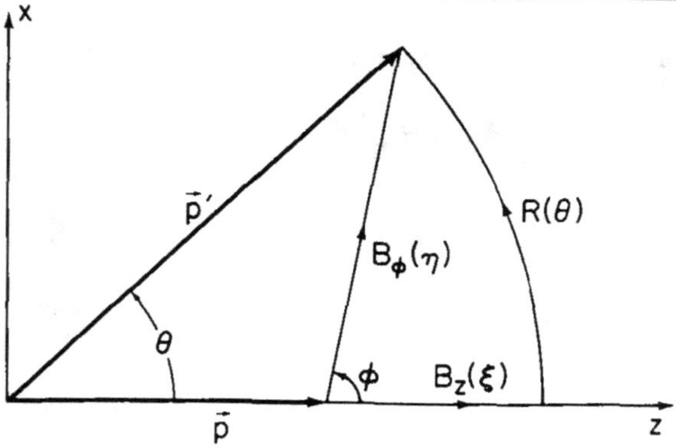


Figure 7.1: Lorentz boost along the $\phi$ direction. The four-momentum $P$ can be directly boosted to $P'$ by $B_\phi(\eta)$, or through the rotation $R(\theta)$ preceded by $B_z(\zeta)$ along the $z$ direction.

speed (or magnitude of momentum) is the same as that of $P'$, and then rotate this boosted vector until its direction coincides with that of $P'$:

$$P' = (R(\theta)B_z(\zeta))\, P, \tag{7.29}$$

where $\theta$ and $\zeta$ are determined from $\alpha$, $\phi$, and $\eta$. The above transformation is a helicity-preserving transformation.

We have seen above that $P$ can be transformed to $P'$ in two different ways. The best way to see this difference is to construct the closed-loop transformation:

$$D_\phi(\alpha, \eta) = [B_\phi(\eta)]^{-1} R(\phi) B_z(\zeta). \tag{7.30}$$

When applied to the four-momentum $P$, this matrix leaves it invariant:

$$P = D_\phi(\alpha, \eta) P. \tag{7.31}$$

Therefore, $D_\phi(\alpha, \eta)$ is an element of the little group which leaves the four-momentum $P$ invariant (Wigner 1939, Han *et al.* 1986, Vassiliadis 1989). Let us now write $B_\phi(\eta)$ of Eq. 7.28 as

$$B_\phi(\eta) = \left(R(\phi)B_z(\zeta)\,(B_z(\zeta))^{-1}\,[R(\phi)]^{-1}\right) B_\phi(\eta). \tag{7.32}$$

Then the right-hand side of the above equation can be rearranged, and

$$B_\phi(\eta) = (R(\phi)B_z(\zeta))\,(D_\phi(\alpha, \eta))^{-1}. \tag{7.33}$$

The transformation $[R(\phi)B_z(\zeta)]$ is a helicity-preserving transformation, but changes the momentum. $[D_\phi(\alpha, \eta)]^{-1}$ is also an element of the little group, but it can change the helicity. Therefore, $B_\phi(\eta)$ can be decomposed into a helicity-preserving transformation which changes the momentum and a momentum-preserving transformation which changes the helicity.

For a massive particle, the $D$ matrix of Eq. 7.30 is an element of the O(3)-like little group (Wigner 1939, Kim and Noz 1986).

$$p = A_z(\alpha)R(\theta^*)[A_z(\alpha)]^{-1}p, \tag{7.34}$$

where $R(\theta^*)$ is a rotation matrix. According to the above formula, the particle is brought to its rest frame and then is rotated before it is brought back to its original frame.

If the transformations are performed on the $xz$ plane, $R(\theta^*)$ represents a rotation matrix around the $y$ axis. This rotation leaves the four-momentum invariant in the Lorentz frame in which the particle is at rest. This rotation changes the direction of spin, and performs a transformation of the O(3)-like little group which leaves the momentum invariant in this Lorentz frame.

Since the transformation of Eq. 7.28 is equivalent to that of Eq. 7.33, we should be able to write (Han *et al.* 1986)

$$D_\phi(\alpha, \eta) = A_z(\alpha)R(\theta^*)\,[A_z(\alpha)]^{-1}, \tag{7.35}$$

and

$$R(\theta^*) = [A_z(\alpha)]^{-1}\, D_\phi(\alpha, \eta)A_z(\alpha). \tag{7.36}$$

The $D$ transformation is therefore a Lorentz-boosted rotation.

For the case of massive particles, every Lorentz transformation can be decomposed into (Han *et al.* 1987a, Vassiliadis 1989)

$$B_\phi(\eta) = R(\phi)\,(B_z(\zeta)A_z(\alpha))\,R(-\theta^*)A_z^{-1}(\alpha). \tag{7.37}$$

The transformation $[B_z(\zeta)A_z(\alpha)]$ is a boost along the direction of momentum. $A_z^{-1}(\alpha)$ is also a boost along the same direction.

## 7.5 Analytic Continuation to the Little Groups for Massless and Imaginary-Mass particles

The calculation of the $D$ matrix is in general cumbersome. However, calculations in simplified situations give some illuminating results. We shall consider here the case in which $\theta$ and $\eta$ are such that $B_z(\zeta)$ is an identity matrix (Kupersztych 1976). The kinematics of this transformation is illustrated in Figure 7.2. The Lorentz boost $B_\phi(\eta)$ in this case is an energy-preserving transformation. The explicit expression for the $D$ matrix is (Han *et al.* 1986):

$$D(\alpha, \theta) = \begin{pmatrix} 1-(1-\alpha^2)\sigma^2/2T & 0 & -\sigma/T & \alpha\sigma/T \\ 0 & 1 & 0 & 0 \\ \sigma/T & 0 & 1-\sigma^2/2T & \alpha\sigma^2/2T \\ \alpha\sigma/T & 0 & -\alpha\sigma^2/2T & 1+\alpha\sigma^2/2T \end{pmatrix}. \tag{7.38}$$

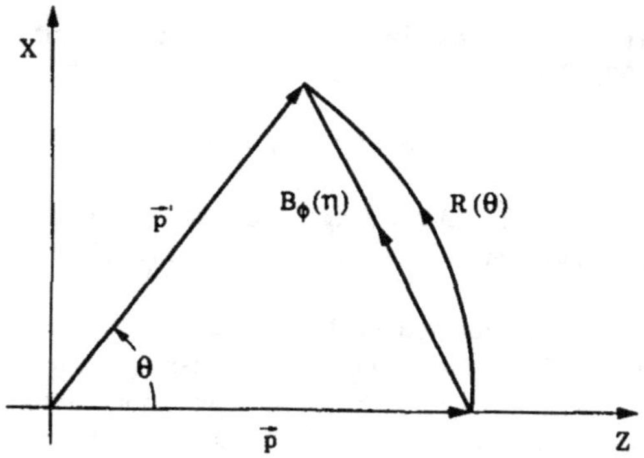


Figure 7.2: Rotation and boost. The momentum $P$ can be rotated by $R(\theta)$. The same effect can be achieved through the boost $B_\phi(\eta)$. This kinematical framework is a special case of the Lorentz kinematics of Figure 7.1.

where

$$\sigma = -2\left(\tan\frac{\theta}{2}\right), \quad \text{and} \quad T = 1 + (1-\alpha^2)\left(\tan\frac{\theta}{2}\right)^2 .$$

Using the expression of Eq. 7.36 for $R(\theta^*)$, we can calculate $\theta^*$, and the result is (Han *et al.* 1986)

$$\theta^* = 2\left[\tan^{-1}\left\{\left(1-\alpha^2\right)^{1/2}\tan\frac{\theta}{2}\right\}\right] . \tag{7.39}$$

According to this expression, $\theta^* = \theta$ for $\theta = 0$. $\theta^*$ is nearly equal to $\theta$ for moderate values of $\alpha$. It becomes vanishingly small as $\alpha$ approaches 1 (Han *et al.* 1986). For massless particles, we can take the $\alpha = 1$ limit, so that the expression for $D(\alpha = 1, \theta)$ becomes

$$D(\theta) = \begin{pmatrix} 1 & 0 & -\sigma & \sigma \\ 0 & 1 & 0 & 0 \\ \sigma & 0 & 1-\sigma^2/2 & \sigma^2/2 \\ \sigma & 0 & 1-\sigma^2/2 & 1+\sigma^2/2 \end{pmatrix} . \tag{7.40}$$

This matrix performs a gauge transformation when applied to the four- potential for a free photon (Weinberg 1964, Kupersztych 1977, Han *et al.* 1982, Kim and Wigner 1987a).

It is remarkable that the expression for the $D(\alpha, \theta)$ is analytic in $\alpha$ at $\alpha = 1$. This allows us to make analytic continuation of $D(\alpha, \theta)$ from $\alpha < 1$ to $\alpha > 1$. If $\alpha$ is greater than 1, we are interested in constructing a set of Lorentz transformations which leave a space-like four-momentum invariant. The momentum four-vector can be written as

$$P = m\left(0, 0, \alpha/(\alpha^2-1)^{1/2}, 1/(\alpha^2-1)^{1/2}\right) . \tag{7.41}$$

We can obtain this space-like four-vector by making the Lorentz boost of $A_z(1/\alpha)$ on the simpler four-vector $(0, 0, m, 0)$. The $A_z(1/\alpha)$ matrix is

$$A_z(1/\alpha) = \begin{pmatrix} 1 & 0 & 0 & 0 \\ 0 & 1 & 0 & 0 \\ 0 & 0 & \alpha/(\alpha^2-1)^{1/2} & 1/(\alpha^2-1)^{1/2} \\ 0 & 0 & 1/(\alpha^2-1)^{1/2} & \alpha/(\alpha^2-1)^{1/2} \end{pmatrix}. \tag{7.42}$$

Then the $D(\alpha, \theta)$ matrix can be written as

$$D(\alpha, \theta) = A_z(1/\alpha) F(\alpha, \theta) [A_z(1/\alpha)]^{-1}, \tag{7.43}$$

where

$$F(\alpha, \theta) = \begin{pmatrix} \cosh\lambda & 0 & 0 & \sinh\lambda \\ 0 & 1 & 0 & 0 \\ 0 & 0 & 1 & 0 \\ \sinh\lambda & 0 & 0 & \cosh\lambda \end{pmatrix},$$

with

$$\cosh\lambda = \frac{1 + (\alpha^2 - 1)\tan^2(\theta/2)}{1 - (\alpha^2 - 1)\tan^2(\theta/2)},$$

$$\sinh\lambda = \frac{-2(\alpha^2 - 1)^{1/2}\tan(\theta/2)}{1 - (\alpha^2 - 1)\tan^2(\theta/2)}.$$

The matrix $F(\alpha, \theta)$ performs a Lorentz transformation along the $x$ direction. This confirms the fact that the little group for an imaginary particle is the (2 + 1)-dimensional Lorentz group.

In Chapter 10, we shall discuss the transition from one little group to another using the theory of group contractions. It will be seen there that the group contraction is a singular procedure. Thus it is quite remarkable that the velocity parameter $\alpha$ enables us to make an analytic continuation from the O(3)-like little group to the (2 + 1)-like little group for imaginary particles via the E(2)-like little group for massless particles.

## 7.6 Light-Cone Coordinate System

We are now interested in how the Lorentz transformation leads to squeezed space-time. Let us start again with the Lorentz transformation

$$x' = x, \quad y' = y,$$

$$z' = (z + \beta t)/(1 - \beta^2)^{1/2},$$

$$t' = (t + \beta z)/(1 - \beta^2)^{1/2}. \tag{7.44}$$

The cumbersome aspect of this transformation is that the coordinate system does not remain orthogonal. One way to avoid this inconvenience is to introduce the light-cone variables (Dirac 1949):

$$u = (z + t)/\sqrt{2}, \quad v = (z - t)/\sqrt{2}. \tag{7.45}$$

Then they are transformed as

$$u' = \left(\frac{1+\beta}{1-\beta}\right)^{1/2} u, \quad v' = \left(\frac{1-\beta}{1+\beta}\right)^{1/2} v. \tag{7.46}$$

Under this transformation, one axis becomes elongated while the other goes through a contraction so that the product $uv$ will stay constant (Kim and Noz 1986):

$$uv = u'v' = (t^2 - z^2)/2 = [(t')^2 - (z')^2]/2. \tag{7.47}$$

This transformation property is illustrated in Figure 7.3.

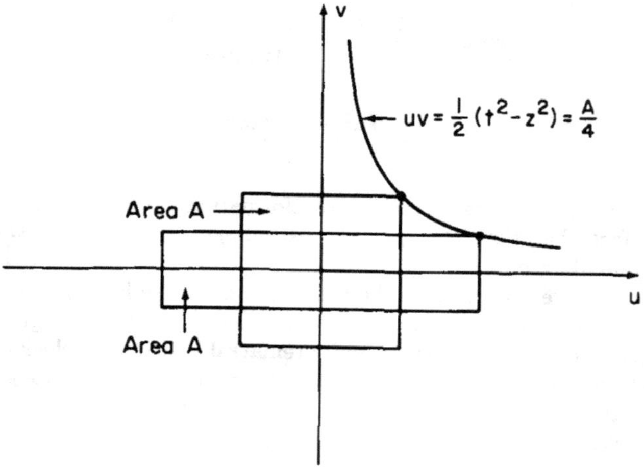


Figure 7.3: Lorentz boost in the light-cone coordinate system. Under the boost, the product $uv$ remains invariant. The product $uv$ is the area of the rectangle in this figure.

Since the momentum and energy form a four-vector, their transformation property is exactly the same as the space-time variables. If we define $p_+$ and $p_-$ as

$$p_+ = (p_z + p_0)/\sqrt{2}, \quad p_- = (p_z - p_0)/\sqrt{2}, \tag{7.48}$$

where $p_0$ and $p_z$ are the energy and the $z$ component of momentum, they will be transformed as

$$p'_+ = \left(\frac{1-\beta}{1+\beta}\right)^{1/2} p_+, \quad p'_- = \left(\frac{1+\beta}{1-\beta}\right)^{1/2} p_-. \tag{7.49}$$

This means that the products $p_-u$ and $p_+v$ are Lorentz-invariant quantities.

In order to see whether they form conjugate pairs, let us write down the Fourier transformation of a function of position-time variables:

$$\phi(p) = \left(\frac{1}{2\pi}\right)^2 \int e^{-ip\cdot x}\psi(x)d^4x, \tag{7.50}$$

where $p \cdot x$ is the scalar product $p^\mu x_\mu$ of the momentum-energy and position-time four-vectors. We are using here the four vector notation:

$$x^\mu = (x, y, z, t), \quad \text{and} \quad x_\mu = (x, y, x, -t). \tag{7.51}$$

This quantity can be written as

$$p \cdot x = \frac{1}{2}(z+t)(p_z - p_0) + \frac{1}{2}(z-t)(p_z + p_0) + xp_x + yp_y. \tag{7.52}$$

Therefore, $p_+$ and $p_-$ are the "Fourier conjugates" of $v$ and $u$ respectively, and they can now be written as

$$p_u = p_-, \quad \text{and} \quad p_v = p_+. \tag{7.53}$$

The area element in the phase space consisting of $p_u$ and $u$ or $p_v$ and $v$ will be invariant under Lorentz boosts along the $z$ direction. This is the kinematical basis for the covariant formulation of phase space.

We can now consider a four-dimensional space whose coordinate variables are $(u, v, p_u, p_v)$. Then the Lorentz transformation matrix applicable to this vector is

$$\begin{pmatrix} e^\lambda & 0 & 0 & 0 \\ 0 & e^{-\lambda} & 0 & 0 \\ 0 & 0 & e^{-\lambda} & 0 \\ 0 & 0 & 0 & e^\lambda \end{pmatrix}, \tag{7.54}$$

where

$$e^\lambda = \left(\frac{1+\beta}{1-\beta}\right)^{1/2}.$$

This matrix indicates that the Lorentz boost is a squeeze operation applicable to the pairs $(u, v)$ and $(p_v, p_u)$ or to the pairs $(u, p_u)$ and $(p_v, v)$. We shall study whether the Wigner function for relativistic problems has this squeeze property in Section 7.8 and also in Chapter 8.

## 7.7 Localized Light Waves

For light waves, the Fourier relation $(\Delta t)(\Delta\omega)$ was known before the present form of quantum mechanics was formulated (Heitler 1954). However, the question of whether this is a Lorentz-invariant relation has not yet been properly addressed. Let us consider a blinking traffic light. A stationary observer will insist on $(\Delta t)(\Delta\omega) \cong 1$. An observer in an automobile moving toward the light will see the same blinking light. This observer will also insist on $(\Delta t')(\Delta\omega') \cong 1$ in his/her coordinate system. However, these observers may not agree with each other, because neither $\Delta t$ nor $\Delta\omega$ is a Lorentz-invariant variable. The product of two non-invariant quantities does not always lead to an invariant quantity.

Let us assume that the automobile is moving in the negative $z$ direction with velocity parameter $\beta$. Since both $t$ and $\omega$ are time-like components of four-vectors $(\mathbf{x}, t)$ and $(\mathbf{k}, \omega)$ respectively, a Lorentz boost along the $z$ direction will lead to new variables:

$$t' = (t + \beta z)/(1 - \beta^2)^{1/2}, \quad \omega' = (\omega + \beta k)/(1 - \beta^2)^{1/2}, \tag{7.55}$$

where the light wave is assumed to travel along the $z$ axis with $k = \omega$. In the above transformation, the light wave is boosted along the positive $z$ direction. If the light passes through the point $z = 0$ at $t = 0$, then $t = z$ on the light front, and the transformations of Eq. 7.55 become

$$t' = \left(\frac{1+\beta}{1-\beta}\right)^{1/2} t, \quad \omega' = \left(\frac{1+\beta}{1-\beta}\right)^{1/2} \omega. \tag{7.56}$$

These equations will formally lead us to

$$(\Delta t')(\Delta\omega') = \left(\frac{1+\beta}{1-\beta}\right)(\Delta t)(\Delta\omega), \tag{7.57}$$

which indicates that the time-energy uncertainty relation is not Lorentz-invariant, and that Planck's constant depends on the Lorentz frame in which the measurement is taken. This is not correct, and we need a better understanding of the transformation properties of $\Delta t$ and $\Delta\omega$.

This is the localization problem for light waves. In approaching the problem of the covariant superposition of light waves, we shall start with the uncertainty relation applicable to nonrelativistic quantum mechanics. The natural starting point for tackling this problem is a free-particle wave packet in nonrelativistic quantum mechanics which we pretend to understand. Let us write down the time-dependent Schrödinger equation for a free particle moving in the $z$ direction:

$$i\frac{\partial}{\partial t}\psi(z,t) = \frac{-1}{2m}\left(\frac{\partial}{\partial z}\right)^2 \psi(z,t). \tag{7.58}$$

The Hamiltonian commutes with the momentum operator. If the momentum is sharply defined, the solution of the above differential equation is

$$\psi(z,t) = \exp\left[i(pz - p^2 t/2m)\right]. \tag{7.59}$$

If the momentum is not sharply defined, we have to take the linear superposition:

$$\psi(z,t) = \int g(k)\exp\left[i(kz - k^2 t/2m)\right] dk. \tag{7.60}$$

The width of the wave function becomes wider as the time variable increases, as is illustrated in Figure 7.4a. This is the wave packet spread discussed in Chapter 3.

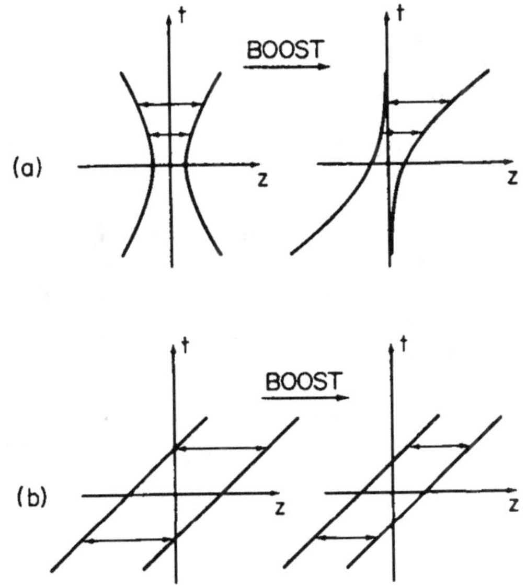


Figure 7.4: The time dependence of the wave packets. (a) shows the spread of the Schrödinger wave function. (b) shows the behavior of the light wave which does not spread. However, for an observer moving in the negative $z$ direction, the Schrödinger wave function is boosted according to the Galilei transformation. The quantum probability interpretation is consistent with the Galilei world. On the other hand, the light wave carries the burden of being consistent with the Lorentzian world.

Let us study the transformation properties of this wave function. The rotation and translation properties are trivial. In order to study the boost property within the framework of the Galilei kinematics, let us imagine an observer moving in the negative $z$ direction. To this observer, the center of the wave function moves along the positive $z$ direction as is specified also in Figure 7.4b. The transformed wave function takes the form

$$\psi_v(z,t) = \exp\left(-im(vz - v^2t/2)\right) g(k) \exp\left(i(kz - k^2t/2m)\right) dk. \tag{7.61}$$

where $v$ is the boost velocity. This expression is different from the usual expression in textbooks by an exponential factor in front of the integral sign (Inonu and Wigner 1952).

In nonrelativistic quantum mechanics, $\psi(z,t)$ has a probability interpretation, and there is no difficulty in giving an interpretation for the transformed wave function in spite of the above-mentioned phase factor. The basic unsolved problem is whether the probabilistic interpretation can be extended into the Lorentzian regime. This has been a fundamental unsolved problem for decades (Newton and Wigner 1949, Ali and Emch 1974, Ali 1985, Han *et al.* 1987c, Kim and Wigner 1987b), and we do not propose to solve all the problems in this Chapter. A reasonable starting point for approaching this problem is to see whether a covariant probability interpretation can be given to light waves.

For light waves, we can start with the usual expression:

$$f(z,t) = \left(\frac{1}{2\pi}\right)^{1/2} \int g(k) e^{i(kz-\omega t)} dk. \tag{7.62}$$

Unlike the case of the Schrödinger wave function, $f(z,t)$ is equal to $k$, and there is no spread of wave packet. The velocity of propagation is always that of light, as is illustrated in Figure 7.4b. We might therefore be led to think that the problem for light waves is simpler than that for nonrelativistic Schrödinger waves. This is not the case for the following reasons.

We like to have a probability interpretation for the light wave given in Eq. 7.62. However, this is not a solution of the Schrödinger equation, but a solution of the relativistic wave equation. On the other hand, the light wave satisfies the Parseval's relation:

$$\int |f(z,t)|^2 dz = \int |g(k)|^2 dk. \tag{7.63}$$

Thus it is possible to carry out a spectral analysis to give a probability interpretation if the above integrals converge. If these integrals converge, we may say that the light wave is localized.

We are then led to the question of covariance. The localized light wave given in Eq. 7.61 and its Parseval's relation are valid in a given Lorentz frame. What form does this equation take for an observer in a different frame?

## 7.8 Covariant Localization of Light Waves

In terms of the light-cone variables, the light wave of Eq. 7.62 can be written as

$$f(q) = \left(\frac{1}{2\pi}\right)^{1/2} \int g(s) e^{isq} ds, \tag{7.64}$$

where

$$q = (z - t), \quad s = (k + \omega)/2.$$

These variables are essentially $v$ and $p_v$ given in Eq. 7.46 and Eq. 7.49 respectively. The variable $q$ is $\sqrt{2}v$, while $s = p_v/\sqrt{2}$. This form is convenient for light waves because s is numerically equal to $\omega$, while $\Delta t$ is equal to $\Delta q$ for a fixed value of $z$ (Han *et al.* 1987c).

Let us now make a Lorentz boost along the $z$ direction. Then the product $s\omega$ remains invariant. Thus the Lorentz-invariant form of the uncertainty relation $(\Delta t)(\Delta\omega)$ should be $(\Delta q)(\Delta s)$. However, $q$ and $s$ in $f(q)$ and $g(s)$ are replaced by $q'$ and $s'$ respectively, where

$$q' = e^{\lambda} q, \quad s' = e^{-\lambda} s, \tag{7.65}$$

where

$$\lambda = \left(\frac{1}{2}\right) \ln \left(\frac{1+\beta}{1-\beta}\right).$$

Furthermore, $dq$ and $ds$ are also replaced by $dq'$ and $ds'$ respectively, where

$$dq' = e^{\lambda} dq, \quad ds' = e^{-\lambda} ds. \tag{7.66}$$

Thus, after the transformation, $f(z', t')$ becomes

$$f(z', t') = \left(\frac{1}{2\pi}\right)^{1/2} \left(\frac{1-\beta}{1+\beta}\right)^{1/2} \int g(k') e^{i(kz-\omega t)} dk',$$

or

$$f'(s) = f(e^{\lambda} q) = \left(\frac{1}{2\pi}\right)^{1/2} e^{-\lambda} \int g(e^{-\lambda} s) e^{iqs} ds, \tag{7.67}$$

and Parseval's relation becomes

$$\int |f(z', t')|^2 dz' = e^{2\lambda} \int |g(k')|^2 dk'. \tag{7.68}$$

This means that the probability is not conserved under Lorentz boosts, unless the transformation generates the factors $e^{\lambda/2}$ and $e^{-\lambda/2}$ in front of $f(e^{\lambda} q)$ and $g(e^{-\lambda} s)$ respectively. Thus the Lorentz transformation of $f(q)$ and $g(s)$ are $f'(q) = e^{\lambda/2} f(e^{\lambda} q)$ and $g'(s) = e^{-\lambda/2} g(e^{-\lambda} q)$ respectively. These multipliers also make the Fourier transformation of Eq. 7.67 consistent with special relativity.

One way to resolve this difficulty is to note that the squeeze transformation on $f(q)$ is generated by $iq\frac{\partial}{\partial q}$. This generator is not Hermitian. The Hermitian form of this squeeze generator is

$$K_q = \left(\frac{i}{2}\right) \left(q\frac{\partial}{\partial q} + \frac{\partial}{\partial q} q\right), \tag{7.69}$$

which can be written as

$$K_q = i \left(q\frac{\partial}{\partial q} + \frac{1}{2}\right). \tag{7.70}$$

Then the transformation operator generated by $K_q$ is

$$\exp(-i\lambda K_q) = e^{\lambda/2} \exp\left(\lambda q \frac{\partial}{\partial q}\right). \tag{7.71}$$

The reason why the transformation operator takes the above form instead of $\exp\left(\lambda q \frac{\partial}{\partial q}\right)$ is simply that we require the generator to be Hermitian.

On the other hand, the generator $K_q$ does not have a dynamical meaning, unlike the case of the generator of rotations which is associated with the angular momentum. There is a physical reason for the generator of rotations to be Hermitian. On the other hand, the boost generator does not carry a dynamical interpretation (Dirac 1949). It is Hermitian only to preserve the unitarity of the representation.

Let us discuss the problem using a concrete example of the Gaussian distribution for $g(k)$. The Gaussian form of the wave packet with the average momentum $p$ is

$$g(k) = \left(\frac{1}{\pi}\right)^{1/4} \exp\left(-(k-p)^2/2\right). \tag{7.72}$$

Then

$$f(z,t) = \left(\frac{1}{\pi}\right)^{1/4} e^{ip(z-t)} \exp\left(-(z-t)^2/2\right). \tag{7.73}$$

Under the Lorentz boost along the $z$ direction, $f(q)$ and $g(k)$ become

$$f'(q) = e^{\lambda/2} f(e^{\lambda} q)$$

$$= \left(\frac{1}{\pi}\right)^{1/4} e^{\lambda/2} e^{ipq} \exp\left(-e^{2\lambda} q^2/2\right),$$

and

$$g'(s) = e^{-\lambda/2} g(e^{-\lambda} s)$$

$$= \left(\frac{1}{\pi}\right)^{1/4} e^{-\lambda/2} \exp\left\{-e^{-2\lambda}\left(s - e^{\lambda}p\right)^2/2\right\}. \tag{7.74}$$

These functions are normalized. The width of $|f'(q)|^2$ is reduced by $e^{\lambda}$, while that for $|g'(s)|^2$ is expanded by $e^{-\lambda}$. Thus the uncertainty product $(\Delta q)(\Delta s)$ remains invariant. In Section 7.9, we shall study the origin of the multiplier $e^{\pm\lambda/2}$ on $f(q)$ and $g(s)$ using the phase-space picture of quantum mechanics.

## 7.9 Covariant Phase-Space Picture of Localized Light Waves

Let us study the localized wave of Section 7.8 using the phase-space picture of quantum mechanics. We can construct the Wigner function in the light-cone coordinate system as

$$W(q,s) = \frac{1}{\pi}\int f^*(q+y) f(q-y) e^{2isy} dy. \tag{7.75}$$

We can develop a theory starting from this form without resorting to an explicit form for $f(q)$. But, it is more efficient to use a Gaussian form for $f(q)$ and $g(s)$.

The Wigner function then takes the form

$$W(q,s) = \frac{1}{\pi}\exp\left\{-\left(q^2 + (s-p)^2\right)\right\}, \tag{7.76}$$

which is localized within a circular region:

$$q^2 + (s-p)^2 < 1, \tag{7.77}$$

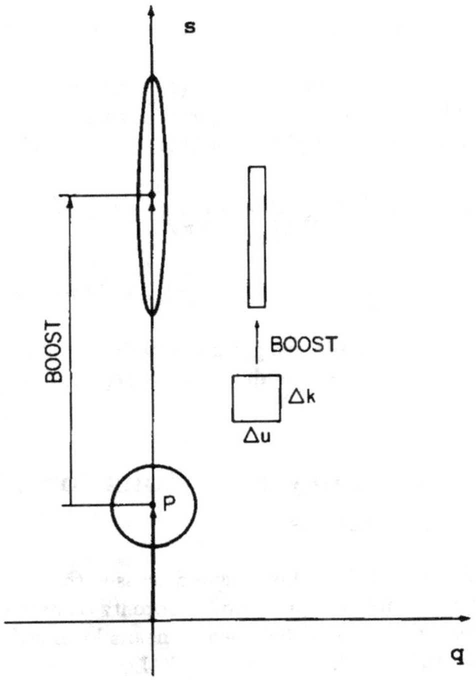


Figure 7.5: The phase space distribution of the localized light wave in the light-cone coordinate system. The Lorentz boost along the $z$ direction contracts (elongates) the $q$ axis while it elongates (contracts) the $s$ axis in such a way that the area element in the phase space is conserved. This represents the Lorentz invariance of Planck's constant.

in the phase space of $q$ and $s$ as is illustrated in Figure 7.5. This Wigner function of Eq. 7.76 is normalized as

$$\int W(q,s)dqds = 1. \tag{7.78}$$

The important difference between the Lorentz boost of the wave functions $f(q)$ and $g(s)$ and that of the above Wigner function is that the integral measure for $W(q,s)$ is $dsdq$, which is invariant under the boost. Thus $W(q,s)$ does not need a multiplier when it is Lorentz boosted (Kim and Wigner 1987b). The boosted Wigner function is

$$\begin{aligned} W'(q,s) &= W\left(e^{\lambda}q, e^{-\lambda}s\right) \\ &= \frac{1}{\pi}\, exp\left\{-\left(e^{2\lambda}q^2 + e^{-2\lambda}(s - e^{\lambda}p)^2\right)\right\}. \end{aligned} \tag{7.79}$$

This function is localized within the elliptic region specified in Figure 7.5. The area of the ellipse is the same as that of the circle, indicating that the boost is a canonical transformation.

The Wigner function reproduces $|f(q)|^2$ and $|g(s)|^2$ after the integral over the $s$ and $q$ variables respectively. The question then is whether this process produces the multipliers on $f(q)$ and $g(s)$. The integrations lead to

$$|f(q)|^2 = \int W'(q,s)ds = \left(\frac{1}{\pi}\right)^{1/2} e^{\lambda} \exp\left(-e^{2\lambda}q^2\right),$$

$$|g(s)|^2 = \int W'(q,s)dq = \left(\frac{1}{\pi}\right)^{1/2} e^{-\lambda} \exp\left(-e^{-2\lambda}(s - e^{\lambda}p)^2\right). \qquad (7.80)$$

Indeed, the phase-space picture gives the desired multipliers on $|f(q)|^2$ and $|g(s)|^2$. From this, the magnitude of multipliers on $f(q)$ and $g(s)$ can be determined, but not their phases.

## 7.10 Uncertainty Relations for Light Waves and for Photons

Throughout Sections 7.7 - 7.9, we have discussed the uncertainty relation for localized light waves and its invariance under Lorentz transformations. The key question then is whether the localized light wave means localized photons. The present answer to this question is "not necessarily." Let us examine the question of photon localization.

It is important to note that the mathematical apparatus for light waves is basically different from the formalism of quantum electrodynamics where photons are created and annihilated. In QED, we use the four-potential $A^{\mu}(x)$, which takes the spectral form

$$A(x) = \left(\frac{1}{2\pi}\right)^{1/2} \int (1/\sqrt{k}) a(k) e^{i(kz-\omega t)} dk, \qquad (7.81)$$

where the superscript $\mu$ has been deleted for simplicity. We consider only the longitudinal direction which is affected by the Lorentz boost.

The photon field $A(x)$ is identified with the amplitude of the light wave $f(q)$ given in Section 7.7. However, for the second quantization for $a(k)$:

$$[a(k), a^{\dagger}(k')] = \delta(k - k'), \qquad (7.82)$$

where $a(k)$ and $a^{\dagger}(k)$ are the annihilation and creation operators respectively, the Lorentz-invariant norm is

$$i \int \left\{ A^{*}(q) \frac{\partial}{\partial t} A(q) - A(q) \frac{\partial}{\partial t} A^{*}(q) \right\} dz. \qquad (7.83)$$

If we use this as the norm, the probability is not always positive. This is due to the fact that there is the $\sqrt{k}$ factor in the integrand of Eq. 7.81. Therefore, it it not possible to associate the concept of localized probability with second-quantized photon fields (Newton and Wigner 1949). At the same time, the formalism for light waves without the $\sqrt{k}$ factor is not suitable for dealing with second quantization. The present limitations on our theoretical tools are summarized in Table 7.1.

Table 7.1: Covariant light waves and photons. They are different. The light waves are localizable but cannot have a particle interpretation. The particle interpretation of free photons is given in terms of the Fock space representation in quantum electrodynamics. However, the photons in field theory are not localizable.

| | Localizable | Particle Interpretation |
|---|---|---|
| Light Waves | Yes | No |
| Photons | No | Yes |

The question then is what uncertainty relations photons obey. The answer to this question is very simple. It is the relation governing the creation and annihilation of photons in Fock space. The uncertainty relation in Fock space has been discussed in Chapter 6. Indeed, the study of coherent and squeezed states of light is the study of the uncertainty principle in Fock space. Our common sense dictates that the position-momentum uncertainty relation for localized light waves and the Fock space uncertainty relations are two-different manifestations of the same principle. We are not yet able to present a mathematical formalism to prove this.

# Chapter 8

# COVARIANT HARMONIC OSCILLATORS

Since wave functions play a central role in nonrelativistic quantum mechanics, one method of combining quantum mechanics and special relativity is to construct relativistic wave functions with an appropriate probability interpretation. Because of its mathematical simplicity, the harmonic oscillator has served as the first concrete solution to many new physical theories. Thus it is quite natural that the first covariant wave function be a covariant harmonic oscillator wave function.

Indeed, the covariant harmonic oscillator wave function has a long history. As early as 1945, Dirac suggested the use of normalizable relativistic oscillator wave functions to construct representations of the Lorentz group (Dirac 1945).

In connection with relativistic particles with internal spacetime structure, Yukawa attempted to construct relativistic oscillator wave functions in 1953. Yukawa observed that an attempt to solve a relativistic oscillator wave equation in general leads to infinite-component wave functions, and that finite-component wave functions may be chosen if a subsidiary condition involving the four-momentum of the particle is considered. This proposal of Yukawa was further developed by Markov (1956), Takabayasi (1965 and 1979), Ginzburg and Man'ko (1965), Sogami (1969), and Ishida (1971).

The effectiveness of Yukawa's oscillator wave function in the relativistic quark model was first demonstrated by Fujimura *et al.* (Fujimura *et al.* 1970) who showed that the Yukawa wave function leads to the correct high-energy asymptotic behavior of the nucleon form factor. In 1971, Feynman, Kislinger and Ravndal noted that the use of relativistic oscillators might be more practical than Feynman diagrams in dealing with hadrons believed to be bound states of quarks (Feynman *et al.* 1971).

In this Chapter, we will construct first the physical representation of the Poincaré group from the harmonic oscillators. It is shown that there are solutions of the Lorentz-invariant oscillator equation of Feynman *et al.* which form the basis for unitary irreducible representations of the O(3)-like little group for massive particles. It is, also shown that this formalism of covariant harmonic oscillators constitutes a solution of the equations for Dirac's "instant-form" relativistic dynamics (Dirac

1949).

Dirac's plan to construct a "relativistic dynamics of atom" using "Poisson brackets" is contained in his 1949 paper entitled "Forms of Relativistic Dynamics." Here Dirac emphasizes that the task of constructing a relativistic dynamics is equivalent to constructing a representation of the inhomogeneous Lorentz group. His approach to relativistic quantum mechanics allows us to construct relativistic bound-state wave functions which can be Lorentz-transformed.

Another advantage of the harmonic oscillator formalism is that it is the basic language for phase space. Indeed, it is possible to construct a phase-space picture of the covariant harmonic oscillator formalism. In this picture, we can see clearly how the time-energy uncertaintyindexuncertainty relation relation can be incorporated into the position-momentum relation in a covariant manner. The Lorentz transformation in this case is a canonical transformation in phase space.

## 8.1 Theory of the Poincaré Group

The Poincaré group is the group of inhomogeneous Lorentz transformations, namely Lorentz transformations followed by space-time translations. We shall use throughout this book the four-vector notation:

$$x^{\mu} = (x, y, z, t) \quad \text{and} \quad x_{\mu} = (x, y, z, -t). \tag{8.1}$$

The superscript and subscript will denote the row and column indices respectively. Therefore, $x^{\mu}$ and $x_{\mu}$ are column and row vectors respectively. The conversion from $x^{\mu}$ to $x_{\mu}$ is achieved through the relation:

$$x_{\mu} = g_{\mu\nu} x^{\nu}, \tag{8.2}$$

where $g_{11} = g_{22} = g_{33} = -g_{44} = 1$ while all other $g_{\mu\nu}$ are zero.

We shall use the notation $x_i$ to cover the first three components of the four-vector and use $x_0$ for the fourth component.

$$x^{i} = x_i = (x, y, z), \qquad x^{0} = -x_4 = t = x_0. \tag{8.3}$$

The lower case Greek and Roman letters will be used for the four-vector and three-component spatial vectors respectively.

The general (homogeneous) Lorentz group consists of all real four-by-four matrices $A$ satisfying the condition:

$$a^{\rho}_{\mu} a^{\sigma}_{\nu} g_{\rho\sigma} = g_{\mu\nu}, \tag{8.4}$$

where $a^{\rho}_{\mu}$ are the elements of $A$, while $\rho$ and $\mu$ are the row and column indices respectively. The $A$ matrix leaves the quantity $(x^2 + y^2 + z^2 - t^2)$ invariant. This expression contains ten conditions on the sixteen elements of the four-by-four matrix $A$. Therefore, the Lorentz transformation matrix contains six independent parameters. The property of this Lorentz transformation matrix is widely discussed in the literature (Wigner 1939, Kim and Noz 1986), and its generators were discussed in Chapter 7.

The study of the Poincaré group starts with the group of inhomogeneous Lorentz transformations on the four-dimensional Minkowski space in the form

$$x'^{\mu} = a^{\mu}_{\nu} x^{\nu} + b^{\mu}. \tag{8.5}$$

This is a Lorentz transformation by $A$ followed by a translation which results in the addition of the four-vector $b^{\mu}$. From our experience in constructing representations of the two-dimensional Euclidean group, we can consider the 5-element column vector whose elements are $(x, y, z, t, 1)$. The matrix performing the Lorentz transformation $A$ can be written as

$$A(a) = \begin{pmatrix} & 4-\text{by}-4 & & \vdots & 0 \\ & \text{Lorentz Trans.} & & \vdots & 0 \\ & \text{Matrix} & & \vdots & 0 \\ \cdots & \cdots & \cdots & \cdots & 0 \\ 0 & 0 & & 0 & 1 \end{pmatrix} \tag{8.6}$$

The matrix which performs the translation takes the form:

$$B(b) = \begin{pmatrix} 1 & 0 & 0 & 0 & b_1 \\ 0 & 1 & 0 & 0 & b_2 \\ 0 & 0 & 1 & 0 & b_3 \\ 0 & 0 & 0 & 1 & b_0 \\ 0 & 0 & 0 & 0 & 1 \end{pmatrix}. \tag{8.7}$$

The matrix which performs the inhomogeneous linear transformation of Eq. 8.5 is $B(b)A(a)$ applicable to the five-element column vector $(x, y, z, t, 1)$. Then the inhomogeneous Lorentz transformation of Eq. 8.1 can be written as

$$x' = B(b)A(a)x, \tag{8.8}$$

where $x$, $x'$, $a$ and $b$ are abbreviations of the their respective quantities with indices. This matrix is a representation of the inhomogeneous Lorentz transformation.

The $A$ matrix has the same properties as the four-by-four Lorentz transformation matrices which we discussed in Chapter 7. The $B$ matrix representing translations in the four-dimensional Minkowski space is Abelian and is an invariant subgroup of the inhomogeneous Lorentz group:

$$\begin{aligned} B(b')B(b) &= B(b' + b), \\ A(a)B(b) &= B(b'')A(a), \end{aligned} \tag{8.9}$$

where

$$b''^{\mu} = a^{\mu}_{\nu} b^{\nu}.$$

The transformation group $BA$ therefore has the algebraic property:

$$(B(b')A(a'))\,(B(b)A(a)) = B(b' + b'')A(a'a). \tag{8.10}$$

Certainly $A(a)$ and $B(b)$ are subgroups of the group $BA$. The translation subgroup $B$ is commutative and is invariant:

$$A(a)B(b)[A(a)]^{-1} = B(b''). \tag{8.11}$$

This invariance property was discussed in Chapter 6. Thus the most general form of the inhomogeneous Lorentz transformation can be decomposed into the form $AB$ (Wigner 1939). This decomposition is particularly convenient when the system is an eigenstate of the translation operator. In this case, $B$ is replaced by its eigenvalue, and the rest of the group is the subgroup $A$ which is the group of homogeneous Lorentz transformations.

From this expression and also from the five-by-five matrix form for $B$, it is clear that the translation applicable to a function of $x^\mu$ is generated by

$$P_\mu = -i\frac{\partial}{\partial x^\mu}, \tag{8.12}$$

From the matrices given in Chapter 7, we can write the most general form for the generators of Lorentz transformations as

$$M_{\mu\nu} = L_{\mu\nu} + S_{\mu\nu}, \tag{8.13}$$

where

$$L_{\mu\nu} = -i\left\{x_\mu\frac{\partial}{\partial x^\nu} - x_\nu\frac{\partial}{\partial x^\mu}\right\}.$$

where the $L_{\mu\nu}$ part is applicable to coordinate variables, while $S_{\mu\nu}$ is applicable to all other variables which are not affected by $P_\mu$ and $L_{\mu\nu}$. In the case of massive Dirac particles, the $S$ operators generate Lorentz transformations on the Dirac spinor. In the case of photons, they generate rotations around the momentum and gauge transformations, as we discussed in Chapter 7. In the case of particles with a finite size and with internal motion of constituent particles, the $S$ operator has to generate Lorentz transformations on the internal coordinate variables. The purpose of this Chapter is to discuss the covariant harmonic oscillator as an example of a system with an internal space-time extension.

The commutation relations among the generators of the Poincaré group are

$$[P_\mu, P_\nu] = 0,$$

$$[M_{\mu\nu}, P_\rho] = i\left(g_{\mu\rho}P_\nu - g_{\nu\rho}P_\mu\right),$$

$$[M_{\mu\nu}, M_{\rho\sigma}] = i\left(g_{\mu\rho}M_{\nu\sigma} - g_{\nu\rho}M_{\mu\sigma} + g_{\mu\sigma}M_{\rho\nu} - g_{\nu\sigma}M_{\rho\mu}\right), \tag{8.14}$$

In constructing irreducible representations of the group, it is convenient to construct, from the generators of the group, the maximal set of operators which commute with all the generators. These are called the Casimir operators (Bargmann and Wigner 1948). In the case of the Poincaré group, the Casimir operators are

$$P^2 = P_\mu P^\mu,$$

$$W^2 = W_\mu W^\mu, \tag{8.15}$$

where

$$W_\mu = \frac{1}{2}\varepsilon_{\mu\nu\alpha\beta}P^\nu M^{\alpha\beta}. \tag{8.16}$$

The transformation property of $W_\mu$ is like that of a four vector. This four vector is called the Pauli-Lyubanski vector in the literature (Bargmann and Wigner 1948). It is constrained to satisfy

$$P_\mu W^\mu = 0. \tag{8.17}$$

The Pauli-Lyubanski vector has only three independent components, and each component of this four-vector commutes with $P^\mu$:

$$[P^\mu, W^\nu] = 0. \tag{8.18}$$

For a free-particle state, the operator $P^2$ determines the (mass)$^2$ of the particle. It is interesting to note that the three independent components of the Pauli-Lyubanski vector become the generators of the little group (Kim and Noz 1986). Indeed, for a massive particle at rest, $W_0 = 0$, and

$$W_i = (\text{mass})\varepsilon_{ijk}S^{jk}. \tag{8.19}$$

We shall continue the discussion of this operator in Section 8.3.

## 8.2 Covariant Harmonic Oscillators

We are considering here a relativistic hadron consisting of two quarks bound together by a harmonic oscillator force. Let us start with the Lorentz-invariant differential equation of Feynman *et al.* (1971) for a hadron of two quarks bound together by a harmonic oscillator potential of unit strength:

$$\left\{-2\left[\left(\frac{\partial}{\partial x_a^\mu}\right)^2 + \left(\frac{\partial}{\partial x_b^\mu}\right)^2\right] + \left(\frac{1}{16}\right)(x_a^\mu - x_b^\mu)^2 + m_0^2\right\}\phi(x_a, x_b) = 0, \tag{8.20}$$

where $x_a$ and $x_b$ are space-time coordinates for the first and second quarks respectively. Then, the following variables are convenient.

$$X = (x_a + x_b)/2, \quad x = (x_a - x_b)/2\sqrt{2}. \tag{8.21}$$

The four-vector $X$ specifies where the hadron is located in space-time, while the variable $x$ measures the space-time separation between the quarks. In terms of these variables, $\phi(x_a, x_b)$ can be separated into

$$\phi(X, x) = f(X)\psi(x), \tag{8.22}$$

and $f(X)$ and $\psi(x)$ satisfy the following differential equations respectively:

$$\left\{\frac{\partial^2}{\partial X_\mu^2} - m_0^2 - (\lambda + 1)\right\} f(X) = 0, \tag{8.23}$$

$$\frac{1}{2}\left\{\frac{\partial^2}{\partial x_\mu^2} - x_\mu^2 + (\lambda + 1)\right\} \psi(x) = 0. \tag{8.24}$$

Eq. 8.23 is a Klein-Gordon equation, and its solution takes the form

$$f(X) = \exp\left(\pm i P_\mu X^\mu\right), \tag{8.25}$$

with

$$-P^2 = -P_\mu P^\mu = M^2 = m_0^2 + (\lambda + 1),$$

where $M$ and $P$ are the mass and four-momentum of the hadron respectively. The eigenvalue $\lambda$ is determined from the solution of Eq. 8.24.

As for the four-momenta of the quarks $p_a$ and $p_b$, we can combine them into the total four-momentum and momentum-energy separation between the quarks:

$$P = p_a + p_b, \qquad q = \sqrt{2}(p_a - p_b), \tag{8.26}$$

where $P$ is the hadronic four-momentum conjugate to $X$. The internal momentum-energy separation $q$ is conjugate to $x$ provided that there exist wave functions which can be Fourier-transformed. If the momentum-energy wave functions can be obtained from the Fourier transformation of the space-time wave function, the differential equation in the $q$ space is the same as the harmonic oscillator equation for the $x$ space given in Eq. 8.24.

There are many different forms of solutions of the Lorentz-invariant oscillator equation of Eq. 8.24. There are many papers in the literature reflecting the view that the solution of this equation has to be Lorentz- invariant and thus has to be a function only of the invariant quantity $(x^2+y^2+z^2-t^2)$ (Feynman *et al.* 1971). This condition leads to wave functions which are not normalizable. In this Chapter, we shall discuss Dirac's normalizable solutions which are covariant but not invariant (1945). These wave functions appear different to observers in different Lorentz frames (Kim and Noz 1973 and 1986).

If the hadron moves along the $Z$ direction which is also the $z$ direction, then the hadronic factor $f(X)$ of Eq. 8.25 is Lorentz-transformed in the same manner as the scalar particles are transformed. The Lorentz transformation of the internal coordinates from the laboratory frame to the hadronic rest frame takes the form

$$x' = x, \quad y' = y,$$

$$z' = (z - \beta t)/(1-\beta^2)^{1/2},$$

$$t' = (t - \beta z)/(1-\beta^2)^{1/2}, \tag{8.27}$$

where $\beta$ is the velocity of the hadron moving along the $z$ direction. This transformation is the same as that given in Eq. 7.44 even though it appears as its inverse. The transformation given here is applicable to the argument of a function, and is therefore a passive transformation (Kim and Noz 1986). The primed quantities are the coordinate variables in the hadronic rest frame.

In terms of the primed variables, the oscillator differential equation is

$$\frac{1}{2}\left\{-\nabla'^2 + \frac{\partial^2}{\partial t'^2}\left((\vec{x}')^2 - t'^2\right)\right\}\psi(x) = (\lambda+1)\psi(x). \tag{8.28}$$

The solution of this equation is a product of four one-dimensional oscillator wave functions.

$$\psi_\beta(x) = \left(\frac{1}{\pi}\right)^2 \left(\frac{1}{2}\right)^{(a+b+n+k)/2} \left(\frac{1}{a!b!n!k!}\right)^{1/2} H_a(x')H_b(y')$$

$$\times H_n(z')H_k(t')\exp\left\{-\left(\frac{1}{2}\right)\left(x'^2 + y'^2 + z'^2 + t'^2\right)\right\}, \tag{8.29}$$

where $a$, $b$, $n$, and $k$ are integers, and $H_a(x')$, $H_b(y')$, ... are the Hermite polynomials. This wave function is normalizable, but the eigenvalue is

$$\lambda = (a + b + n - k). \tag{8.30}$$

Thus, for a given finite value of $\lambda$, there are infinitely many possible combinations of $a$, $b$, $n$ and $k$. The most general solution of the oscillator differential equation is infinitely degenerate (Yukawa 1953).

Because the wave functions are normalizable, all the generators of the Lorentz transformations given Section 8.1 are Hermitian operators. The Lorentz transformation applicable to this function space is therefore a unitary transformation. Indeed, we can write any function of the coordinate variables $x$, $y$, $z$ and $t$ as a linear combination of the above solutions. In particular, a solution of the oscillator equation with a given set of quantum numbers in the hadronic rest frame can be written as a linear sum of infinitely many solutions in the hadronic rest frame as we shall see in Section 8.6.

It is very difficult, if not impossible, to give physical interpretations to infinite-component wave functions. For this reason, it is quite natural to seek a finite set from the infinite number of wave functions at least in one Lorentz frame. The simplest way to obtain such a finite set of wave functions is to invoke the restriction that there be no time-like oscillations in the Lorentz frame in which the hadron is at rest by setting $k = 0$ in Eq. 8.29 and Eq. 8.30. In doing so, we are led to:

**(a).** Is it possible to give physical interpretations to the wave functions belonging to the resulting finite set?

**(b).** Is it still possible to maintain Lorentz covariance with this condition?

We shall find physical interpretation for the first question in Sections 8.4 and 8.5. Let us first examine question (b).

When the hadron moves along the $z$ axis, the $k = 0$ condition is equivalent to

$$\left(\frac{\partial}{\partial t'} + t'\right) \psi_\beta(x') = 0. \tag{8.31}$$

The most general form of the above condition is

$$P_\mu \left(x^\mu - \frac{\partial}{\partial x_\mu}\right) \psi_\beta(x) = 0. \tag{8.32}$$

Thus the $k = 0$ condition is covariant (Kim and Noz 1973, Hussar *et al.* 1985). Once this condition is set, we can write the wave function belonging to this finite set as

$$\psi_\beta^{abn}(x) = \left(\frac{1}{\pi}\right)^2 \left(\frac{1}{2}\right)^{(a+b+n)/2} \left(\frac{1}{a!b!n!}\right)^{1/2} H_a(x')H_b(y')H_n(z')$$
$$\times \exp\left\{-\left(\frac{1}{2}\right)\left(x'^2 + y'^2 + z'^2 + t'^2\right)\right\}, \tag{8.33}$$

with the eigenvalue $\lambda = a + b + n$.

Since the above oscillator wave functions are separable in the Cartesian coordinate system, and since the transverse coordinate variables are not affected by the boost along the $z$ direction, we can omit the factors depending on the $x$ and $y$ variables when studying their Lorentz transformation properties. The essential part of the covariant wave function is

$$\psi_\beta^n(z,t) = \left(\frac{1}{\pi}\right)^{1/2} \left(\frac{1}{2}\right)^{n/2} \left(\frac{1}{n!}\right)^{1/2} H_n(z') \exp\left\{-\left(\frac{1}{2}\right)\left(z'^2 + t'^2\right)\right\}. \tag{8.34}$$

Let us next study the orthogonality relations of the wave functions. Since the volume element is Lorentz-invariant:

$$dzdt = dz'dt', \tag{8.35}$$

there is no difficulty in understanding the orthogonality relation:

$$\int \psi_\beta^n(z,t)\psi_\beta^m(z,t)dzdt = \int \psi_0^n(z,t)\psi_0^m(z,t)dzdt = \delta_{mn}. \tag{8.36}$$

However, a more interesting problem is the inner product of two wave functions belonging to different Lorentz frames. The orthogonality relation is (Ruiz 1974)

$$\int \psi_\beta^n(z,t)\psi_0^m(z,t)dzdt = \left(1 - \beta^2\right)^{(n+1)/2} \delta_{nm}. \tag{8.37}$$

The remarkable fact is that the orthogonality in the quantum number $n$ is still preserved, because of the Lorentz invariance of the harmonic oscillator differential equation. The oscillator equation does not depend on the velocity parameter $\beta$.

As for the factor $(1-\beta^2)^{(n+1)/2}$ in Eq. 8.37, we note first that, when the oscillator is in the ground state, it becomes like a Lorentz contraction of a rigid rod by $(1-\beta^2)^{1/2}$. Excited-state wave functions are obtained from the ground state wave function through repeated applications of the step operator:

$$\psi_\beta^n(x) = \left(\frac{1}{n!}\right)^{1/2} \left(z' - \frac{\partial}{\partial z'}\right)^n \psi_\beta^0(x). \tag{8.38}$$

The transformation property of each step-up operator is like that of $z$. Therefore, if the ground-state wave function is like a rigid rod along the $z$ direction, the n[th] excited state should behave like a multiplication of $(n+1)$ rigid rods (Kim and Noz 1986). This contraction property is summarized in Figure 8.1.

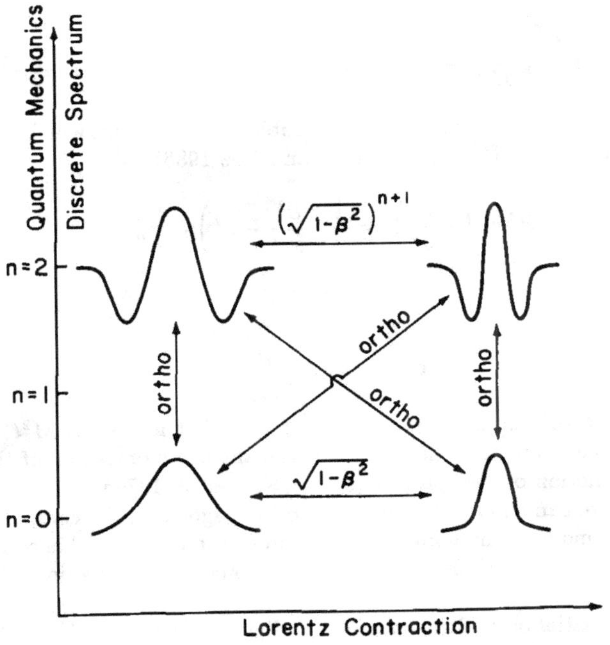


Figure 8.1: Orthogonality and Lorentz contraction properties of the normalizable harmonic oscillator wave functions. The ground-state wave function is contracted like a rigid rod. The n[th] excited state is contracted like a product of $(n+1)$ rigid rods (Kim and Noz 1986).

## 8.3 Irreducible Unitary Representations of the Poincaré Group

The Poincaré transformation consists of space-time translations and Lorentz transformations. Let us go back to the quarks coordinates $x_a$ and $x_b$ in Eq. 8.21, and consider performing inhomogeneous Lorentz transformations on the quarks. The same Lorentz transformation matrix is applicable to $x_a$, $x_b$, $x$ as well as $X$. However, under the space-time translation which changes $x_a$ and $x_b$ to $x_a + a$ and $x_b + a$, $X$ becomes $X + a$, and $x$ remains invariant. The quarks separation coordinate $x$ is not affected by translations. For this reason, the generators of translations for this system are

$$P_\mu = -i\frac{\partial}{\partial X^\mu}, \tag{8.39}$$

while the generators of Lorentz transformations are (Kim *et al.* 1979a)

$$M_{\mu\nu} = L_{\mu\nu} + S_{\mu\nu}, \tag{8.40}$$

where

$$L_{\mu\nu} = -i\left(X_\mu\frac{\partial}{\partial X^\nu} - X_\nu\frac{\partial}{\partial X^\mu}\right), \quad S_{\mu\nu} = -i\left(x_\mu\frac{\partial}{\partial x^\nu} - x_\nu\frac{\partial}{\partial x^\mu}\right).$$

We are interested in constructing normalizable wave functions which are diagonal in the Casimir operators $P^2$ and $W^2$ (Kim and Noz 1986):

$$P^2 = \left(\frac{\partial}{\partial X^\mu}\right)^2 = \frac{1}{2}\left(\frac{\partial^2}{\partial x_\mu^2} - x_\mu^2\right) - m_0^2, \tag{8.41}$$

$$W^2 = M^2\left(\mathbf{L}\right)^2, \tag{8.42}$$

where

$$\mathbf{L}_i = -i\epsilon_{ijk}x_j\frac{\partial}{\partial x_k}.$$

The eigenvalue of $-P^2$ is $M^2 = m_0^2 + (\lambda + 1)$, and that for $W^2$ is $M^2\ell(\ell + 1)$. $M$ is the hadronic mass, and $\ell$ is the total intrinsic angular momentum of the hadron due to internal motion of the spinless quarks (Kim *et al.* 1979a).

In addition, we can choose the solutions to be diagonal in the component of the intrinsic angular momentum along the direction of the motion. This component is often called the helicity. If the hadron moves along the $Z$ direction, the helicity operator is $S_3$.

Because the spatial part of the harmonic oscillator equation in Eq. 8.28 is separable also in the spherical coordinate system, we can write its solution using spherical variables in the hadronic rest frame space spanned by $x'$, $y'$ and $z'$. The most general form of the solution is

$$\psi_{\beta\lambda\ell}^{k,m}(x) = R_\mu^\ell(r')Y_\ell^m(\theta', \phi')\left(1/\sqrt{\pi}2^k k!\right)^{1/2}H_k(t')e^{-t'^2/2}, \tag{8.43}$$

where

$$r' = \left(x'^2 + y'^2 + z'^2\right)^{1/2}, \quad \theta' = \cos^{-1}\left(\frac{z'}{r'}\right), \quad \phi' = \tan^{-1}\left(\frac{y'}{x'}\right),$$

and

$$\lambda = 2\mu + \ell - k. \tag{8.44}$$

$R^\ell_\mu(r')$ is the normalized radial wave function for the three-dimensional harmonic oscillator:

$$R^\ell_\mu(r) = \left(2(\mu!)/(\mu + \ell + 3/2)^3\right)^{1/2} r^\ell L^{\ell+1/2}_\mu(r^2) e^{-r^2/2}, \tag{8.45}$$

where $L^{\ell+1/2}_\mu(r^2)$ is the associated Laguerrefunction. The spherical form given in Eq. 8.43 can of course be expressed as a linear combination of the wave functions in the Cartesian coordinate system given in Eq. 8.33.

The wave function of Eq. 8.43 is diagonal in the Casimir operators of Eqs. 8.41 and 8.42, as well as in $S_3$. It indeed forms a vector space for the O(3)-like little group. However, the system is infinitely degenerate due to excitations along the $t'$ axis. As we did in Section 8.1, we can suppress the time-like oscillation by imposing the subsidiary condition of Eq. 8.32. The solution then takes the form

$$\psi^m_{\beta\lambda\ell}(x) = R^\ell_\mu(r') Y^m_\ell(\theta', \phi') \left\{(1/\pi)^{1/4} e^{-t'^2/2}\right\}, \tag{8.46}$$

with $\lambda = 2\mu + \ell$. Thus for a given $\lambda$, there are only a finite number of solutions. The above spherical form can be expressed as a linear combination of the solutions without time-like excitations in the Cartesian coordinate system given in Eq. 8.33.

We can now write the solution of the differential equation of Eq. 8.20 as

$$\phi(X, x) = e^{\pm iP\cdot X} \psi^m_{\beta\lambda\ell}(x). \tag{8.47}$$

This wave function describes a free hadron with a definite four-momentum having an internal space-time structure which can be described by an irreducible unitary representation of the Poincaré group. The representation is unitary because the portion of the wave function depending on the internal variable $x$ is square-integrable, and all the generators of Lorentz transformations are Hermitian operators.

## 8.4 C-number Time-Energy Uncertainty Relation

As early as 1927, Dirac observed that the uncertaintyindexuncertainty relation relation applicable to the time and energy variables is different from Heisenberg's uncertaintyindexuncertainty relation relations applicable to the position and momentum variables, and that this space-time asymmetry is one of the problems we have to face in the process of making quantum mechanics relativistic. In 1945, Dirac considered the possibility of using the four-dimensional harmonic oscillator with normalizable time-like wave functions in connection with relativistic Fock space. The covariant harmonic oscillator wave functions discussed in Sections 8.2 and 8.3 are of Dirac's type. Indeed, the use of the covariant oscillator formalism is perfectly consistent with Dirac's overall plan to construct a relativistic quantum mechanics.

The time-energy uncertaintyindexuncertainty relation relation in the form of $(\Delta t)(\Delta E) \simeq 1$ was known to exist even before the present form of quantum mechanics was formulated (Wigner 1972). However, we are still debating over the following issues.

**(a).** While there exists the time-energy uncertaintyindexuncertainty relation relation in the real world, possibly with the form $[t, H] = -i$ (Heisenberg 1927), this commutator is zero in the case of Schrödinger quantum mechanics. As was noted by Dirac in 1927, the time variable is a c-number. Then, is the c-number time-energy uncertainty relation universal, or true only in nonrelativistic quantum mechanics?

**(b).** If the time variable is a c-number and the position variables are q-numbers, then the coordinate variables in different Lorentz frames are mixtures of $c$ and $q$ numbers. This cannot be consistent with special relativity, as was also pointed out by Dirac (1927).

We are concerned here with the time-energy uncertaintyindexuncertainty relation relation applicable to the time-separation between the quarks. The covariant harmonic oscillator will enable us to answer these questions. The subsidiary condition of Eq. 8.32 forbids time-like excitations, while allowing the minimal time-energy uncertaintyindexuncertainty relation relation. This is consistent with the commutation relation $[t, H] = 0$. Classically, this corresponds to the fact that $t$ and $H$ are not canonically conjugate variables. In quantum mechanics, this means that there is no Hilbert space in which $t$ and $i\partial/\partial t$ act as operators. However, it is important to note that there still exists a "Fourier" relation between time and energy which limits the precision to $(\Delta t)(\Delta E) \simeq 1$ (Dirac 1927, Weisskopf and Wigner 1930a and 1930b, Heitler 1954, Blanchard 1982).

In discussing the uncertainty relations, we need momentum-energy variables in addition to the space-time coordinates introduced in Eq. 8.21. They are given in Eq. 8.26. The spatial and time-like components of the four-vector $P$ is the sum of the momenta and the energies of the two quarks respectively. We assume here that the system is in an eigenstate of this four-momentum, and that every component of $P$ is a sharply defined number. The spatial component of $q$ measures the momentum difference between the quarks. The time-like component of $q$ is the energy difference between the quarks. The concept of this energy separation does not exist in nonrelativistic quantum mechanics. If the wave functions can be Fourier-transformed, $P$ and $q$ are conjugate to $X$ and $x$ respectively. We are particularly interested in the uncertainty relation applicable between the time-like components of four-vectors $x$ and $q$.

If the hadron has a definite four-momentum and moves along the $z$ direction with velocity parameter $\beta$, it is possible to find the Lorentz frame in which the hadron is at rest. We shall use $x'$, $y'$, $z'$, and $t'$ to denote the space-time separations in this frame, and $q'_x$, $q'_y$, $q'_z$, $q'_0$ for momentum-energy separations. In the Lorentz frame where the hadron is at rest, the uncertainty principle applicable to the space-time separation of quarks is expected to be the same as the presently accepted form. The usual Heisenberg uncertainty relation holds for each of the three spatial coordinates:

$$[x', q'_x] = i, \quad [y', q'_y] = i, \quad [z', q'_z] = i. \tag{8.48}$$

On the other hand, the time-separation variable is a c-number and therefore does not cause quantum excitations. Thus

$$[t', q'_0] = 0. \tag{8.49}$$

However, this commutator allows the existence of the "Fourier relation" between time and energy variables (Blanchard 1982, Hussar *et al.* 1985) resulting in

$$(\Delta t')(\Delta q'_0) \simeq 1. \tag{8.50}$$

The crucial question is how these uncertainty relations appear to an observer in the laboratory frame with the space-time separation variables $x$, $y$, $z$ and $t$.

We assume here that the hadron moves along the $z$ axis. Thus the $x$ and $y$ coordinates are not affected by boosts, and the first two commutation relations of Eq. 8.48 remain invariant. If we consider only the ground state, the third and longitudinal commutator of Eq. 8.48 can be quantified as (Hussar *et al.* 1985)

$$<\Delta z'><\Delta q'_z>\simeq 1. \tag{8.51}$$

Then the uncertainty relations associated with both the longitudinal and time-separation variables will lead to a distribution centered around the origin in the $z't'$ coordinate system.

Next, we should examine this localization region in the laboratory frame where the hadron moves along the $z$ direction with the velocity parameter $\beta$. This is why we need a covariant formulation of the phase-space picture. We shall return to this problem in Section 8.7.

## 8.5 Dirac's Form of Relativistic Theory of "Atom"

One of the most fruitful and still promising approaches to incorporating special relativity into quantum mechanics is the covariant formulation of quantum field theory. However, quantum field theory is not an efficient tool for handling relativistic bound-state problems (Feynman *et al.* 1971). Indeed, in 1949, Dirac proposed a relativistic theory of "atom" based on the representations of the Poincaré group. Dirac's atom in modern language is a relativistic bound state of quarks. Dirac's form of relativistic quantum mechanics is therefore the quantum mechanics of bound states.

In an attempt to construct a relativistic theory of bound states, Dirac considered several constraint conditions, each of which reduces the four-dimensional Minkowskian space-time into a three-dimensional Euclidean space. It will be shown in this Section that Dirac's instant-form quantum mechanics is a representation of Wigner's O(3)-like little group. Thus, the covariant harmonic oscillator discussed in Section 8.2 represent also Dirac's instant-form quantum mechanics.

In his "instant form", Dirac considered the condition

$$x_0 \approx 0, \tag{8.52}$$

whose covariant form is

$$P \cdot x \approx 0, \tag{8.53}$$

where $P$ is the total momentum of the hadron. Eq. 8.53 becomes Eq. 8.52 when the hadron is at rest. Dirac avoided using the exact numerical equality in writing down the above constraint in order to allow further physical interpretations consistent with quantum mechanics and special relativity. In particular, Dirac had in mind the possibility of the left-hand side becoming an operator acting on state vectors.

After introducing this constraint in his instant-form dynamics, Dirac introduced the generators of transformations of the "inhomogeneous Lorentz group" which is the Poincaré group. He then emphasized that the construction of a relativistic dynamics is the same as the construction of the representation of the Poincaré group.

The only remaining step in constructing Dirac's dynamical system is therefore to make the constraint condition consistent with the transformations of the Poincaré group. Dirac noted in particular that the constraint condition of Eq. 8.53 can be an operator equation, and that its "Poisson brackets" with other dynamical variables should be zero or become zero in the manner in which the right-hand side of Eq. 8.53 vanishes.

Let us see whether the covariant harmonic oscillator formalism satisfies all these conditions. The question is whether the subsidiary condition of Eq. 8.32 can act as the instant-form constraint. What Dirac wanted from his conditional equality of Eq. 8.53 was to freeze the motion along the time separation variable in a manner consistent with quantum mechanics and relativity. This means that we can allow a time-energy uncertainty along this time-like axis without excitations, in accordance with the c-number time-energy uncertainty relation discussed in Section 8.4. The c-number in matrix language is a one-by-one matrix, and is the ground state with no excitations in the harmonic oscillator system. The subsidiary condition of Eq. 8.32 can be written as

$$P^{\mu} a_{\mu}^{\dagger} \psi_{\beta}(x) = 0, \tag{8.54}$$

with

$$a_{\mu}^{\dagger} = \left(\frac{1}{2}\right)^{1/2} \left(x^{\mu} - \frac{\partial}{\partial x_{\mu}}\right).$$

In order that the dynamical system be completely consistent, the subsidiary condition should commute with the generators of the Poincaré group:

$$[P_\alpha, P^\mu a_\mu^\dagger] = 0, \quad [M_{\alpha\beta}, P^\mu a_\mu^\dagger] = 0. \tag{8.55}$$

The above equations follow immediately from the fact that the operator $P^\mu a_\mu^\dagger$ is invariant under translations and Lorentz transformations.

Since the Casimir operators are constructed from the generators of the Poincaré group, we are tempted to say that the constraint operator commutes with the invariant Casimir operators. However, from Eq. 8.25, we note that the operator $P^2$ also takes the form

$$P^2 = \frac{1}{2}\left\{\frac{\partial^2}{\partial x_\mu^2} - x_\mu^2 - 2m_0^2\right\}. \tag{8.56}$$

The subsidiary condition should commute also with this form of $P^2$. If we compute the commutator using this expression of $P^2$,

$$[P^2, P^\mu a_\mu^\dagger] = P^\mu a_\mu^\dagger. \tag{8.57}$$

The right-hand side of this commutator does not vanish. This however should not alarm us. Because of the subsidiary condition of Eq. 8.32, the right-hand side vanishes in the applicable Hilbert space.

The constraint condition of Eq. 8.32 and its commutator with other operators produce zero either identically or in the manner in which the Eq. 8.32 is zero. Therefore, the subsidiary condition of Eq. 8.32 satisfies all the requirements of the "instant form" constraint of Eq. 8.53. Indeed, the covariant harmonic oscillator formalism, while serving as the basis for the representations of the Poincaré group, is a solution of Dirac's "Poisson bracket" equations consistent with the "instant form" constraint (Kim and Noz 1986).

In his 1949 paper, Dirac notes a "real difficulty" associated with making his dynamical models consistent with the energy-momentum relation for each particle participating in the interacting system. Let us examine why we do not have this difficulty in the harmonic oscillator formalism. The basic difference between Dirac's original paper and the present case is that, for each quarks in the dynamics, Dirac uses the free-particle energy-momentum relation:

$$E_a = \left(\mathbf{p}_a^2 + m_a^2\right)^{1/2}, \tag{8.58}$$

because the concept of off-mass-shell particles was not firmly established in 1949.

In the harmonic oscillator formalism, the Casimir operators of the Poincaré group indicate clearly that the mass of the hadron is a Poincaré-invariant constant, but they do not tell us anything about the masses of the constituent particles. In fact, the (mass)$^2$ operator for an individual quarks $(p_a)^2$ does not commute with $P^2$ of Eq. 8.25 (Kim and Noz 1986):

$$[p_a^2, P^2] \neq 0, \quad [p_b^2, P^2] \neq 0, \tag{8.59}$$

These non-vanishing commutators would have caused very serious difficulties in 1949 when the concept of off-mass-shell particles did not exist.

The above commutators do not cause any difficulty today, because it is by now a well-accepted view that particles in the interacting field or within a bound state are not necessarily on their mass shells. We know also that those particles can become unphysical or virtual particles due to the time-energy uncertainty relation. It is interesting to note that Dirac's 1949 difficulty can be resolved by the c-number time-energy uncertainty relation which Dirac discussed in 1927 (Kim and Noz 1986).

## 8.6 Lorentz Transformations of Harmonic Oscillator Wave functions

Since we now understand the physics of the covariant harmonic oscillator formalism, we can continue our discussion of its covariance properties. Let us go back to the covariant form given in Eq. 8.34.

If $\beta = 0$, the wave function becomes that of the hadron at rest. As $\beta$ increases, it should be a linear combination of the wave functions in the rest frame. It should be of the form

$$\psi_\beta^n(z,t) = \psi_\beta^{n,0}(z,t) = \sum_{m,k} A_{mk}^n(\beta)\psi_0^{mk}(z,t), \tag{8.60}$$

where

$$\psi_\beta^{m,k}(z,t) = \left(\frac{1}{\pi}\right)^{1/2} \left(\frac{1}{2}\right)^{(m+k)/2} \left(\frac{1}{m!k!}\right)^{1/2} H_m(z')H_k(t')$$
$$\times \exp\left\{-\left(\frac{1}{2}\right)\left(z'^2 + t'^2\right)\right\}. \tag{8.61}$$

We have concluded in Section 8.5 that there are no time-like oscillations, and therefore that the physical wave functions should have $k = 0$. Indeed, in Eq. 8.60, the left-hand side is a physical wave function. However, this does not keep us from expanding in a complete set of orthonormal wave functions which include time-like excitations. Each term on the right-hand side may not be physical, but the summation may still result is a physical wave function.

Because the oscillator differential equation is Lorentz invariant, the eigenvalue $n$ for $\psi_\beta^n(z,t)$ of Eq. 8.60 remains invariant, and only the terms which satisfy the condition

$$n = (m - k) \tag{8.62}$$

make non-zero contributions in the sum. Thus the above expression can be simplified to

$$\psi_\beta^n(z,t) = \sum_k A_k^n(\beta)\psi_0^{n+k,k}(z,t). \tag{8.63}$$

This is indeed an infinite-dimensional unitary representation of the Lorentz group (Kim *et al.*, 1979b).

The remaining task is to determine the coefficient $A_k^n(\beta)$. Using the orthogonality relation of the Hermite polynomials, we can write

$$A_k^n(\beta) = \frac{1}{\pi}\left(\frac{1}{2}\right)^{(n+k)}\left(\frac{1}{n!(n+k)!k!}\right)^{1/2}\int H_m(z)H_k(t)H_n(z')$$
$$\times \exp\left\{-\left(\frac{1}{2}\right)\left(z^2+z'^2+t^2+t'^2\right)\right\}dzdt. \tag{8.64}$$

In this integral, the Hermite polynomials and the Gaussian form are mixed with the kinematics of Lorentz transformation. However, if we use the generating function for the Hermite polynomial, the evaluation of the integral is straightforward (Ruiz 1974, Kim and Noz 1986), and the result is

$$A_\beta^n(\beta) = \left(1-\beta^2\right)^{(n+1)/2}\left(\frac{(n+k)!}{n!k!}\right)^{1/2}\beta^k. \tag{8.65}$$

The binomial expansion of $(1-\beta^2)^{-(n+1)}$ is

$$\left(1-\beta^2\right)^{-(n+1)} = \sum_k \frac{(n+k)!}{n!k!}\beta^{2k}, \tag{8.66}$$

which leads to

$$\sum_k |A_k^n(\beta)|^2 = 1. \tag{8.67}$$

This relation assures us that the expansion given in Eq. 8.63 is a unitary transformation.

If $n = 0$, the transformation corresponds to the Lorentz boost of the ground-state wave function which is of the Gaussian form. Let us go back to Section 3.5 where the density matrix for thermally excited states was discussed. It is interesting to note that the thermal excitation of the oscillator shares the same mathematics as the Lorentz boost of the ground-state wave function.

Let us next consider the Lorentz-boosted oscillator wave function in the light-cone coordinate system. As was noted in Section 7.6, the basic advantage of the light-cone coordinate system is that the Lorentz boost can be formulated in terms of a squeeze in the $zt$ plane. In the case of the covariant harmonic oscillators, the Gaussian factor determines the localization property of the wave function. The Gaussian factor is localized within the region

$$\left(z'^2+t'^2\right) < 1. \tag{8.68}$$

If the hadron is at rest, this becomes a circular region in the $zt$ plane with $(z^2+t^2) < 1$. In terms of the light-cone variables, this circular region is

$$\left(u^2+v^2\right) < 1, \tag{8.69}$$

where the $u$ and $v$ variables are defined in Eq. 7.45.

When the hadron moves, the region becomes

$$\left(\frac{1-\beta}{1+\beta}\right) u^2 + \left(\frac{1+\beta}{1-\beta}\right) v^2 < 1. \tag{8.70}$$

Certainly this is a squeezed circle. In terms of the $zt$ variables, this elliptic region can be expressed as

$$\frac{1}{2}\left(\frac{1-\beta}{1+\beta}\right)(z+t)^2 + \frac{1}{2}\left(\frac{1+\beta}{1-\beta}\right)(z-t)^2 < 1. \tag{8.71}$$

The Lorentz boost on the harmonic oscillator wave function results in a squeeze in the $zt$ plane. It was also noted in Section 7.6 that the squeeze property is applicable to the energy-momentum plane. We shall examine the squeeze property of the covariant oscillators in phase space in Section 8.7.

## 8.7 Covariant Phase-Space Picture of Harmonic Oscillators

We are now ready to give a phase-space picture of the covariant harmonic oscillator formalism in the light-cone coordinate system. The harmonic oscillator wave function consists of a Gaussian factor and Hermite polynomials.

Since the Gaussian factor determines the localization property of the wave function, let us study first the ground state wave function, whose form is

$$\psi_0^0(z,t) = \left(\frac{1}{\pi}\right)^{1/2} \exp\left(-(z^2+t^2)/2\right). \tag{8.72}$$

In terms of the light-cone variables, this wave function can be written as

$$\psi_0^0(z,t) = \psi_0^0(u,v) = \left(\frac{1}{\pi}\right)^{1/2} \exp\left(-(u^2+v^2)/2\right). \tag{8.73}$$

If the system is boosted, the wave function becomes

$$\psi_\beta^0(z,t) = \left(\frac{1}{\pi}\right)^{1/2} \exp\left\{-\left(\frac{1}{2}\right)\left(\frac{1-\beta}{1+\beta}u^2 + \frac{1+\beta}{1-\beta}v^2\right)\right\}. \tag{8.74}$$

This wave function undergoes a Lorentz deformation as $\beta$ increases.

The momentum-energy wave function is

$$\phi_\beta^0(q_u, q_v) = \left(\frac{1}{2\pi}\right)\int \psi_\beta^0(x,t)e^{-i(q_z z - q_0 t)}dz dt, \tag{8.75}$$

where the momentum-energy separation variables $q_z$ and $q_0$ are given in Eq. 8.26. The evaluation of this integral leads to

$$\phi_\beta^0(q_u, q_v) = \left(\frac{1}{\pi}\right)^{1/2} \exp\left\{-\left(\frac{1}{2}\right)\left(\frac{1+\beta}{1-\beta}q_u^2 + \frac{1-\beta}{1+\beta}q_v^2\right)\right\}, \tag{8.76}$$

where

$$q_u = (q_z - q_0)/\sqrt{2}, \quad q_v = (q_z + q_0)/\sqrt{2}. \tag{8.77}$$

Both of these wave functions have the same Lorentz deformation property.

For the ground state, the Wigner function can now be defined as

$$W_\beta^0(u, q_u; v, q_v) = \left(\frac{1}{\pi}\right) \int \left(\psi_\beta^0(u + x, v + y)\right)^* \psi_\beta^0(u - x, v - y)$$
$$\times \exp\{2i(q_u x + q_v y)\}\, dx dy. \tag{8.78}$$

After the evaluation of this integral, the Wigner function becomes

$$W_\beta^0(u, q_u; v, q_v) = \left(\frac{1}{\pi}\right)^2 \exp\left\{-\left(\frac{1}{2}\right)\left(\frac{1-\beta}{1+\beta}u^2 + \frac{1+\beta}{1-\beta}q_u^2\right)\right\}$$
$$\times \exp\left\{-\left(\frac{1}{2}\right)\left(\frac{1+\beta}{1-\beta}v^2 + \frac{1-\beta}{1+\beta}q_v^2\right)\right\}. \tag{8.79}$$

The above Wigner function is defined in two independent phase spaces consisting of $(u, q_u)$ and $(v, q_v)$ respectively. When the hadron moves with velocity $\beta$, the Wigner function is localized in the regions:

$$\left(\frac{1-\beta}{1+\beta}u^2 + \frac{1+\beta}{1-\beta}q_u^2\right) < 1,$$
$$\left(\frac{1+\beta}{1-\beta}v^2 + \frac{1-\beta}{1+\beta}q_v^2\right) < 1. \tag{8.80}$$

These localization regions are described in Figure 8.2. When the hadron moves, these regions become squeezed. It is interesting to note that this squeeze property is the same as one of the six possible squeezes in the four-dimensional phase space of two-mode squeezed states of photons discussed in Section 6.8.

This Wigner function reproduces the distributions $|\psi_\beta^0(u, v)|^2$ and $|\phi_\beta^0(q_u, q_v)|^2$ after the appropriate integrals:

$$|\psi_\beta^0(u, v)|^2 = \int W_\beta^0(u, v; q_u, q_v) dq_u dq_v,$$
$$|\phi_\beta^0(q_u, q_v)|^2 = \int W_\beta^0(u, v; q_u, q_v) du dv. \tag{8.81}$$

As for the excited states, there are no time-like oscillations in the hadronic rest frame, and the oscillations in the transverse direction are not affected. Therefore, the only factor we have to consider is the Hermite polynomial $H_n(z')$ multiplied by the ground-state wave function. In terms of the $u'$ and $v'$ variables, $H_n(z')$ can be written as (Magnus and Oberhettinger 1949)

$$H_n(z') = H_n([u' + v']/\sqrt{2}) = \left(\frac{1}{2}\right)^{n/2} \sum_{m=0}^{n} \binom{n}{m} H_{n-m}(u') H_m(v'). \tag{8.82}$$

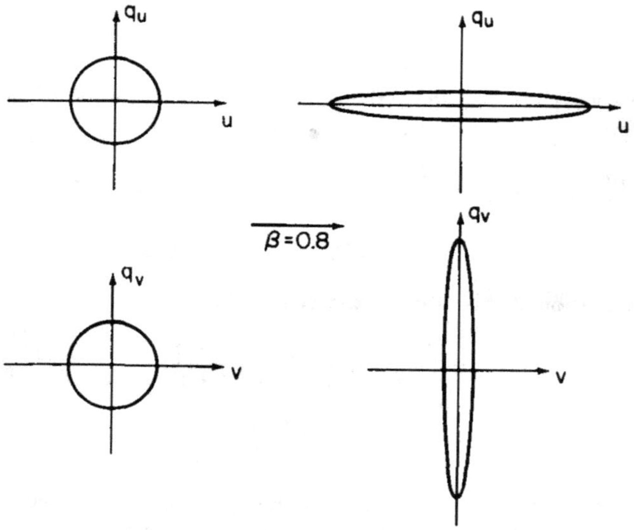


Figure 8.2: Lorentz squeezes in the light-cone phase space consisting of two pairs of conjugate variables. The major (minor) axis in the $uv$ coordinate system is conjugate to the minor (major) axis in the $q_u q_v$ coordinate system. The Lorentz boost results in a canonical transformation of the Wigner function in this four-dimensional phase space.

Thus the explicit form of the physical wave function becomes

$$\psi_\beta^n(u,v) = \left(\frac{1}{2}\right)^{n/2}\left(\frac{1}{\pi n!}\right)^{1/2} \times \left\{\sum_{m=0}^{n}\binom{n}{m} H_{n-m}(u')H_m(v')\right\} e^{-(u'^2+v'^2)}. \tag{8.83}$$

This means that we need off-diagonal Wigner functions for the one-dimensional harmonic oscillator such as

$$W_{nm}(x,p) = \frac{1}{\pi}\int \psi_n^*(x+y)\psi_m(x-y)e^{2ipy}dy \tag{8.84}$$

to evaluate the Wigner function for the covariant harmonic oscillator (Carruthers and Zachariasen 1983). If we evaluate this integral,

$$W_{nm}(x,p) = \frac{(n!m!)^{1/2}}{\pi} \times \left\{\sum_{k=0}^{s}\frac{(-1)^k[\sqrt{2}(x+ip)]^{n-k}[\sqrt{2}(x-ip)]^{m-k}}{k!(n-k)!(m-k)!}\right\} e^{-(x^2+p^2)}, \tag{8.85}$$

where $s$ is $n$ or $m$ whichever is smaller. We can then go back to the original definition to complete the evaluation of the Wigner function. The localization and deformation properties of the Wigner function for excited states are essentially the same as those of the ground-state oscillator.

Let us go back to the localization problem, and compare the present case with the localization of light waves discussed in Section 7.9. Unlike the present case illustrated in Figure 8.2, there is only one pair of conjugate variables for light waves as is illustrated in Figure 7.5. The squeeze given in Figure 7.5 is essentially that for the $vq_v$ pair. The lack of the $uq_u$ pair complicated the question of unitarity (Han *et al.* 1987c).

# Chapter 9

# LORENTZ-SQUEEZED HADRONS

We have shown in Chapter 8 that it is possible to construct a covariant phase-space picture of quantum mechanics for harmonic oscillators, and that the localized region in phase space undergoes Lorentz squeeze. The question then is whether this Lorentz deformation manifests itself in the real world. We shall examine some high-energy hadronic phenomena using the quark model.

In the quark model, the hadron is a quantum bound state of quarks with a localized wave function with probability interpretation. This bound state consists of a ground state and excited states. Indeed, the quark model with a harmonic oscillator potential gives a reasonable description of the observed hadronic mass spectra. This is one of the strongest experimental supports for the quark model.

However, the above-mentioned picture of bound states is applicable only to observers in the Lorentz frame in which the hadron is at rest. How would hadrons appear to observers in other Lorentz frames? More specifically, can we use the picture of Lorentz squeeze described in Chapter 8 to tackle this problem. The purpose of this Chapter is to examine whether this picture is consistent with what we observe in high-energy laboratories. We are particularly interested in studying the nucleon form factors and Feynman's parton picture.

Since the size of the proton is $10^{-5}$ of that of the hydrogen atom, it has long been believed that the proton has a point charge in atomic physics. However, while carrying out experiments on electron scattering from proton targets, Hofstadter in 1955 observed that the proton charge is spread out. In this experiment an electron emits a virtual photon, which then interacts with the proton. If the proton consists of quarks distributed within a finite space-time region, the virtual photon will interact with quarks which carry fractional charges. The scattering amplitude will depend on the way in which quarks are distributed within the proton.

The portion of the scattering amplitude which describes the interaction between the virtual photon and the proton is called the form factor. Although there have been many attempts to explain this phenomenon within the framework of quantum field theory, it is quite natural to expect that the wave function in the quark model will describe the charge distribution. In high-energy experiments, we are dealing with the situation in which the momentum transfer in the scattering process is large. We expect that the Lorentz-squeezed oscillator wave function will give a reasonable description of rapidly moving hadrons.

While the form factor is the quantity which can be extracted from the elastic scattering, it is important to realize that in high-energy processes, many particles are produced in the final state. They are called inelastic processes. While the elastic process is described by the total energy and momentum transfer in the center-of-mass coordinate system, there is, in addition, the energy transfer in inelastic scattering. Therefore, we would expect that the scattering cross section would depend on the energy, momentum transfer, and energy transfer. However, one prominent feature in inelastic scattering is that the cross section remains nearly constant for a fixed value of the momentum-transfer/energy-transfer ratio. This phenomenon is called "scaling."

In order to explain the scaling behavior in inelastic scattering, Feynman in 1969 observed that a fast-moving hadron can be regarded as a collection of many "partons" whose properties do not appear to be identical to those of quarks. For example, the number of quarks inside a static proton is three, while the number of partons in a rapidly moving proton appears to be infinite. The question then is how the proton looking like a bound state of quarks to one observer can appear different to an observer in a different Lorentz frame? Can the covariant harmonic oscillator discussed in Chapter 8 explain this puzzle? We shall deal with this problem in the present Chapter.

It is remarkable that the mathematics of squeezed states of light is directly applicable to the basic features of high-energy hadronic physics. The Lorentz squeeze to be discussed in this Chapter has already been studied in Chapters 5 and 6 in connection with two-mode squeezed states. This allows us to import some the physical concepts from the squeezed states of light. We shall study the hadronic temperature and entropy derivable from the squeeze in Chapter 10.

## 9.1 Quark Model

Historically quantum mechanics was developed for the purpose of explaining discrete spectra. The first discrete spectrum was that of the hydrogen atom. The second discrete spectrum was that of nuclei which are quantum bound states of nucleons. Quantum mechanics has been very successful in explaining these two traditional spectra.

The third discrete spectrum is the mass spectrum of strongly interacting "elementary particles" commonly called "hadrons." It is by now firmly established that hadrons are quantum bound states of quarks. One of the wonders of modern high-energy physics is that the use of nonrelativistic quantum mechanics leads to a reasonable understanding of hadronic mass spectra. Then what is the theoretical basis for using a nonrelativistic bound-state picture for this highly relativistic situation?

The answer to the above question is very simple. When we study hadronic mass spectra, we often employ the language of the direct product of O(3), SU(2) and SU(3) (Greenberg and Resnikoff 1967, Feynman *et al.* 1971), where SU(2) describes quark spins, and SU(3) is used for other internal quantum numbers such as "flavor" and "color." The O(3) symmetry, the three- dimensional rotation group, means that nonrelativistic quantum mechanics with rotational symmetry is used. Because O(3) is not Lorentz-invariant, it is very easy to make a hasty conclusion that the quark model is inherently nonrelativistic. This is not correct. As we studied in Chapters 7 and 8, the little group for massive particles is locally isomorphic to O(3). We shall see in this Chapter that the O(3) group associated with the quark model has the same space-time symmetry as Wigner's O(3)-like little group for massive particles.

In the quark model, hadrons are quantum bound states of quarks and/or antiquarks. Baryons such as the proton and neutron are bound states of three quarks, and mesons such as the $\pi$ and $K$ mesons are bound states of a quark and antiquark. Like electrons, quarks are spin-1/2 fermions and have negligible size. Unlike electrons quarks carry two additional quantum numbers commonly called flavor and color. The traditional name for flavor is unitary spin.

In spite of the importance of understanding relativistic bound-state problems, the quark model was originally developed to explain selection rules in hadronic processes. The idea started from the concept of isotopic spins in the system of mesons and nucleons. Let us look at the nucleons. The proton and neutron have approximately the same mass, and further study of their properties had led us to believe that they belong to the same isotopic multiplet, and that the only difference between them is the electromagnetic property. For example, it is by now firmly believed that the neutron and the proton will have the same mass once the electromagnetic interaction is turned off. This isotopic spin is often called isospin for simplicity. What then is isospin?

The symmetry of electron spin is governed by the group SU(2). The spin can be up or down. Likewise, we can consider a Hilbert space of nucleonic states, and use $I$ and $I_3$ to specify the total isospin and its third component. $I$ for the nucleonic system is 1/2. If the nucleon is proton, its $I_3$ is 1/2. The neutron's $I_3$ is $-1/2$. There are three $\pi$ mesons with approximately the same mass separated only by electromagnetic interaction. The total isospin for this mesonic system is 1, and the eigenvalues of $I_3$ for $\pi^+$, $\pi^0$, $\pi^-$ are 1, 0 and $-1$ respectively.

The nucleonic multiplet and mesonic multiplets are represented by a spinor and a vector in isospin space respectively. From these vectors and spinors, we can construct scalar quantities which are invariant under rotations in isospin space. Indeed, the observed strong-interaction symmetries are consistent with the rotational symmetry in isospin space. This aspect of strong-interaction physics has been widely discussed in textbooks (Frazer 1966).

It was observed that there are, in addition to the nucleons, six more particles which may be put into the same multiplet, and that not all hadronic transitions are due to strong interactions. It was absolutely necessary to add another dimension to the internal symmetry space, called the hypercharge to explain why the selection rules for these eight particles are similar to the nucleon. If we add another dimension to the nucleonic multiplet, the symmetry group has to be enlarged from SU(2) to SU(3). However, since there are eight particles in the multiplet, it is not possible to describe this in terms of the fundamental representation of the SU(3) group whose dimension is three. This line of reasoning led to the concept of quarks (Gell-Mann 1964). Quarks have fractional charges and fractional baryon numbers as listed in Table 9.1.

Table 9.1: Quantum numbers for quarks and antiquarks, B, Q, S, C, B, T are baryon number, charge, strangeness, charm, bottom and top respectively. I is the total isospin quantum number, and $I_3$ is its third component. The u and d quarks form a doublet in isospin space, while the s, c, b, and t quarks are isospin singlets.

| Quantum Nos. | | B | Q | I | $I_3$ | S | C | B | T |
|---|---|---|---|---|---|---|---|---|---|
| Quarks | u | 1/3 | 2/3 | 1/2 | 1/2 | 0 | 0 | 0 | 0 |
| | d | 1/3 | -1/3 | 1/2 | -1/2 | 0 | 0 | 0 | 0 |
| | s | 1/3 | -1/3 | 0 | 0 | -1 | 0 | 0 | 0 |
| | c | 1/3 | 2/3 | 0 | 0 | 0 | 1 | 0 | 0 |
| | b | 1/3 | -1/3 | 0 | 0 | 0 | 0 | 1 | 0 |
| | t | 1/3 | 2/3 | 0 | 0 | 0 | 0 | 0 | 1 |
| Antiquarks | | -1/3 | -2/3 | 1/2 | -1/2 | 0 | 0 | 0 | 0 |
| | | -1/3 | 1/3 | 1/2 | 1/2 | 0 | 0 | 0 | 0 |
| | | -1/3 | 1/3 | 0 | 0 | 1 | 0 | 0 | 0 |
| | | -1/3 | -2/3 | 0 | 0 | 0 | -1 | 0 | 0 |
| | | -1/3 | 1/3 | 0 | 0 | 0 | 0 | -1 | 0 |
| | | -1/3 | -2/3 | 0 | 0 | 0 | 0 | 0 | -1 |

Three quarks or a quark/antiquark pair can form an integer charge and an integer baryon number to become an observable hadron. For example, the proton consists of two u quarks and one d quark. The way in which the quarks form observable baryonic multiplets is discussed briefly in Section 9.2. It is not necessary to present here a full-fledged discussion of the quark model, because the model has been extensively discussed in the physics literature (Kokkedee 1969, Lichtenberg 1970, Greenberg 1982, Huang 1982, Kim and Noz 1986).

We are interested only in those aspects of the quark model which are needed in studying the bound-state property of hadrons. The key question is whether quarks can be regarded as constituent particles within quantum bound states. For instance, the proton and electron are clearly constituent particles in the hydrogen atom. The consequence of this bound-state picture is that the hydrogen atom has a localized probability of the electron around the proton with the radius determined by the electron mass and by the strength of the interaction between the proton and electron. The localization condition imposed on the hydrogen wave function is responsible for the discreteness of the energy spectrum. The electron can sometimes be separated from the proton to become a free particle.

If the hadron is a bound state of quarks, it should have a non-zero radius determined by the interaction between the quarks, and should show evidence for the existence of a discrete mass spectra due to this localization condition. Since quarks have fractional charges, it is not advisable to consider unbound or free quarks. Therefore, unlike the case of the hydrogen atom, quarks in hadrons cannot be separated. The force between quarks should therefore be of the harmonic-oscillator type.

## 9.2 Hadronic Mass Spectra

We do not yet know the exact form of the force between quarks. However, from both phenomenological and field theoretic approaches, the present indications are that the force is very weak at short distances and becomes very strong at large distances. The potential governing the mass is expected to be linear in distance between the quarks. It is therefore reasonable to study the harmonic oscillator potential for the $(\text{mass})^2$ spectrum (Critchfield 1976). This means that we should see only equal-spaced $(\text{mass})^2$ spectra in high-energy laboratories. This is not the case.

What we see in the real world is the perturbed mass spectra. Quarks have spin and unitary spin. In addition to the approximate harmonic oscillator force, their spin and unitary spin can remove the degeneracy of the harmonic oscillator system. Precisely for this reason, we have to know how to construct unperturbed symmetric wave functions and then perform perturbations on them.

Since mesons are two-body states of two different particles, there are no problems connected with identical particles, and the two-body problem should not cause any mathematical complications. However, baryons are bound states of three quarks obeying the Pauli exclusion principle. It is now firmly experimentally established that the baryonic wave functions be totally symmetric under the exchange of quarks. This exchange degeneracy is analogous to the case of many electrons. As Dirac pointed out in Sections 55 and 56 of his classic book on quantum mechanics (Dirac 1958), the Hamiltonian should be invariant under the exchange of quarks, and physical states should therefore be eigenstates of the permutation operators whose eigenvalues correspond to constants of motion.

When we attempt to construct a totally symmetric wave function from spin 1/2 particles, we are led to the question of the Pauli exclusion principle. This is in fact a very serious question, and there have been many attempts to rectify the situation. At present, the prevailing view is that there is an additional quarks quantum number called "color" (Greenberg 1964). There are three colors forming a unitary multiplet. The observable three-quark system always manifests itself in a color singlet or a totally antisymmetric state of this quantum number. For this reason, the hadronic wave functions are totally symmetric in all other quantum numbers.

The hypothesis of the color space leads to very rich experimental and theoretical consequences (Greenberg and Nelson 1977, Marciano and Pagels 1978). However, since it does not have a direct relation to what we plan to establish in this Chapter, we shall not discuss this subject further.

Once the quarks are identified, it is not difficult to add their quantum numbers to calculate the resulting quantum number for the hadron. The non-trivial aspect of constructing wave functions is to make symmetric combinations forming irreducible representations. Indeed, this is also a widely discussed subject (Greenberg and Resnikoff 1967, Kim and Noz 1986). The remarkable fact is that this symmetry problem was considered in depth by Dirac in his book on quantum mechanics (1958), many years before the formulation of the quark model in 1964. The method used by Feynman *et al.* (1971) is along the line suggested by Dirac.

In constructing representations of three-quark states, we have to consider totally symmetric and totally antisymmetric states for a given set of quantum numbers. If the quark carries more than one quantum number, there are cases where a combination of partially symmetric states for two different quantum numbers give a totally symmetric or antisymmetric state. Let us use $|S>$ and $|A>$ for the totally symmetric and antisymmetric states, and $|\alpha>$ and $|\beta>$ for the states symmetric and antisymmetric under the exchange of the first two quarks.

For three quantum states $x$, $y$, and $z$, the totally symmetric and antisymmetric states are

$$|xyz>_S = (1/\sqrt{6})(|xyz> + |yzx> + |zxy> + |yxz> + |zyx> + |xzy>),$$

$$|xyz>_A = (1/\sqrt{6})(|xyz> + |yzx> + |zxy> - |yxz> - |zyx> - |xzy>). \quad (9.1)$$

There are two states symmetric under the exchange of the first two quarks. They are

$$|xyz>_{\alpha 1} = (1/2\sqrt{3})(|xyz> + |xzy> + |yxz> + |yzx> - 2|zxy> - 2|zyx>),$$

$$|xyz>_{\alpha 2} = (1/2)(|xyz> - |yzx> + |xzy> - |yxz>). \quad (9.2)$$

The states antisymmetric under the exchange of the first two particles are

$$|xyz>_{\beta 1} = (1/2)(|xyz> - |xzy> - |yxz> + |yzx>),$$

$$|xyz>_{\beta 2} = (1/2\sqrt{3})(|xyz> - |yxz> - |xzy> + |yzx> + 2|zyx> - 2|zxy>). \quad (9.3)$$

Quarks can have spin and unitary spin quantum numbers. Thus the spin-unitary spin wave function is a combination of the two wave functions. The obvious combinations of the $\alpha$ and $\beta$ sets of quantum numbers are

$$|a>_S\,|b>_S = |ab>_S, \quad |a>_S\,|b>_\alpha = |ab>_\alpha,$$

$$|a>_S\,|b>_\beta = |ab>_\beta, \quad |a>_S\,|b>_A = |ab>_A,$$

$$|a>_A |b>_S = |ab>_A, \quad |a>_A |b>_\alpha = |ab>_\beta,$$

$$|a>_A |b>_\beta = |ab>_\alpha, \quad |a>_A |b>_A = |ab>_S \,. \tag{9.4}$$

In addition, there are combinations consisting of $\alpha$ and $\beta$ states (Feynman *et al.* 1971). They are

$$|ab>_S = (1/\sqrt{2})(|a>_\alpha |b>_\alpha + |a>_\beta |b>_\beta),$$

$$|ab>_\alpha = (1/\sqrt{2})(-|a>_\alpha |b>_\alpha + |a>_\beta |b>_\beta),$$

$$|ab>_\beta = (1/\sqrt{2})(|a>_\alpha |b>_\beta + |a>_\beta |b>_\alpha),$$

$$|ab>_A = (1/\sqrt{2})(-|a>_\alpha |b>_\beta + |a>_\beta |b>_\alpha). \tag{9.5}$$

After combining the spin and unitary spin wave functions, we have to combine with the space-time wave functions, by repeating the same procedure.

If $x_a$, $x_b$, and $x_c$ are the space-time coordinates of the quarks, it is more convenient to use the variables (Feynman *et al.* 1971):

$$X = \frac{1}{3}(x_a + x_b + x_c),$$

$$r = \frac{1}{6}(x_b + x_c - 2x_a),$$

$$s = (1/2\sqrt{3}(x_c - x_b), \tag{9.6}$$

and their conjugate variables:

$$P = p_a + p_b + p_c,$$

$$q = p_b + p_c - 2p_a,$$

$$k = \sqrt{3}(p_c - p_b). \tag{9.7}$$

In terms of these variables, the ground-state oscillator wave function for the three-particle bound system takes the form (Kim and Noz 1986)

$$\psi_\beta(r,s) = \left(\Omega^2/\pi\right)^2 \exp\left\{-\left(\frac{\Omega}{2}\right)\left(\mathbf{r}^2 + r_0^2 + \mathbf{s}^2 + s_0^2\right)\right\}, \tag{9.8}$$

where $\Omega$ is the spring constant between the quarks. This constant can be estimated from the mass spectrum. Since

$$r^2 + s^2 = \left(\frac{1}{16}\right)\left(x_a^2 + x_b^2 + x_c^2\right) - \left(\frac{1}{9}\right)\left(x_a x_b + x_b x_c + x_c x_a\right), \tag{9.9}$$

the ground-state wave function is totally symmetric. The harmonic oscillator becomes excited through the multiplication of the Hermite polynomials in **r** and **s**. The symmetry of each component of the three-vector **r** is of the $\alpha$ type, while that of **s** is of the $\beta$ type. We thus have enough tools to construct the totally symmetric wave functions for the three-quarks system.

In studying the excitation levels of the oscillator system, we can start with the total energy. Within a given energy level, there are degeneracies. These degeneracies are then removed by perturbation. Indeed, the study of mass spectra within the framework of nonrelativistic quantum mechanics generates very respectable numerical results. For instance, the use of the nonrelativistic harmonic oscillator describes very accurately the degeneracy of the spectrum (Van Royen and Weisskopf 1967, Feynman *et al.* 1971) and gives a reasonably accurate mass formula for nonstrange hadrons (Hussar *et al.* 1980). Hadron spectroscopy is one of the essential branches of high-energy physics (De *et al.* 1973, Critchfield 1976, Isgur and Karl 1979, Maltman and Isgur 1984, Oneda 1985). At present, the agreement of the harmonic oscillator model with experimental data may be summarized in Table 9.2.

Table 9.2: Summary of the present status of the multiplet scheme in the harmonic oscillator quark model. A means "excellent," B means "good,", etc.

| N | Baryons | | Mesons | |
|---|---|---|---|---|
| Level | Nonstrange | Strange | Nonstrange | Strange |
| N = 0 | A | A | A | A |
| N = 1 | A | A⁻ | B | C |
| N = 2 | A⁻ | C | D | D |

## 9.3 Hadrons in the Relativistic Quark Model

In order that the hadron be a quantum bound state, it has to have a space-time extension. This is consistent with the earlier observation by Hofstadter and McAllister (1955), derived from the elastic scattering of electrons from the proton target, that the proton is not a point particle, and that the charge distribution of the proton has a non-zero radius. If the electron energy becomes sufficiently high, the recoil effect of the target proton becomes substantial.

This recoil effect manifests itself in the electromagnetic form factor of the nucleon. In order to treat this recoil effect relativistically, we need a hadronic wave function which can be Lorentz-boosted. Indeed, the covariant harmonic oscillator model is a good candidate for this purpose.

While the observed form factor decreases as $(1/(\text{momentum transfer}))^4$ for large momentum transfer, the use of the nonrelativistic harmonic oscillator wave function leads to an incorrect asymptotic behavior. On the other hand, if we take into account the Lorentz-squeezed wave function of the covariant harmonic oscillator formalism, the calculation gives the correct behavior for the proton form factor (Fujimura *et al.* 1970, Ishida 1971, Lipes 1972).

Another hadronic phenomenon which can be regarded as a manifestation of the Lorentz-squeezed wave function is Feynman's parton picture (Feynman 1969). If a proton is sitting quietly on the desk, it appears as a bound state of three quarks to a stationary observer. The same proton appears as a collection of partons to an observer who is moving very fast. Therefore, these two observers will quarrel over the structure of the proton. The Lorentz-squeezed wave function in the covariant harmonic oscillator wave function settles this quarrel (Kim and Noz 1977).

The relativistic quark model based on the covariant harmonic oscillator formalism explains other basic features in hadronic physics, such as $g_A/g_V$ (Ruiz 1975), electroproduction processes (Lipes 1972), and the jet phenomenon (Kitazoe and Hama 1979, Kim *et al.* 1979b, Kitazoe and Morii 1980a and 1980b). This model also forms the basis for a quantum field theory of relativistic extended hadrons (Sogami 1973, Kim 1976, Karr 1976; Capri and Chiang 1976, 1977, and 1978, Pocsik 1978, Mita 1978, Das 1983 and 1988).

It should be noted that the above-mentioned Lorentz-Dirac deformation property is not restricted to the harmonic oscillator wave functions, and that it is not difficult to find the same deformation property in other theoretical and phenomenological models (Drell and Yan 1971, Preparata and Craigie 1976). One important advantage of using the harmonic oscillator formalism is that it is mathematically simple. According to Dirac (1970), a theory possessing mathematical beauty is more likely to survive than one which does not. The covariant oscillator has another important strength. It can be easily converted to the phase-space picture which often gives a clearer interpretation of the uncertainty principle.

In the rest of this Chapter we shall discuss in detail Feynman's parton picture and the proton form factor within the framework of the covariant oscillator formalism, and also in the phase-space picture of relativistic quantum mechanics. Let us start with the form factor.

The form factor is caused by the charge distribution of the target in the scattering of electrons. If we scatter electrons from a fixed charge distribution whose density is $e\rho(r)$, the scattering amplitude is

$$f(\theta) = -\frac{me^2}{2\pi}\int e^{-i\mathbf{K}\cdot\mathbf{x}}\frac{\rho(r')}{R}d^3x d^3x', \tag{9.10}$$

where $r = |\mathbf{x}|$, $R = |\mathbf{x} - \mathbf{x}'|$, and $\mathbf{K} = \mathbf{K}_f - \mathbf{K}_i$ which is the momentum transfer. This amplitude can be reduced to

$$f(\theta) = \left(2me^2/K^2\right) F(K^2). \tag{9.11}$$

$F(K^2)$ is the Fourier transform of the density function

$$F(K^2) = \int \rho(r) e^{-\mathbf{K}\cdot\mathbf{x}} d^3x. \tag{9.12}$$

The above quantity is called the form factor. It describes the charge distribution in terms of the momentum transfer. The charge density is normalized:

$$\int \rho(r) d^3x = 1. \tag{9.13}$$

Then $F(0) = 1$ from Eq. 9.12. If the density function is a delta function corresponding to a point charge, $F(K^2) = 1$ for all values of $K^2$, and the scattering amplitude of Eq. 9.11 becomes the Rutherford formula for Coulomb scattering. The deviations from Rutherford scattering for increasing values of $K^2$ give a measure of the charge distribution (Hofstadter and McAllister 1955, Hofstadter 1963). This was precisely what Hofstadter found in the scattering of electrons from a proton target.

As the electron energy increases, the relativistic effect on the target proton becomes important. We then have to use quantum field theory and Feynman diagrams to formulate this problem. The Feynman diagram for this elastic scattering is given in Figure 9.1. When the electron approaches the target, it emits a virtual photon, which interacts with the proton. The proton is not a point particle, but is a bound state of quarks with a charge distribution dictated by the Lorentz-squeezed bound-state wave function of the quarks. We shall return to this problem in detail in Sections 9.4 and 9.5.

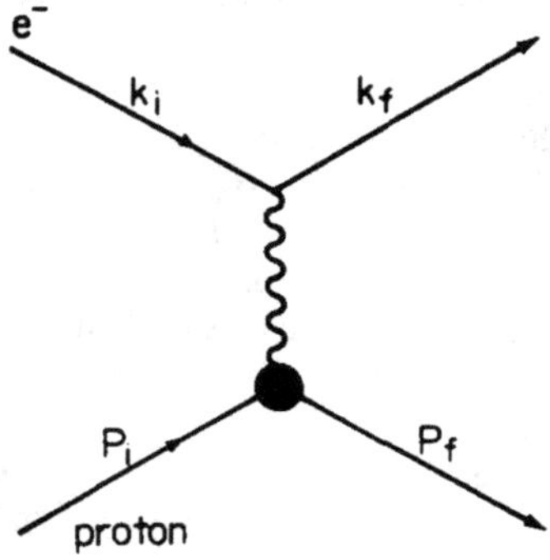


Figure 9.1: Elastic electron-proton scattering. The electron behaves like a point charge. However, the proton has its hadronic structure, and has a spread-out charge distribution.

Another high energy phenomenon which exhibits the squeeze property of the hadronic wave function is the scaling behavior in high-energy inelastic scattering (Bjorken 1969). In order to explain this phenomenon, Feynman developed the the parton picture of the hadron (Feynman 1969). Let us consider the process of inelastic scattering of an electron on a proton target, as is described in Figure 9.2.

In the usual notation of the Dirac matrices, the matrix element for this process can be written as (Kim and Noz 1986)

$$M_{fi} = \left(e^2/k^2\right) \overline{U}(K_f)\gamma_\mu U(K_i) < P', \quad \text{all}|J^\mu|P>, \tag{9.14}$$

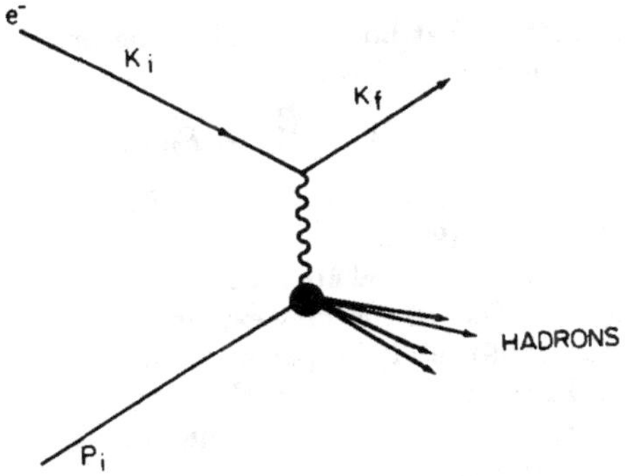


Figure 9.2: Inelastic electron-proton scattering. The electron emits a virtual photon. After absorbing the virtual photon, the proton becomes may hadrons. Unlike the case of elastic scattering described in Figure 9.1, there is an energy transfer through the virtual photon, in addition to the momentum transfer. If the energy transfer is large, the sum over the final states results in an approximate sum over the complete states which simplifies the problem.

where $K = K_i - K_f$ is the four-momentum transfer of the electron. $P$ is the four-momentum of the initial nucleon, and $P'$ is the total four-momentum of the final-state hadrons. $U(K_i)$ and $U(K_f)$ are the Dirac spinors for the initial and final electrons respectively. Here again, the electron-photon vertexindexvertex is point-like. What matters most is the hadronic part. In the calculation of the inelastic cross section, the hadronic part will appear as

$$W_{\mu\nu} = (2\pi)^3 \sum_{\text{all}} <P|J_\mu|\text{all}><\text{all}|J_\nu|P> \delta(P' - P + K). \tag{9.15}$$

The summation is over all hadrons in the final state. Considering all relevant four-vectors in the system, we can reduce this to

$$W_{\mu\nu} = W_1\left(K^2, \varepsilon\right)\left(g_{\mu\nu} - K_\mu K_\nu / K^2\right)$$
$$+ W_2\left(K^2, \varepsilon\right)\left(\frac{1}{M}\right)^2 \left(P_\mu - K_\mu K \cdot P/K^2\right)\left(P_\nu - K_\nu K \cdot P/K^2\right), \tag{9.16}$$

where $\varepsilon = -K \cdot P/M$, and $M$ is the proton mass. $W_1$ and $W_2$ are now the Lorentz-invariant functions of the two Lorentz-invariant variables. In the Lorentz frame in which the target proton is at rest,

$$\varepsilon = E - E', \quad K^2 = (4EE')\sin\frac{\theta}{2}, \tag{9.17}$$

where $E$ and $E'$ are the energies of the initial and final electrons, and $\theta$ is the scattering angle of the electron.

We call the above inelastic scattering deep inelastic scattering when both the momentum and energy transfer variables become very large. In this case, we can define the ratio:

$$\omega = -K^2/2K \cdot P. \tag{9.18}$$

It was observed (Bjorken 1969) that both $W_1$ and $W_2$ become functions of only one variable in the deep inelastic limit:

$$W_1 \to F_1(\omega), \quad \varepsilon W_2 \to F_2(\omega). \tag{9.19}$$

This is called the scaling phenomenon. $F_1(\omega)$ and $F_2(\omega)$ are called the structure functions (Bjorken and Paschos 1969).

There are several different theoretical approaches to explain this scaling behavior in deep inelastic scattering. One of the first explanations was that of Feynman's parton picture (Feynman 1969), and the parton model still plays the major role in high-energy physics. In the parton model, the structure functions are directly related to the parton distribution function. We shall discuss the parton picture of high-energy hadrons in Section 9.5.

## 9.4 Form Factors of Nucleons

As we noted in Section 9.3, in studying form factors, we need a matrix element of the form

$$M_{fi} = (\phi, e^{ikx}\psi). \tag{9.20}$$

This is the inner product of the wave functions $\phi$ and $\psi'$ with $\psi' = e^{ikx}\psi$. Thus, we have to construct the Wigner function for $\psi'$.

$$W_{\psi'}(x,p) = \frac{1}{\pi}\int \psi^*(x+y)\psi(x-y)e^{2i(p-k)y}dy. \tag{9.21}$$

This leads to

$$W_{\psi'}(x,p) = W_{\psi}(x,(p-k)). \tag{9.22}$$

Therefore,

$$|(\phi, e^{ikx}\psi)|^2 = 2\pi \int W_{\phi}(x,p)W_{\psi}(x,(p-k))dx\,dp. \tag{9.23}$$

This formalism can be generalized to three- or four-dimensional space.

Let us return to the problem of form factors. If electrons are scattered by a charged point particle, the scattering amplitude in the Born approximation is $f(\theta) = 2me^2/(\mathbf{k}_f - \mathbf{k}_i)^2)$. On the other hand, if the electron is scattered by a spread-out charge due to quantum probability distribution, the scattering amplitude is

$$f(\theta) = (2me^2/K^2)F(K^2), \tag{9.24}$$

where $\mathbf{K} = \mathbf{k}_f - \mathbf{k}_i$, and $K^2 = (\mathbf{k}_f - \mathbf{k}_i)^2$. $F(K^2)$ is called the form factor and takes the form

$$F(K^2) = (\psi_f, e^{-i\mathbf{K}\cdot\mathbf{x}}\psi_i) = \int [\psi_f(\mathbf{x})]^\dagger \psi_i(\mathbf{x})e^{-i\mathbf{K}\cdot\mathbf{x}}d^3x, \tag{9.25}$$

with $F(0) = 1$, if the initial and final state wave functions are the same. If $\{[\psi_f(\mathbf{x})]^\dagger\psi_i(\mathbf{x})\}$ describes a point charge distribution with $\delta(\mathbf{x})$, then $F(K^2) = 1$ for all values of $K^2$. According to Eq. 9.23, the form factor should take the form

$$|F(K^2)|^2 = \int W_f(\mathbf{x},\mathbf{p})W_i(\mathbf{x},(\mathbf{p}-\mathbf{K}))d^3x\,d^3p. \tag{9.26}$$

This is a generalization of Eq. 9.12 to three-dimensional space.

As the energy of incoming electrons becomes higher for the fixed nucleon target, $K^2$ becomes very large, and the problem becomes relativistic. For the electromagnetic interactions of point particles, we have to use quantum electrodynamics, where the scattering amplitude is expanded in a power series of the fine structure constant $\alpha = e^2/4\pi$ in the Lorentz-Heaviside unit. The lowest non-trivial term in this expansion is essentially a relativistic version of the Born approximation.

For the lowest order in $\alpha$, we can describe the scattering of an electron by a proton using the diagram given in Figure 9.1. The corresponding matrix element is given in many textbooks on elementary particle physics (Frazer 1966). It is proportional to

$$e^2\left(1/K^2\right)\left\{\overline{U}(P_f)\Gamma_\mu(P_f,P_i)U(P_i)\right\}\left\{\overline{U}(k_f)\gamma^\mu U(k_i)\right\}, \tag{9.27}$$

where $P_i$, $P_f$, $K_i$ and $K_f$ are the initial and final four-momenta of the proton and electron respectively. $U(P_i)$ is the Dirac spinor for the initial proton. $K^2$ is the (four-momentum transfer)$^2$ given by

$$K^2 = \mathbf{K}^2 - K_0^2 = (P_f - P_i)^2 = (K_i - K_f)^2. \tag{9.28}$$

The $(1/K^2)$ factor in Eq. 9.27 comes from the virtual photon being exchanged between the electron and the proton. In the metric we use, this quantity is positive for physical values of the four-momenta for the particles involved in the scattering process.

The function $\Gamma_\mu(P_f, P_i)$ in Eq. 9.27 represents the shaded circle in Figure 9.1 and carries the effect of the nucleon structure. If the proton were a point charge, we would have $\Gamma_\mu = \gamma_\mu$. If the proton has an extended charge structure, we will be inclined to write it as $\Gamma_\mu = \gamma_\mu F(K^2)$. However, the proton and neutron have anomalous magnetic moments whose values are 2.79 and $-1.91$ in units of $e/2M$ for the proton and neutron respectively, where $M$ is the nucleon mass. If we include these observed anomalous magnetic moments, $\Gamma_\mu$ should be written as

$$\Gamma_\mu = \gamma_\mu F_1(K^2) + i(\sigma_{\mu\nu}K^\nu/2M)F_2(K^2). \tag{9.29}$$

The above form factors are scalar functions in the Lorentz-invariant variable $K^2$. When we compare $F_1(K^2)$ and $F(K^2)$ with experimental data, it is more convenient to use the following linear combinations:

$$G_M(K^2) = F_1(K^2) + F_2(K^2),$$

$$G_E(K^2) = F_1(K^2) + (K^2/4M^2)F_2(K^2). \tag{9.30}$$

The above form should be written for the proton and neutron separately. We may use the superscripts $p$ and $n$ to distinguish them. When $K^2 = 0$,

$$G_M^p(0) = \mu_p = 2.79, \qquad G_E^p(0) = 1 \qquad \text{for proton,}$$

$$G_M^n(0) = \mu_n = -1.91, \qquad G_E^n(0) = 0 \qquad \text{for neutron.} \tag{9.31}$$

These numbers are the magnetic moments and electric charges of the proton and neutron respectively.

In the past, many attempts have been made to understand the form factors (Frazer and Fulco 1960), we are interested here in treating this problem within the framework of the quark model. In this model, the proton is a bound state of two $u$ quarks and one $d$ quark, while the neutron consists of one $u$ and two $d$ quarks. Indeed, one of the early successes of the quark model was the calculation of the magnetic moment ratio $\mu_n/\mu_p = -2/3$ (Beg *et al.* 1964). However, it is even more challenging to calculate the form factors for increasing values of $K^2$ (Fujimura *et al.* 1970, Lipes 1972, Haberman 1984, Kim and Wigner 1989). At present, we can make the following experimental observation. For the four form factors in the nucleonic system given in Eq. 9.30, the neutron charge form factor is zero at $K^2 = 0$, and remains small (not zero) for all values of $K^2$ (Haberman and Hussar 1988). The three remaining form factors decrease like $1/(K^2)^2$ as $K^2$ increases beyond the value of the (nucleon mass)$^2$ (Stanley and Robson 1982). This behavior is usually called the dipole fit. We are interested in the question of whether each of the form factors can be written in terms of a single form factor $G(K^2)$, multiplied by a constant, where $G(K^2)$ is normalized as $G(0) = 1$, and is proportional to $1/(K^2)^2$ for large values $K^2$, as is illustrated in Figure 9.3. In the case of the neutron charge form factor, we have to multiply $G(K^2)$ by zero within the framework of the model in which the spin, unitary spin and spatial wave functions are factorized (Kim and Noz 1986). The question then is whether it is possible to calculate the above-mentioned dipole behavior of $G(K^2)$ using the wave functions obtained from the quark model. The success of the quark model in the hadronic mass spectra forces us to consider the ground-state harmonic oscillator wave function for the proton. However, the Gaussian distribution gives an exponential cut-off in $K^2$. This contradicts experimental observation. It is therefore interesting to see whether the effect of Lorentz squeeze could transform this exponential decrease into a polynomial cut-off.

## 9.5 Phase-Space Picture of Overlapping Wave Functions

Let us now go back to the expression for the form factor in Eq. 9.25. For a two-body bound state, the nonrelativistic calculation without the Lorentz deformation effect gives a form factor of the form

$$g(K^2) = \exp(-K^2/4\Omega). \tag{9.32}$$

We use $g(K^2)$, instead of $G(K^2)$, for the two-body bound state. This expression does not lead to a polynomial cut-off for large values of $K^2$, and therefore is not consistent with the real world as is described in Figure 9.3. The story is the same for $G(K^2)$ for the three-body bound state.

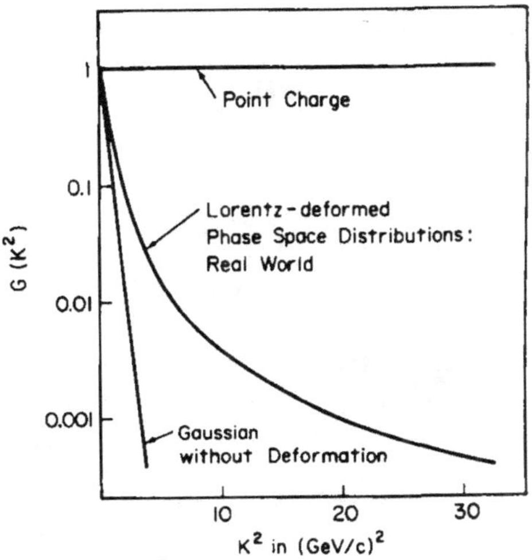


Figure 9.3: Form factor behaviors for increasing values of $K^2$. If the proton is a point charge, the form factor should be independent of $K^2$, as is illustrated by the horizontal line. If the charge distribution is Gaussian, the nonrelativistic calculation leads to an exponential cut-off in $K^2$. The relativistic calculation gives a reasonably accurate description of the real world. At present, the experimental data are available for $G_M^p(K^2)$, $G_E^p(K^2)$ and $G_M^p(K^2)$ from $K^2 = 0$ to 25, 15 and $7(\text{Gev/c})^2$ respectively. They are all consistent with the relativistic calculation with Lorentz deformation.

We are interested in the question of whether the Lorentz effect on the Gaussian distribution will lead to a dipole fit. For this purpose, let us go to the Lorentz frame in which the momenta of the incoming and outgoing nucleons have equal magnitude but opposite signs. In this frame (Kim and Noz 1986),

$$\mathbf{P}_f + \mathbf{P}_i = 0. \tag{9.33}$$

The Lorentz frame in which the above condition holds is usually called the Breit frame. We assume that the proton comes in along the $z$ direction and goes out along the negative $z$ direction after the scattering process. In this frame, the four-vector $K = (k_f - k_i) = (P_i - P_f)$ has no time-like component. Thus the exponential factor $\exp(-i\mathbf{K}\cdot\mathbf{x})$ can be replaced by the Lorentz-invariant form $\exp(-iK\cdot x)$.

We can use the covariant harmonic oscillator wave function discussed in Chapter 8 for the proton. If we assume for simplicity that the proton is a bound state of two quarks, the form factor should take the form:

$$g(K^2) = \int \psi_f(x)\psi_i(x)e^{-iK\cdot x}d^4x. \tag{9.34}$$

The only difference between the above form and the nonrelativistic cases is that the integral of Eq. 9.34 requires an integration over the time-like variable. This time-separation variable has been thoroughly discussed in Chapter 7. If $\beta$ is the velocity parameter for the incoming proton, $\psi_i$ and $\psi_f$ in Eq. 9.34 should be replaced by $\psi_\beta$ and $\psi_{-\beta}$ which correspond to the protons moving in the positive and negative directions respectively. The form factor integral in the Breit frame takes the form:

$$g(K^2) = \int \psi^*_{-\beta}(z,t)\psi_\beta(z,t)e^{-iK_z}dzdt. \tag{9.35}$$

In terms of the light-cone variables,

$$g(K^2) = \int \psi^*_{-\beta}(u,v)\psi_\beta(u,v)e^{-iK(u+v)/\sqrt{2}}dudv, \tag{9.36}$$

where $K$ is the magnitude of the vector $\mathbf{K}$. The form factor can then be computed from the overlap integral of two Lorentz-squeezed Wigner functions (Kim and Wigner 1989):

$$|g(K^2)|^2 = (2\pi)^2 \int W_{-\beta}(u,q_u;v,q_v)$$
$$\times W_\beta(u,(q_u - K/\sqrt{2});v,(q_v - K/\sqrt{2}))dudq_udvdq_v. \tag{9.37}$$

Then $g(K^2)$ takes the simpler form

$$g(K^2) = 2\pi \int J_{-\beta}(u,(q_u J_\beta))(u,(q_u - K/\sqrt{2}))dudq_u, \tag{9.38}$$

where

$$J_\beta(u,q_u) = \left(\frac{1}{\pi}\right)\exp\left\{-\left(\frac{1-\beta}{1+\beta}u^2 + \frac{1+\beta}{1-\beta}q_u^2\right)\right\}.$$

This overlap integral is illustrated in Figure 9.4.

The evaluation of this integral in Eq. 9.38 is straightforward, and the result is

$$g(K^2) = \left\{2M^2/\left(2M^2+K^2\right)\right\}\exp\left\{-M^2K^2/\left(2\Omega\left(2M^2+K^2\right)\right)\right\}, \tag{9.39}$$

where $\Omega$ is the spring constant. The expression of Eq. 9.38 becomes the nonrelativistic form of Eq. 9.25 for small values of $K^2$ and becomes 1 for $K^2 = 0$. It decreases like $1/K^2$ as $K^2$ becomes large, but does not decrease like $1/(K^2)^2$. How are we going to get an extra $1/K^2$ factor?

The $1/K^2$ factor comes from the two-quark system with one oscillator mode. We can therefore expect that each mode will contribute a $1/(K^2)^2$ factor to give the net decrease of $1/(K^2)^2$ as $K^2$ becomes very large. There are indeed two oscillator modes in the nucleon. The generalization of the form factor calculation to this three-quark system is straightforward for the harmonic oscillator wave functions (Fujimura *et al.* 1970, Feynman *et al.* 1971, Kim and Noz 1986). Since the oscillator wave functions are separable, the construction of the Wigner function is also straightforward. The result of the calculation is

$$G(K^2) = \left\{2M^2/\left(2M^2+K^2\right)\right\}^2\exp\left\{-M^2K^2/\left(\Omega\left(2M^2+K^2\right)\right)\right\}, \tag{9.40}$$

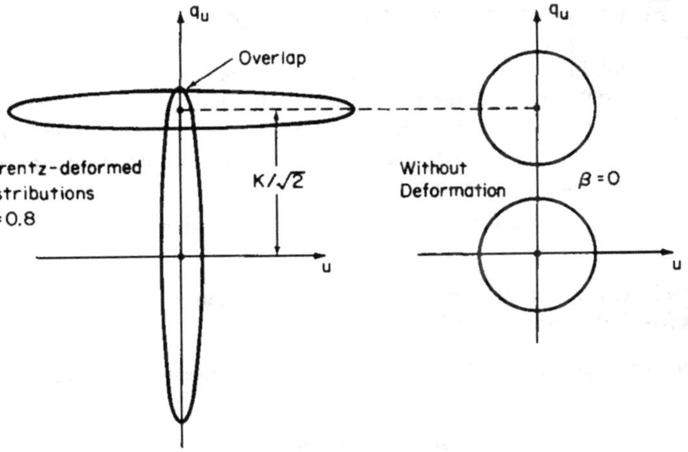


Figure 9.4: Lorentz-squeezed Wigner functions and their overlaps. The $J_\beta$ function is localized within a circular or elliptic region. As the momentum transfer increases, the Wigner functions become separated. Without the squeeze, the Wigner functions completely separated in the overlap integral of Eq. 9.38. This lack of overlap is the cause of the unacceptable exponential cutoff in $K^2$. However, the Lorentz-squeezed Wigner functions maintain a small overlapping region as $K^2$ increases. This leads to a polynomial decrease of the form factor.

which is 1 at $K^2 = 0$, and decreases as $1/(K^2)^2$. The behavior of this function is illustrated in Figure 9.3.

The experimental curves are nicely summarized in the paper of Stanley and Robson 1982). For protons, the data for the electric and magnetic form factors are available from $K^2 = 0$ to 15 and $25(\mathrm{GeV}/c)^2$ respectively. For neutrons, the data for the magnetic form factor is available from $K^2 = 0$ to $7(\mathrm{GeV}/c)^2$. All data are consistent with the form given in Eq. 9.40 and illustrated in Figure 9.3.

As for the charge form factor of neutrons, in the harmonic oscillator model for which only the ground state is taken into account, the coefficient which multiplies the $G(K^2)$ of Eq. 9.40 is zero. However, the observed neutron charge form factor is not zero for non-zero values of $K^2$. This is a clear indication that excited oscillator states should also be taken into account. This point has been discussed by Hussar and Haberman in their recent paper in the conventional harmonic oscillator formalism (Haberman and Hussar 1988).

Throughout this Chapter, we have ignored the effect of spin and assumed that the nucleon form factor can be decomposed into the form of Eq. 9.29 in the quark model. Indeed, there are models in which the quark spins can be combined for the nucleon to give the form of Eq. 9.40 (Henrique *et al.* 1975, Ishida *et al.* 1977 and 1979). On the other hand, we still do not know how to treat the spins in the covariant phase-space formalism. This is a challenging future problem.

## 9.6 Feynman's Parton Picture

In order to explain the scaling behavior in high-energy inelastic electron scattering, Feynman in 1969 proposed the parton model for high-energy hadrons. Feynman observed that a rapidly moving proton, which is now believed to be a bound state of three quarks, can be regarded as a collection of particles called partons which exhibit the following peculiar features.

**(a).** The picture is valid only for hadrons moving with velocity close to that of light.

**(b).** The interaction time between the quarks becomes dilated, and partons behave as free independent particles.

**(c).** The momentum distribution of partons becomes widespread as the hadron moves fast.

**(d).** The number of partons seems to be much larger than that of quarks, as is illustrated in Figure 9.5.

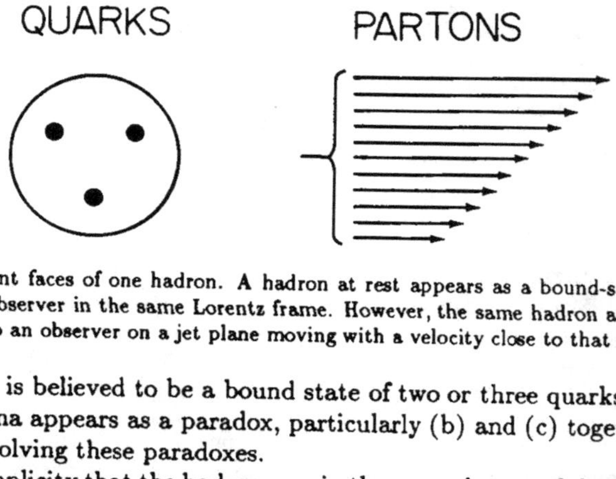


Figure 9.5: Two different faces of one hadron. A hadron at rest appears as a bound-state of two or three quarks to an observer in the same Lorentz frame. However, the same hadron appears as a collections of partons to an observer on a jet plane moving with a velocity close to that of light.

Because the hadron is believed to be a bound state of two or three quarks, each of the above phenomena appears as a paradox, particularly (b) and (c) together. We are interested in resolving these paradoxes.

Assuming for simplicity that the hadrons are in the ground state of the covariant harmonic oscillator, let us go back to the Wigner function of Eq. 8.79. As the hadron moves very fast along the $z$ direction and the velocity parameter $\beta$ approaches one, the distribution along the $v$ and $q_u$ axes becomes very sharp, and behaves like a delta function at the origin. If we integrate the Wigner function over these variables, the result is a wide spread distribution over the $u$ and $q_v$ variables (Kim and Wigner 1988):

$$Q^0_\beta(u;q_v) = \int W^0_\beta(u,q_u;v,q_v)dv\,dq_u$$
$$= \frac{1}{\pi}\left(\frac{1-\beta}{1+\beta}\right)\exp\left\{-\left(\frac{1-\beta}{1+\beta}\right)\left(u^2+q_v^2\right)\right\}. \qquad (9.41)$$

This is not a phase-space distribution function, because there is no conjugate relation between $u$ and $q_v$. These variables are in principle simultaneously measurable.

As $\beta$ approaches 1, because of the delta function behavior in the distributions of $v$ and $q_u$, $(1+\beta)$ becomes 2, and

$$u = \sqrt{2}z = \sqrt{2}t, \quad q_v = \sqrt{2}q_z = \sqrt{2}q_0. \tag{9.42}$$

Then $Q^0_\beta(u; q_v)$ of Eq. 9.41 becomes

$$Q^0_\beta(z; q_z) = \left(\frac{1-\beta}{2\pi}\right) \exp\left\{-\left(\frac{1-\beta}{2}\right)\left(u^2 + q_v^2\right)\right\}. \tag{9.43}$$

This function is widespread in both $z$ and $q_z$. From Eq. 8.21, we can write the longitudinal coordinate for each quarks as

$$z_a = Z + \sqrt{2}z, \quad z_b = Z - \sqrt{2}z. \tag{9.44}$$

where $Z$ is the longitudinal coordinate of the hadron. The distribution in $z$ given in Eq. 9.43 tells us that the position of each quarks appears widespread to observers in the laboratory frame, and that the quarks appear like free particles as the hadronic velocity parameter $\beta$ approaches 1. This effect, first noted by Feynman (1969), is universally observed in high-energy hadronic experiments.

Because the quarks appear almost free, we would normally expect that the momentum of each quark will appear sharply defined. However, this is not the case. We can write the momentum of each quark as

$$p_{az} = P_z/2 + q_z/2\sqrt{2}, \quad p_{bz} = P_z/2 - q_z/2\sqrt{2}. \tag{9.45}$$

When the hadron moves very fast, $(1-\beta)$ becomes $(M/P_0)^2/2$ in the expression given in Eq. 9.43, where $M$ and $P_0$ are the hadronic mass and energy respectively. Then $q_z$ is spread over the region:

$$-2(P_0/M) < q_z < 2(P_0/M), \tag{9.46}$$

which becomes widespread as $P_0$ becomes large. Consequently, the longitudinal momentum of the first quark given in Eq. 9.48 mostly lies within the interval between

$$p_z^{max} = P_0\left\{\frac{1}{2} + \left(\Omega/2M^2\right)^{1/2}\right\} \quad \text{and} \quad p_z^{min} = P_0\left\{\frac{1}{2} - \left(\Omega/2M^2\right)^{1/2}\right\}, \tag{9.47}$$

If we use the proton mass for $M$, and use the value $\Omega = 0.5(\mathrm{GeV})^2$ (Hussar *et al.* 1980), which is commonly used for the quark model analysis, then the quantity $(\Omega/2M^2)^{1/2}$ is of the same order of magnitude as 1/2. For this reason, the (almost) light-like four-momentum $p_a$ can be written as

$$p_a = xP, \tag{9.48}$$

where the parameter $x$ ranges approximately between zero and 1. This type of distribution, which was proposed first by Feynman (1969), is routinely observed in high-energy hadronic experiments. The parameter $x$ is known as Feynman's $x$ (Feynman 1969, Bjorken and Paschos 1969).

As for the time-separation variable, the light-cone variable $u$ can be replaced by $\sqrt{2}t$ in the infinite-momentum limit, in accordance with Eq. 9.42. Thus, when $\beta$ approaches 1, the relative quark motion is an oscillation along the positive light-cone axis. Classically, the quarks make two "collisions" with each other within one period of oscillation. Therefore, the interquark collision time is one half of the oscillation time. If the hadron is at rest with $\beta = 1$, the collision time is $1/2\sqrt{\Omega}$. If the hadronic velocity is close to that of light, the collision time becomes $1/2\sqrt{\Omega(1-\beta)/2}$ to an observer in the laboratory frame. The ratio of this collision time to that in the hadronic rest frame becomes

$$\frac{\text{Collision time for } \beta \to 1}{\text{Collision time for } \beta = 0} = P_0/2M. \tag{9.49}$$

As the hadron becomes very fast, the above ratio becomes very large. This is consistent with the observation that quarks appear as almost free particles in the laboratory frame. This time dilation allows us to make an incoherent sum of cross sections due to each parton (Feynman 1969).

The widespread momentum distribution appears to contradict our initial expectation that, based upon the widespread spatial distribution, the quark be a free particle with a sharply defined momentum. This apparent contradiction presents to us the following two fundamental questions.

**(a).** If both the spatial and momentum distributions become widespread as the hadron moves, and if we insist on Heisenberg's uncertainty relation, is Planck's constant dependent on the hadronic velocity?

**(b).** Is this apparent contradiction related to another apparent contradiction that the number of partons is infinite while there are only two or three quarks inside the hadron?

The answer to the first question is "No", and that for the second question is "Yes". The answer to the first question is contained in Figure 8.2, in which $u$ does not form a phase space with $q_v$, but with $q_u$. The area of the $(u, q_u)$ phase space remains constant.

In order to resolve the puzzle of why the number of partons appears to be infinite while there are only a finite number of quarks inside the hadron, let us translate the above analysis based on the Wigner function into the space-time and momentum-energy coordinate systems. Figure 9.6 contains the essential points of the Lorentz-squeezed space-time and momentum energy distributions (Kim and Noz 1986). It is not difficult to see from Figure 9.6 that both the $x$ and $q$ distributions become concentrated along the positive light-cone axis, as the hadronic velocity approaches the speed of light. This means that the quarks also move with velocity very close to that of light. Quarks in this case behave like massless particles.

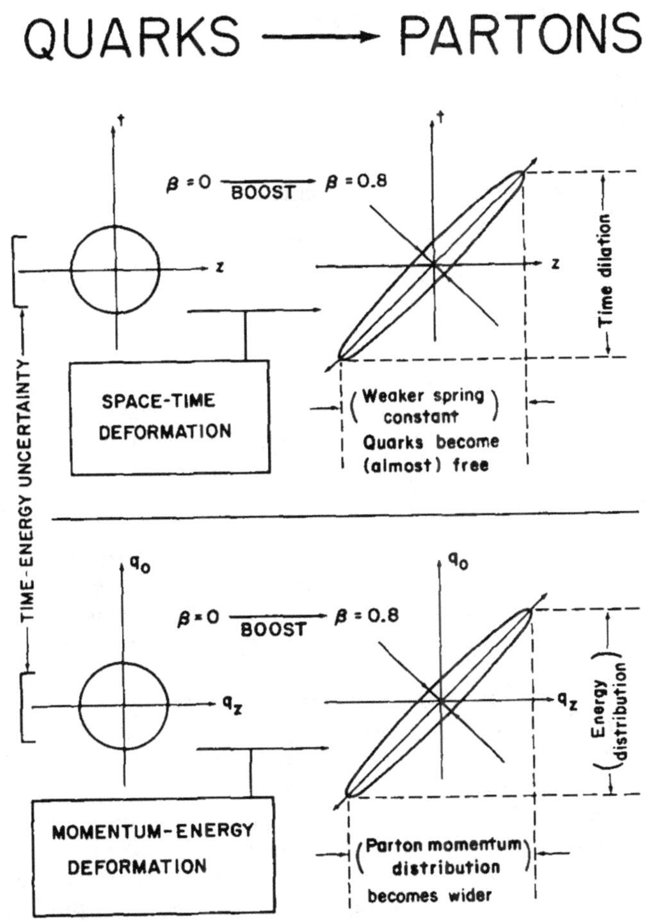


Figure 9.6: Lorentz-squeezed hadron. The upper half of this figure describes the space-time squeeze, while the lower half is for the squeeze in the momentum-energy plane, where $q_z$ and $q_0$ measure the longitudinal and time-like momentum separations respectively. When the hadron is at rest, it appears as a bound state of quarks. When it moves with its velocity close to that of light, the hadronic matter becomes concentrated along one of the light-cones, with wide-spread distributions in both space-time and momentum-energy. As a consequence, the hadron appears as a collection of free partons.

We know from statistical mechanics that the number of massless particles is not a conserved quantity. For instance, in black-body radiation, free light-like particles have a widespread momentum distribution. However, this does not contradict the known principles of quantum mechanics, because the massless photons can be divided into infinitely many massless particles with a continuous momentum distribution (Kim and Noz 1977).

Likewise, in the parton picture, massless free quarks have a wide-spread momentum distribution. They can appear as a distribution of an infinite-number of free particles. These free massless particles are the partons.

The next crucial question is whether this form of distribution is observable in laboratories. The answer to this question is YES.

## 9.7 Experimental Observation of the Parton Distribution

We studied in Section 9.6 the kinematics through which the peculiarities of Feynman's parton picture can be derived from the Lorentz-squeezed hadronic wave functions. A more interesting problem is to calculate the parton momentum distribution function in the proton and compare it with the distribution measured in high-energy experiments.

Before carrying out this program, let us see what we do in the Lamb shift calculation. Because calculations in QED (quantum electrodynamics) are much more complicated than those in nonrelativistic quantum mechanics, we often forget the fact that the starting point in the Lamb shift calculation is the Rydberg formula for unperturbed hydrogen energy levels. The Rydberg formula is obtained from the solution of the Schrödinger equation for the hydrogen atom. The discreteness of the energy levels is a consequence of the boundary condition on the wave function requiring that the probability distribution be localized around the proton. QED then makes corrections to the Rydberg formula by taking into account the detailed interaction mechanism between photons and electrons.

Likewise, in calculating the parton distribution function, we have to know first how the partons are distributed. After that, we may make corrections due to the detailed interaction between the free partons and probing particles. Like QED in the Lamb shift calculation, there is at present a field theoretic method called quantum chromodynamics (QCD) for calculating the corrections to the distribution function.

We discuss here a calculation of the parton distribution function using the covariant harmonic oscillator wave function. The nucleon in its rest frame is regarded as the ground state of the three-quark bound state. As before, we name the three quarks $a$, $b$, $c$, respectively, and use three independent four-momentum variables $P$, $q$ and $k$ given in Eq. 9.7. $P$ represents the four-momentum of the hadron, and is on its mass shell. $q$ and $k$ are the relative internal momenta. Here again, we ignore the transverse components. In terms of the $q$ and $k$ variables, the ground-state wave function takes the form

$$\phi_\beta(q_z, q_0, k_z, k_0) = \left(\frac{1}{\pi\Omega}\right) \exp\left\{-\left(\frac{1}{\Omega}\right)\left[\left(\frac{1-\beta}{1+\beta}\right) q_+^2 + \left(\frac{1+\beta}{1-\beta}\right) q_-^2\right.\right.$$

$$+\left(\frac{1-\beta}{1+\beta}\right)k_+^2+\left(\frac{1+\beta}{1-\beta}\right)k_-^2\Big]\Big\} \tag{9.50}$$

in the momentum-energy space, with

$$q_\pm=(q_z\pm q_0)/\sqrt{2}, \quad k_\pm=(k_z\pm k_0)/\sqrt{2}. \tag{9.51}$$

The distribution function is the square of the wave function.

In the parton model calculation, only one of the quarks interacts with the virtual photon. Since the wave function is totally symmetric under the exchange of quarks, we can pick quark $a$ as the one interacting with the virtual photon. As can be seen from Eq. 9.7, the four-momentum of this quark does not depend on $k$. Thus we can integrate over the $k$ variables. After this integration, the probability distribution function for the ground-state takes the form

$$\rho(q_z,q_0)=\left(\frac{1}{\pi\Omega}\right)\exp\left\{-\left(\frac{1}{\Omega}\right)\left[\left(\frac{1-\beta}{1+\beta}\right)q_+^2+\left(\frac{1+\beta}{1-\beta}\right)q_-^2\right]\right\}. \tag{9.52}$$

In the limit $\beta\to 1$, the $q_-$ integral can be performed as in the case of the two-quark system. As in the case of the two-particle system, it is possible to introduce the $x$ variable defined as

$$p_{a0}=xP_0, \tag{9.53}$$

for this three-quark system. Then the parton distribution function becomes

$$\rho(x)=3M\left(\frac{1}{2\pi\Omega}\right)^{1/2}\exp\left\{-\left(M^2/2\Omega\right)(3x-1)^2\right\} \tag{9.54}$$

The above distribution function is normalized as

$$\int_{-\infty}^{\infty}\rho(x)dx=1. \tag{9.55}$$

This normalization is of course that of the three-quark wave function in the hadronic rest frame.

Before we compare the above form of the distribution function, we have to consider whether the interaction of the quark with the virtual photon is a point interaction or is enhanced by an effect of other interactions. The answer to this question is very simple. There is a significant perturbation between the exact harmonic oscillator and the real world. The distribution function given in Eq. 9.54 does not take into account the effects of perturbation. We therefore have to retreat to the parton distribution function of Eq. 9.54 and consider all necessary perturbative effects before making an attempt to compare quantitatively with the experimental curve.

The existence of this perturbative effect manifests itself in the experimental result that the exact scaling is violated and that the structure function, in addition to $x$, depends on the momentum-transfer variable $K^2$. Fortunately there is an effective theoretical tool to calculate the perturbative corrections to the distribution

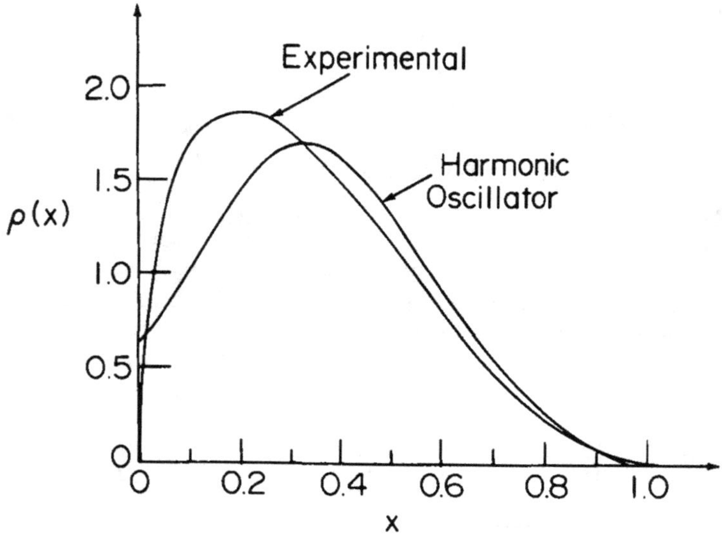


Figure 9.7: Parton distribution inside a rapidly moving proton in the harmonic oscillator model. The $x$ variable is the ratio of the parton momentum to that of the proton. The experimental distribution is obtained from the measured structure function after the QCD correction. For $x > 0.3$, the calculated curve is very close to the experimental distribution. For $x < 0.3$, the agreement does not appear to be impressive. However, it should be noted that the error bars are still very large in this region, and that strong-interaction dynamics is not yet completely understood for small values of $x$. See Hussar (1981) for a detailed discussion.

or structure function. This method is based on a field theoretical approach called "quantum chromodynamics" or simply QCD. Since it will require a separate book to explain QCD, and since perturbative QCD is discussed exhaustively in the literature (Altarelli and Parisi 1977, Marciano and Pagels 1978, Buras 1980), we shall not go into this subject.

The strength of perturbative QCD is its ability to calculate the departure from the exact scaling. On the other hand, The weakness of QCD is its inability to produce the parton distribution function. For this reason, there have been models of QCD which start from an assumed parton distribution which does not depend on $K^2$. The model of Hwa is such a model (Hwa 1980, Hwa and Zahir 1981). Hwa's distribution function has been compared with the expression given in Eq. 9.54 (Hussar 1981), as is shown in Figure 9.7. The agreement is not good enough for us to say that the above-mentioned calculation constitutes the final word on this subject. On the other hand, the agreement is reasonable enough to continue research along this line.

There are of course many related problems. For simplicity, we have discussed only one of the structure functions for electron proton scattering. As we can see, there are two independent functions in Eq. 9.19. If we use heavy vector mesons, instead of photons, in inelastic neutrino-nucleon scattering, there are three structure functions. We can also measure the neutron structure function using a deutron target (Hanlon *et al.* 1980).

The analysis given in this section indicates that the neutron structure function, if appropriately renormalized, would be identical to that of the proton structure function. But this is not the case in the real world. The ratio of the neutron structure function to that of the proton is not constant (Bodek *et al.* 1973). There has been an attempt to explain this ratio using the departure from the exact SU(6) scheme and the accompanying level mixing in the harmonic oscillator system and to relate this mixing parameter to the observed neutron form factor. According to the published result, the sign of the mixing parameter is not consistent with the neutron form factor (Le Yaouanc *et al.* 1978). This discrepancy in sign could be due to the fact that we do not yet completely understand quark spins.

The distribution of Eq. 9.54 is normalized to be one when integrated from $x = -\infty$ to $\infty$. However, the experimental observation is made only from $x = 0$ to $x = 1$. Thus the integral of $\rho(x)$ from $x = 0$ to $x = 1$ should be less than 1. According to recent experimental data, there are indications that this number is smaller than one. The analyses given by Gluck *et al.* (1982) and by Duke and Owens (1984) are consistent with the assertion that this number is 1. The numbers given by Allasia *et al.* (1984) and Cole *et al.*(1988) are 1.01 and 0.97 with error bars respectively. According to Abbot *et al.* (1980), and to Aubert *et al.* (1983), this number should be 0.610 and 0.72 respectively. While we are not able to compare these numbers critically, we can take the arithmetic average. The average value is 0.90. This number is quite consistent with the Gaussian curve given in Eq. 9.22 (Hussar 1981). This "statistical" agreement encourages future investigation along this line.

The determination of the structure functions, both experimental and theoretical, will remain as one of the main branches of high energy physics for many years to come.

# Chapter 10

# SPACE-TIME GEOMETRY OF EXTENDED PARTICLES

In Chapters 8 and 9, within the framework of the phase-space picture of quantum mechanics, we were able to state the uncertainty principle applicable to extended hadrons in a Lorentz-invariant manner. The hadronic deformation of phase space due to a Lorentz boost is a volume-preserving squeeze. Thus the physical basis for Lorentz-squeezed hadrons is in the phase-space picture of quantum mechanics. With this point in mind, we can now construct a space-time geometry of relativistic extended hadrons, and compare with that of point particles with spin.

The internal space-time symmetries of relativistic particles are governed by the little groups of the Poincaré group (Wigner 1932, Kim and Noz 1986). It was shown in Chapter 7 that the internal space-time symmetry groups for massive and massless particles are isomorphic to the three-dimensional rotation group and the two-dimensional Euclidean group respectively. It is shown further in Chapter 7 that the E(2)-like little group for a massless particle is the infinite-momentum and/or zero-mass limit of the O(3)-like little group for a massive particle. This limiting process is illustrated in the second row of Figure 10.1.

It was Einstein who unified the energy-momentum relation for massive and massless particles. Wigner's little groups unify the internal space-time symmetries of massive and massless particles. In this Chapter, we are concerned with relativistic extended particles. We have shown in Chapter 8 that the covariant harmonic oscillator formalism can be framed into Wigner's little groups. In Chapter 9, we discussed the physical basis for the same oscillator formalism. The question then is whether the process of taking the infinite-momentum limit can also be incorporated into the covariant oscillator formalism. In order to accomplish this, we need a better understanding of the space-time geometry of relativistic particles.

It is straight-forward to give a geometrical interpretation of the O(3)-like little group for a massive particle. However, we still have to construct such a geometry for massless particles starting from the local isomorphism between the little group and the two-dimensional Euclidean group. In this Chapter we shall start with the three-dimensional geometry which will describe the symmetry of the E(2)-like little

| | Massive Slow | $\longleftarrow$  between  $\longrightarrow$ | Massless Fast |
|---|---|---|---|
| Energy Momentum | $E = \dfrac{p^2}{2m}$ | $\longleftarrow \quad E = \sqrt{m^2 + p^2} \quad \longrightarrow$ | $E = p$ |
| Spin, Gauge Helicity | $S_3$<br>$S_1 \qquad S_2$ | $\longleftarrow$  Little Groups  $\longrightarrow$ | $S_3$<br>Gauge Trans. |
| Quarks Partons | Quark Model $\longleftarrow$ | $\left(\begin{array}{c}\text{Covariant}\\\text{Phase Space}\end{array}\right)$ $\longrightarrow$ | Parton Model |

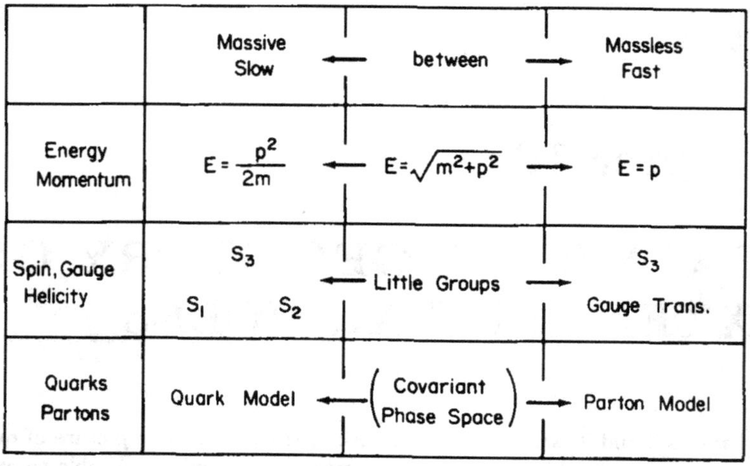

Figure 10.1: Space-time symmetries of relativistic particles. It was Einstein who unified the energy-momentum relations for massive and massless particles. In addition to the energy and momentum, the particle can have internal space-time structures. A massive particle which can be brought to its rest frame has spin degrees of freedom. A massless particle has helicity and gauge degrees of freedom. Wigner's little group unifies this aspect of internal space-time symmetries. As for space-time extensions, the concept of covariant phase space provides a unification of the quark model for slow hadrons and the parton model for fast hadrons.

group for massless particles.

It will be shown that the three-parameter cylindrical group is locally isomorphic to the two-dimensional Euclidean group. It will then be shown that both E(2) and the cylindrical group are limiting cases of the three-dimensional rotation group. It will further be shown that both groups as well as the limiting process are needed for the relativistic description of point and composite particles.

Let us go back to the question of extended particles. If an electron and a hydrogen atom are at rest, we know how to rotate them. The electron is a point particle with spin, while the hydrogen atom is a composite particle with orbital angular momentum as well as the spin of its electron. We use different representations of the same rotation group for these two different particles. The correspondence between these two different representations, namely the Pauli spinors and spherical harmonics, is well understood (Wigner 1959). Does such a correspondence exist between a light-like point particle with its spin and helicity variables and a composite particle with its space-time extension?

As was pointed out before, the internal space-time symmetry of massless particles is dictated by the cylindrical group (Boya and de Azcarraga 1967, Kim and Wigner 1987a). It is of interest to see whether the covariant harmonic oscillator formalism provides a representation space for this symmetry.

Within the framework of the phase-space picture, another interesting problem is to study the effect of incomplete measurement. The time-separation variable which plays the pivotal role in the covariant oscillator formalism is not a measurable variable in the sense that the present form of quantum measurement theory does not deal with this variable. It is therefore worthwhile to study the effect of this non-measurement. It will be shown in this Chapter that this leads to a rising temperature and an increasing entropy. This concept is derivable from the space-time geometry.

## 10.1 Two-Dimensional Euclidean Group and Cylindrical Group

Let us start with the three-dimensional coordinate system with the coordinate variables $(x, y, z)$. We are quite familiar with rotations and translations in this space. In Chapter 4, we studied the two-dimensional Euclidean group, and its applications to coherent states of light in Chapter 6.

The most general form of the two-dimensional Euclidean transformation is

$$\begin{pmatrix} x' \\ y' \\ z' \end{pmatrix} = \begin{pmatrix} \cos\alpha & -\sin\alpha & \xi \\ \sin\alpha & \cos\alpha & \eta \\ 0 & 0 & 1 \end{pmatrix} \begin{pmatrix} x \\ y \\ z \end{pmatrix}, \tag{10.1}$$

with a fixed value of $z$ which remains invariant. In Chapters 4 and 6, $z$ was set to be 1 for convenience.

The three-by-three matrix in the above expression can be exponentiated as

$$E(\xi, \eta, \alpha) = \exp\left(-i\left(\xi P_1 + \eta P_2\right)\right)\exp\left(-i\alpha L_3\right), \tag{10.2}$$

where $L_3$ is the generator of rotations, and $P_1$ and $P_2$ generate translations. These generators take the form:

$$L_3 = \begin{pmatrix} 0 & -i & 0 \\ i & 0 & 0 \\ 0 & 0 & 0 \end{pmatrix},$$

$$P_1 = \begin{pmatrix} 0 & 0 & i \\ 0 & 0 & 0 \\ 0 & 0 & 0 \end{pmatrix}, \quad P_2 = \begin{pmatrix} 0 & 0 & 0 \\ 0 & 0 & i \\ 0 & 0 & 0, \end{pmatrix} \tag{10.3}$$

and satisfy the commutation relations:

$$[P_1, P_2] = 0, \quad [L_3, P_1] = iP_2, \quad [L_3, P_2] = -iP_1. \tag{10.4}$$

$P_1$, $P_2$, and $L_3$ are the generators of the group E(2), which was discussed in Chapter 4. $P_1$ and $P_2$ take the same form as $N_1$ and $N_2$ given in Eq. 4.9. $L_3$ is proportional to $L$ of Eq. 4.11.

The above commutation relations are invariant under the sign change in $P_1$ and $P_2$. They are also invariant under Hermitian conjugation. Since $L_3$ is Hermitian, we can replace $P_1$ and $P_2$ by

$$Q_1 = -(P_1)^\dagger, \quad Q_2 = -(P_2)^\dagger, \tag{10.5}$$

respectively to obtain

$$[Q_1, Q_2] = 0, \quad [L_3, Q_1] = iQ_2, \quad [L_3, Q_2] = -iQ_1. \tag{10.6}$$

These above commutation relations are identical to those for E(2) given in Eq. 10.4. However, $Q_1$ and $Q_2$ are not the generators of translations in the two-dimensional space. Let us write down their matrix forms:

$$Q = \begin{pmatrix} 0 & 0 & 0 \\ 0 & 0 & 0 \\ i & 0 & 0 \end{pmatrix}, \quad Q_2 = \begin{pmatrix} 0 & 0 & 0 \\ 0 & 0 & 0 \\ 0 & i & 0 \end{pmatrix}. \tag{10.7}$$

$L_3$ is given in Eq. 10.3. As in the case of E(2), we can consider the transformation matrix:

$$C(\xi, \eta, \alpha) = C(0, 0, \alpha) C(\xi, \eta, 0), \tag{10.8}$$

where $C(0, 0, \alpha)$ is the rotation matrix and takes the form

$$C(0, 0, \alpha) = \exp(-i\alpha L_3) = \begin{pmatrix} \cos\alpha & -\sin\alpha & 0 \\ \sin\alpha & \cos\alpha & 0 \\ 0 & 0 & 1 \end{pmatrix}, \tag{10.9}$$

$$C(\xi, \eta, 0) = \exp(-i(\xi Q_1 + \eta Q_2)) = \begin{pmatrix} 1 & 0 & 0 \\ 0 & 1 & 0 \\ \xi & \eta & 1 \end{pmatrix}. \tag{10.10}$$

It is seen that $Q_1$ and $Q_2$ generate translations along the $z$ axis. The multiplication of the above two matrices results in the most general form of $C(\xi, \eta, \alpha)$. If this matrix is applied to the column vector $(x, y, z)$, the result is

$$\begin{pmatrix} \cos\alpha & -\sin\alpha & 0 \\ \sin\alpha & \cos\alpha & 0 \\ \xi & \eta & 1 \end{pmatrix} \begin{pmatrix} x \\ y \\ z \end{pmatrix} = \begin{pmatrix} x\cos\alpha - y\sin\alpha \\ x\sin\alpha + y\cos\alpha \\ z + \xi x + \eta y \end{pmatrix}. \tag{10.11}$$

This transformation leaves $(x^2 + y^2)$ invariant, while $z$ can vary from $-\infty$ to $+\infty$. For this reason, it is quite appropriate to call the group of the above linear transformation the cylindrical group. This group is locally isomorphic to E(2).

If, for convenience, we set the radius of the cylinder to be unity:

$$x^2 + y^2 = 1, \tag{10.12}$$

then $x$ and $y$ can be written as

$$x = \cos\phi, \quad y = sin\phi, \tag{10.13}$$

and the transformation of Eq. 10.11 takes the form

$$\begin{pmatrix} \cos\alpha & -\sin\alpha & 0 \\ \sin\alpha & \cos\alpha & 0 \\ \xi & \eta & 1 \end{pmatrix} \begin{pmatrix} \cos\phi \\ \sin\phi \\ z \end{pmatrix} = \begin{pmatrix} \cos(\phi+\alpha) \\ \sin(\phi+\alpha) \\ z + \xi(\cos\phi) + \eta(\sin\phi) \end{pmatrix}. \tag{10.14}$$

The generators of the Euclidean and cylindrical group can also take differential forms. The generator of rotations takes the familiar form

$$L_3 = -i\left(x\frac{\partial}{\partial y} - y\frac{\partial}{\partial x}\right). \tag{10.15}$$

The generators of translations are

$$P_1 = -i\frac{\partial}{\partial x}, \quad P_2 = -i\frac{\partial}{\partial y}. \tag{10.16}$$

The $Q_1$ and $Q_2$ generators are

$$Q_1 = -i\left(\frac{x}{R}\right)\frac{\partial}{\partial z}, \quad Q_2 = -i\left(\frac{y}{R}\right)\frac{\partial}{\partial z}, \tag{10.17}$$

with $x^2 + y^2 = R^2$. In Section 10.2, we shall study the E(2) and the cylindrical groups as the limiting cases of the rotation group.

## 10.2 Contractions of the Three-Dimensional Rotation Group

Let us start with a sphere with a large radius, like the earth. A small potion of the earth's surface is like a flat plane. This is an indication that the two-dimensional Euclidean group can be a small-area approximation of a sphere of large radius. For convenience, we can visualize a flat tangential plane at the north pole.

With this point in mind, we can start with the commutation relations for the generators of the rotation group satisfying the commutation relations:

$$[L_i, L_j] = ie_{ijk}L_k. \tag{10.18}$$

Transformations applicable to the coordinate variables $x$, $y$, and $z$ are generated by $L_3$ given in Eq. 10.3 and

$$L_1 = \begin{pmatrix} 0 & 0 & 0 \\ 0 & 0 & -i \\ 0 & i & 0 \end{pmatrix}, \quad L_2 = \begin{pmatrix} 0 & 0 & i \\ 0 & 0 & 0 \\ -i & 0 & 0 \end{pmatrix}. \tag{10.19}$$

In order to take into account the effect of a large radius, we introduce the matrix

$$B(R) = \begin{pmatrix} 1 & 0 & 0 \\ 0 & 1 & 0 \\ 0 & 0 & R \end{pmatrix}. \tag{10.20}$$

Near the north pole, $z = R$. Thus it is more convenient to introduce the vector

$$\begin{pmatrix} x \\ y \\ 1 \end{pmatrix} = \begin{pmatrix} 1 & 0 & 0 \\ 0 & 1 & 0 \\ 0 & 0 & 1/R \end{pmatrix} \begin{pmatrix} x \\ y \\ z \end{pmatrix}, \tag{10.21}$$

where $z/R = 1$ in the large radius limit, while $x$ and $y$ take moderate values.

The generator of rotations $L_3$ around the $z$ axis remains invariant under this transformation. As for $L_1$ and $L_2$, we can consider

$$W_1(R) = \left(\frac{1}{R}\right) B^{-1}(L_2) B, \quad W_2(R) = -\left(\frac{1}{R}\right) B^{-1}(L_1) B. \tag{10.22}$$

Then

$$[L_3, W_1(R)] = iW_2(R), \quad [L_3, W_2(R)] = -iW_1(R),$$

$$[W_1(R), W_2(R)] = i\left(\frac{1}{R}\right)^2 L_3. \tag{10.23}$$

The explicit forms for $W_1(R)$ and $W_2(R)$ are

$$W_1(R) = \begin{pmatrix} 0 & 0 & i \\ 0 & 0 & 0 \\ -i/R^2 & 0 & 0 \end{pmatrix}, \quad W_2(R) = \begin{pmatrix} 0 & 0 & 0 \\ 0 & 0 & i \\ 0 & -i/R & 0 \end{pmatrix} \tag{10.24}$$

It is now clear that $W_1(R)$ and $W_2(R)$ become $P_1$ and $P_2$ respectively in the large-$R$ limit. This is the contraction of the O(3) to E(2) (Inonu and Wigner 1953, Misra and Maharana 1976, Han *et al.* 1983). This contraction procedure is illustrated in Figure 10.2.

Next, let us consider the equatorial belt where $z$ is very small, while $x^2+y^2 = R^2$. Then

$$Q_1 = -\left(\frac{1}{R}\right) B(L_2) B^{-1}, \quad Q_2 = \left(\frac{1}{R}\right) B(L_1) B^{-1}, \tag{10.25}$$

in the large $R$ limit. $Q_1$ and $Q_2$ generate translations along the $z$ direction on the surface of the cylinder tangential to the sphere at the equator. We shall call the group generated by $L_3$, $Q_1$ and $Q_2$ the cylindrical group (Kim and Wigner 1987a). The contraction of O(3) to this cylindrical group is also illustrated in Figure 10.2.

In differential forms, $P_1$ and $P_2$ are given in Eq. 10.16. We can start with

$$L_1 = -i\left(y\frac{\partial}{\partial z} - z\frac{\partial}{\partial y}\right), \quad L_2 = -i\left(z\frac{\partial}{\partial x} - x\frac{\partial}{\partial z}\right), \tag{10.26}$$

applicable functions of $x$, $y$, and $z$. The $B(R)$ transformation applicable to these operators is

$$B(R) = \exp\left(-\rho z\frac{\partial}{\partial z}\right), \tag{10.27}$$

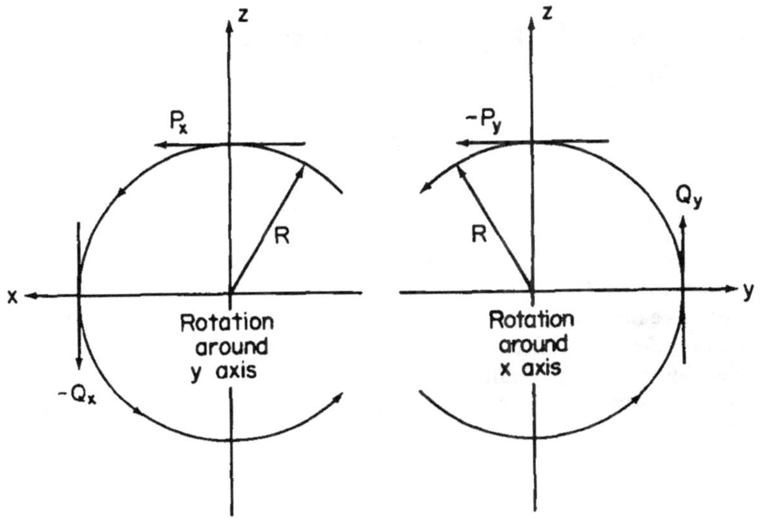


Figure 10.2: Contraction of the three-dimensional rotation group to the two-dimensional Euclidean group and to the cylindrical group. The rotation around the $z$ axis remains unchanged as the radius becomes large. In the case of $E(2)$, rotations around the $y$ and $x$ axes become translations in the $x$ and $-y$ directions respectively within a flat area near the north pole. In the case of the cylindrical group, the rotations around the $y$ and $x$ axes result in translations in the negative and positive $z$ directions respectively within a cylindrical belt around the equator.

where $\rho = \ln(R)$. This operator commutes with $L_3$. The application of this formula to Eq. 10.9 in the large-$R$ limit leads to $Q_1$ and $Q_2$ given in Eq. 10.17. The application of $(B(R))^{-1}$ leads to $P_1$ and $P_2$. On the other hand, a more transparent way of taking the limit is to replace $z$ by $R$ in $L_1$ and $L_2$ to get the generators of E(2), while the replacement of $x$ and $y$ by $(R\cos\phi)$ and $(R\sin\phi)$ leads to the cylindrical group in the large-$R$ limit.

Since $P_1$ commutes with $Q_2$, and $P_2$ commutes with $Q_1$, we can consider the following combination of generators:

$$F_1 = P_1 + Q_1, \quad F_2 = P_2 + Q_2. \tag{10.28}$$

Then these operators also satisfy the commutation relations:

$$[F_1, F_2] = 0, \quad [L_3, F_1] = iF_2, \quad [L_3, F_2] = -iF_1. \tag{10.29}$$

Is it then possible to construct a representation of this group?

## 10.3 Three-Dimensional Geometry of the Little Groups

It is not difficult to construct the geometry of the little group for a massive particle at rest. It is the geometry of the three-dimensional rotation group. As for the geometry of the E(2)-like little group for massless particles, we may continue the discussion of Section 10.2.

Let us go back to the question of adding $P_i$ to $Q_i$ to get $F_i$ as in Eq. 10.28. While both $P_i$ and $Q_i$ are three-by-three matrices, it is not possible to construct a three-by-three matrix representation for $F_i$ . This is due to the fact that the vector spaces are different for the $P_i$ and $Q_i$ representations. We can accommodate this difference by creating two different $z$ coordinates, one with a contracted $z$ and the other with an expanded $z$, namely $(x, y, Rz, z/R)$, as is illustrated in Figure 10.3. Then the generators become

$$P_1 = \begin{pmatrix} 0 & 0 & 0 & i \\ 0 & 0 & 0 & 0 \\ 0 & 0 & 0 & 0 \\ 0 & 0 & 0 & 0 \end{pmatrix}, \quad P_2 = \begin{pmatrix} 0 & 0 & 0 & 0 \\ 0 & 0 & 0 & i \\ 0 & 0 & 0 & 0 \\ 0 & 0 & 0 & 0 \end{pmatrix},$$

$$Q_1 = \begin{pmatrix} 0 & 0 & 0 & 0 \\ 0 & 0 & 0 & 0 \\ i & 0 & 0 & 0 \\ 0 & 0 & 0 & 0 \end{pmatrix}, \quad Q_2 = \begin{pmatrix} 0 & 0 & 0 & 0 \\ 0 & 0 & 0 & 0 \\ 0 & i & 0 & 0 \\ 0 & 0 & 0 & 0 \end{pmatrix}. \qquad (10.30)$$

Then $F_1$ and $F_2$ will take the form:

$$F_1 = \begin{pmatrix} 0 & 0 & 0 & i \\ 0 & 0 & 0 & 0 \\ i & 0 & 0 & 0 \\ 0 & 0 & 0 & 0 \end{pmatrix}, \quad F = \begin{pmatrix} 0 & 0 & 0 & 0 \\ 0 & 0 & 0 & i \\ 0 & i & 0 & 0 \\ 0 & 0 & 0 & 0 \end{pmatrix}. \qquad (10.31)$$

The contraction and expansion of the $z$ axis are illustrated in Figure 10.3.

Next, let us consider the transformation matrix generated by the above matrices. It is easy to visualize the transformations generated by $P_i$ and $Q_i$. It would be easy to visualize the transformation generated by $F_1$ and $F_2$, if $P_i$ commuted with $Q_i$. However, $P_i$ and $Q_i$ do not commute with each other. Thus the transformation matrix takes a somewhat complicated form:

$$\exp\left(-i(\xi F_1 + \eta F_2)\right) = \begin{pmatrix} 1 & 0 & 0 & \xi \\ 0 & 1 & 0 & \eta \\ \xi & \eta & 1 & (\xi^2 + \eta^2)/2 \\ 0 & 0 & 0 & 1 \end{pmatrix}. \qquad (10.32)$$

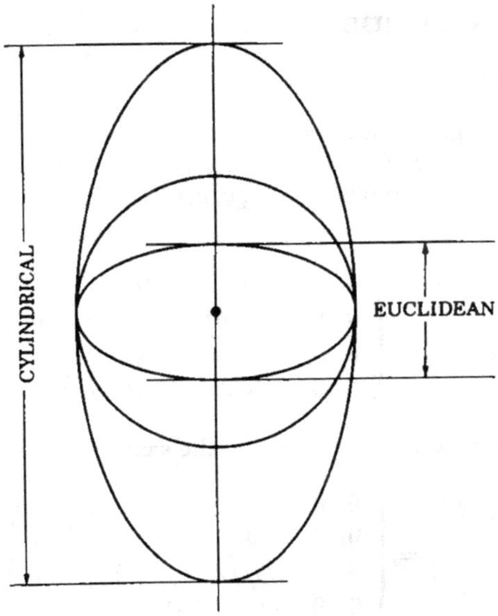


Figure 10.3: Euclidean and cylindrical deformations of the sphere. It is possible to contract the $z$ axis by dividing it by $R$ which becomes large. This is the contraction of $O(3)$ to $E(2)$. It is also possible to expand the $z$ axis by multiplying it by $R$. This is the contraction of $O(3)$ to the cylindrical group.

If we make a similarity transformation on the above form using the matrix,

$$\begin{pmatrix} 1 & 0 & 0 & 0 \\ 0 & 1 & 0 & 0 \\ 0 & 0 & 1/\sqrt{2} & -1/\sqrt{2} \\ 0 & 0 & 1/\sqrt{2} & 1/\sqrt{2} \end{pmatrix}, \tag{10.33}$$

then, $\exp(-i(\xi F_1 + \eta F_2))$ of Eq. 10.32 becomes

$$\begin{pmatrix} 1 & 0 & -\xi/\sqrt{2} & \xi/\sqrt{2} \\ 0 & 1 & -\eta/\sqrt{2} & \eta/\sqrt{2} \\ \xi/\sqrt{2} & \eta/\sqrt{2} & 1-(\xi^2+\eta^2)/4 & (\xi^2+\eta^2)/4 \\ \xi/\sqrt{2} & \eta/\sqrt{2} & -(\xi^2+\eta^2)/4 & 1+(\xi^2+\eta^2)/4 \end{pmatrix}. \tag{10.34}$$

This form is readily available in the literature (Wigner 1939, Weinberg 1964, Kim and Noz 1986) as the translation-like transformation matrix for the little group for massless particles. In this section, we have given a geometrical interpretation to this matrix.

## 10.4 Little Groups in the Light-Cone Coordinate System

Let us now study the group of Lorentz transformations using the light-cone coordinate system. If the space-time metric coordinate is specified by $(x, y, z, t)$, then the light-cone coordinate variables are $(x, y, u, v)$ for a particle moving along the $z$ direction, where

$$\begin{pmatrix} x \\ y \\ u \\ v \end{pmatrix} = \begin{pmatrix} 0 & 0 & 0 & 0 \\ 0 & 0 & 0 & 0 \\ 0 & 0 & 1/\sqrt{/2} & 1/\sqrt{2} \\ 0 & 0 & 1/\sqrt{2} & -1/\sqrt{2} \end{pmatrix} \begin{pmatrix} x \\ y \\ z \\ t \end{pmatrix} . \tag{10.35}$$

The inverse transformation is performed by the same matrix:

$$\begin{pmatrix} x \\ y \\ z \\ t \end{pmatrix} = \begin{pmatrix} 0 & 0 & 0 & 0 \\ 0 & 0 & 0 & 0 \\ 0 & 0 & 1/\sqrt{/2} & 1/\sqrt{2} \\ 0 & 0 & 1/\sqrt{2} & -1/\sqrt{2} \end{pmatrix} \begin{pmatrix} x \\ y \\ u \\ v \end{pmatrix} . \tag{10.36}$$

In this light-cone coordinate system, the generators of Lorentz transformations are

$$J_1 = (1/\sqrt{2}) \begin{pmatrix} 0 & 0 & 0 & 0 \\ 0 & 0 & -i & i \\ 0 & i & 0 & 0 \\ 0 & -i & 0 & 0 \end{pmatrix} , \quad K_1 = (1/\sqrt{2}) \begin{pmatrix} 0 & 0 & 0 & i \\ 0 & 0 & 0 & 0 \\ 0 & 0 & 0 & 0 \\ i & 0 & 0 & 0 \end{pmatrix} ,$$

$$J_2 = (1/\sqrt{2}) \begin{pmatrix} 0 & 0 & i & -i \\ 0 & 0 & 0 & 0 \\ -i & 0 & 0 & \\ i & 0 & 0 & 0 \end{pmatrix} , \quad K_2 = (1/\sqrt{2}) \begin{pmatrix} 0 & 0 & 0 & 0 \\ 0 & 0 & 0 & i \\ 0 & 0 & 0 & 0 \\ 0 & i & 0 & 0 \end{pmatrix} ,$$

$$J_3 = \begin{pmatrix} 0 & -i & 0 & 0 \\ i & 0 & 0 & 0 \\ 0 & 0 & 0 & 0 \\ 0 & 0 & 0 & 0 \end{pmatrix} , \quad K_3 = \begin{pmatrix} 0 & 0 & 0 & 0 \\ 0 & 0 & 0 & 0 \\ 0 & 0 & i & 0 \\ 0 & 0 & 0 & -i \end{pmatrix} . \tag{10.37}$$

For $J_1$, $J_2$, and $J_3$, we can consider the three-by-three matrices consisting of the first three rows and columns. Then they are clearly the generators of the rotation matrix. The set of three-by-three matrices consisting of the first, second and third rows and columns also constitutes the set of rotation generators.

If a massive particle is at rest, its little group is generated by $J_1$, $J_2$ and $J_3$. For a massless particle, the little group is generated by $J_3$, $N_1$ and $N_2$, where

$$N_1 = (K_1 - J_2) , \quad N_2 = (K_2 + J_1) , \tag{10.38}$$

which can be written in the matrix form

$$N_1 = (1/\sqrt{2})\begin{pmatrix} 0 & 0 & 0 & i \\ 0 & 0 & 0 & 0 \\ i & 0 & 0 & 0 \\ 0 & 0 & 0 & 0 \end{pmatrix}, \quad N_2 = (1/\sqrt{2})\begin{pmatrix} 0 & 0 & 0 & 0 \\ 0 & 0 & 0 & i \\ 0 & i & 0 & 0 \\ 0 & 0 & 0 & 0 \end{pmatrix}. \tag{10.39}$$

These generators of course satisfy the commutation relations for the E(2)-like little group given in Eqs. 7.20 and 7.21.

Let us go back to $F_1$ and $F_2$ of Eq. 10.31. Indeed, they are proportional to $N_1$ and $N_2$ respectively:

$$N_1 = (1/\sqrt{2})F_1, \quad N_2 = (1/\sqrt{2})F_2. \tag{10.40}$$

Since $F_2$ and $F_2$ are somewhat simpler than $N_1$ and $N_2$, and since the commutation relations of Eq. 7.8 are invariant under multiplication of $N_1$ and $N_2$ by constant factors, we shall hereafter use $F_1$ and $F_2$ for $N_1$ and $N_2$.

In the light-cone coordinate system, the boost matrix along the $z$ direction takes the form

$$B(R) = \exp(-i\rho K_3) = \begin{pmatrix} 1 & 0 & 0 & 0 \\ 0 & 1 & 0 & 0 \\ 0 & 0 & R & 0 \\ 0 & 0 & 0 & 1/R \end{pmatrix}, \tag{10.41}$$

with $\rho = \ln(R)$, and $R = ((1+\beta)/(1-\beta))^{1/2}$, where $\beta$ is the velocity parameter of the particle. Under this transformation, $x$ and $y$ coordinates are invariant, and the light-cone variables $u$ and $v$ are transformed as

$$u' = Ru, \quad v' = v/R. \tag{10.42}$$

If we boost $J_1$ and $J_2$ and multiply them by $\sqrt{2}/R$, as

$$W_1(R) = -(\sqrt{2}/R)BJ_2B^{-1} = \begin{pmatrix} 0 & 0 & -i/R^2 & i \\ 0 & 0 & 0 & 0 \\ i & 0 & 0 & 0 \\ -i/R^2 & 0 & 0 & 0 \end{pmatrix},$$

$$W_2(R) = (\sqrt{2}/R)BJ_1B^{-1} = \begin{pmatrix} 0 & 0 & 0 & 0 \\ 0 & 0 & -i/R^2 & i \\ 0 & i & 0 & 0 \\ 0 & i/R^2 & 0 & 0 \end{pmatrix}, \tag{10.43}$$

then $W_1(R)$ and $W_2(R)$ become $F_1$ and $F_2$ respectively in the large-$R$ limit (Han *et al.* 1983).

The algebra given in this section is identical with that of Section 9.3 based on the three-dimensional geometry of a sphere going through a contraction/expansion of the $z$ axis. Therefore, it is possible to give a concrete geometrical picture to the little groups of the Poincaré group governing the internal space-time symmetries of relativistic particles.

The most general form of the transformation matrix is

$$D(\xi, \eta, \alpha) = D(\xi, \eta, 0) D(0, 0, \alpha), \tag{10.44}$$

where

$$D(\xi, \eta, 0) = \exp\left(-i(\xi F_1 + \eta F_2)\right), \quad D(0, 0, \alpha) = \exp\left(-i\alpha J_3\right).$$

$D(0,0,\alpha)$ represents a rotation around the $z$ axis, and does not need further explanation. In the light-cone coordinate system, $D(\xi,\eta,0)$ takes the form of Eq. 10.32. It is then possible to decompose it into

$$D(\xi, \eta, 0) = C(\xi, \eta) E(\xi, \eta) S(\xi, \eta), \tag{10.45}$$

where

$$C(\xi, \eta) = \exp\left(-i\xi Q_1 - i\eta Q_2\right) = \begin{pmatrix} 1 & 0 & 0 & 0 \\ 0 & 1 & 0 & 0 \\ \xi & \eta & 1 & 0 \\ 0 & 0 & 0 & 1 \end{pmatrix}, \tag{10.46}$$

$$E(\xi, \eta) = \exp\left(-i\xi P_1 - i\eta P_2\right) = \begin{pmatrix} 1 & 0 & 0 & \xi \\ 0 & 1 & 0 & \eta \\ 0 & 0 & 1 & 0 \\ 0 & 0 & 0 & 1 \end{pmatrix}, \tag{10.47}$$

$$S(\xi, \eta) = I + \frac{1}{2}\left[C(\xi, \eta), E(\xi, \eta)\right] = \begin{pmatrix} 1 & 0 & 0 & 0 \\ 0 & 1 & 0 & 0 \\ 0 & 0 & 1 & (\xi^2 + \eta^2)/2 \\ 0 & 0 & 0 & 1 \end{pmatrix}. \tag{10.48}$$

The matrix $C(\xi,\eta)$ performs a cylindrical transformation on the first, second and third components, while $E(\xi,\eta)$ is for a Euclidean transformation on the first, second and fourth components. The matrix $S(\xi,\eta)$ performs a translation along the third axis and commutes with both $C(\xi,\eta)$ and $E(\xi,\eta)$. Both $E(\xi,\eta)$ and $S(\xi,\eta)$ become identity matrices when applied to four-vectors satisfying the Lorentz condition which have a vanishing fourth component.

## 10.5 Cylindrical Group and Gauge Transformations

In order to illustrate the transformation property of the vector to which the above matrices are applicable, let us consider a particle represented by a four-vector:

$$A^\mu(x) = A^\mu e^{i(kz - \omega t)}, \tag{10.49}$$

where $A^\mu = (A_1, A_2, A_3, A_0)$. In the light-cone coordinate system,

$$A^\mu = (A_1, A_2, A_u, A_v), \tag{10.50}$$

where $A_u = (A_3 + A_0)/\sqrt{2}$, and $A_v = (A_3 - A_0)/\sqrt{2}$. If it is boosted by the matrix of Eq. 10.41, then

$$A'^\mu = (A_1, A_2, RA_u, (1/R)A_v). \tag{10.51}$$

Thus the fourth component will vanish in the large-$R$ limit, while the third component becomes large.

The momentum-energy four-vector is

$$P^\mu = \left(0, 0, (k + w)/\sqrt{2}, (k - w)/\sqrt{2}\right), \tag{10.52}$$

which in the rest frame becomes

$$P^\mu = \left(0, 0, m/\sqrt{2}, -m/\sqrt{2}\right), \tag{10.53}$$

where $m$ is the mass. If we boost this four-momentum using the matrix of Eq. 10.41, then

$$P'^\mu = \left(0, 0, Rm/\sqrt{2}, -m/\sqrt{2}R.\right) \tag{10.54}$$

Here again, the fourth component vanishes for large values of $R$, while the third component becomes large.

Let us go back to $W_1(R)$ and $W_2(R)$ of Eq. 10.43. If $W_1(R)$ is applied to the four-vector $A'^\mu$, the result is

$$i\left((A_u - A_v)/R, 0, A_1, -(1/R)^2 A_1\right), \tag{10.55}$$

which becomes $(0, 0, -iA_1, 0)$. When $W_2(R)$ is applied, the result is $(0, 0, -iA_2, 0)$. Thus, the $i/R^2$ factors in $W_1(R)$ and $W_2(R)$ can be dropped in the large-$R$ limit. We can thus safely apply the transformation matrix generated by $F_1$ and $F_2$.

Since the fourth component of the vector vanishes or becomes vanishingly small, the application of $S(\xi, \eta)$ of Eq. 10.48 on $A'^\mu$ and $P'^\mu$ will produce no effects in the large-$R$ limit. The same is true for $E(\xi, \eta)$ of Eq. 10.47. Thus, among the three factors of the transformation matrix, only the matrix $C(\xi, \eta)$ given in Eq. 10.46 will produce a nontrivial effect. This is the cylindrical transformation discussed in Section 10.1.

During the limiting process, the three-dimensional geometry consisting of the $x$, $y$, and $v$ coordinates describes a pancake-like compression of the sphere in which the $v$ coordinate shrinks to zero, as is indicated in Figure 10.3. Because of this contraction of the $v$ coordinate, the Euclidean component of the little group disappears. This is the content of the Lorentz condition for massive particles in the infinite-momentum limit. The three-dimensional geometry of the $x$, $y$ and $u$ coordinates corresponds to the expanding $z$ coordinate, resulting in the cylindrical symmetry, as is indicated in Figure 10.3.

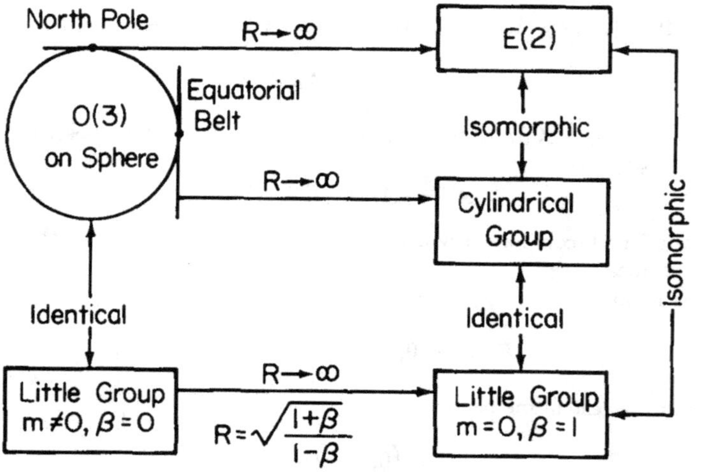


Figure 10.4: $E(2)$, the $E(2)$-like little group for massless particles, and the cylindrical group. The correspondence between $E(2)$ and the $E(2)$-like little group is isomorphic but not identical. The cylindrical group is identical to the $E(2)$-like little group. Both $E(2)$ and the cylindrical group can be regarded as contractions of $O(3)$ in the large-radius limit. The Lorentz boost of the $O(3)$-like little group for a massive particle at rest to the $E(2)$-like little group for a massless particle is exactly the same as the contraction of $O(3)$ to the cylindrical group. The radius of the sphere in this case can be identified as $((1+\beta)/(1-\beta))^{1/2}$.

Let us see the effect of $C(\xi,\eta)$ on the four-vector of Eq. 10.51. If we apply $C(\xi,\eta)$ to the four-vector, then

$$\begin{pmatrix} 1 & 0 & 0 & 0 \\ 0 & 1 & 0 & 0 \\ \xi & \eta & 1 & 0 \\ 0 & 0 & 0 & 1 \end{pmatrix} \begin{pmatrix} A_1 \\ A_2 \\ RA_u \\ (1/R)A_v \end{pmatrix} = \begin{pmatrix} A_1 \\ A_2 \\ RA_u + \xi A_1 + \eta A_2 \\ (1/R)A_v \end{pmatrix}. \qquad (10.56)$$

This is not unlike the $D(\xi,\eta)$ transformation applied to the four-vector satisfying the Lorentz condition $A_v = 0$:

$$\begin{pmatrix} 1 & 0 & 0 & \xi \\ 0 & 1 & 0 & \eta \\ \xi & \eta & 1 & (\xi^2+\eta^2)/2 \\ 0 & 0 & 0 & 1 \end{pmatrix} \begin{pmatrix} A_1 \\ A_2 \\ RA_u \\ 0 \end{pmatrix}$$

$$= \begin{pmatrix} 1 & 0 & 0 & 0 \\ 0 & 1 & 0 & 0 \\ \xi & \eta & 1 & 0 \\ 0 & 0 & 0 & 1 \end{pmatrix} \begin{pmatrix} A_1 \\ A_2 \\ RA_u \\ 0 \end{pmatrix} = \begin{pmatrix} A_1 \\ A_2 \\ RA_u + \xi A_1 + \eta A_2 \\ 0 \end{pmatrix}. \qquad (10.57)$$

As we noted in Section 7.3, the Lorentz condition eliminates the Euclidean component in the $D(\xi,\eta,0)$ matrix. It is remarkable that the matrix in Eq. 10.57 is strikingly similar to Eq. 10.10. The cylindrical transformation is quite independent of the fourth component in both cases, and it produces the same result for the first three components. Thus the elimination of the Euclidean component which led to Eq. 10.57 can thus be regarded as an extension of the Lorentz condition to all four-vectors (Kim and Wigner 1990a).

The fact that the cylindrical group shares the same geometry as that of the E(2)-like little group is illustrated in Figure 10.4.

## 10.6 Little Groups for Relativistic Extended Particles

We are now ready to study the space-time symmetry of relativistic extended particles or hadrons, using the covariant harmonic oscillator formalism. Here again, we consider a hadron consisting of two quarks bound together by an attractive force such as the harmonic oscillator force. We use four-vectors $x_a$ and $x_b$ to specify space-time positions of the two quarks. Then it is more convenient to use the variables $X$ and $x$ defined in Eq. 8.21, which specify the space-time coordinate for the hadron and the space-time separation between the quarks. The oscillator functions are defined in the $x$ space and are normalizable. It is thus possible to generate unitary Lorentz transformations from the Hermitian operators:

$$J_i = -i\epsilon_{ijk} x_j \frac{\partial}{\partial x_k}, \quad K_i = -i\left\{t\frac{\partial}{\partial x_i} + x_i\frac{\partial}{\partial t}\right\}. \tag{10.58}$$

In the light-cone coordinate system, the generators of rotations are

$$\begin{aligned} J_1 &= -(i/\sqrt{2})\left(y\left(\frac{\partial}{\partial u} + \frac{\partial}{\partial v}\right) - (u+v)\frac{\partial}{\partial y}\right), \\ J_2 &= (i/\sqrt{2})\left(x\left(\frac{\partial}{\partial u} + \frac{\partial}{\partial v}\right) - (u+v)\frac{\partial}{\partial x}\right), \\ J_3 &= -i\left(x\frac{\partial}{\partial y} - y\frac{\partial}{\partial z}\right). \end{aligned} \tag{10.59}$$

The boost generators are

$$\begin{aligned} K_1 &= (i/\sqrt{2})\left(x\left(\frac{\partial}{\partial u} - \frac{\partial}{\partial v}\right) + (u-v)\frac{\partial}{\partial x}\right), \\ K_2 &= (i/\sqrt{2})\left(y\left(\frac{\partial}{\partial u} - \frac{\partial}{\partial v}\right) + (u-v)\frac{\partial}{\partial y}\right), \\ K_3 &= -i\left(u\frac{\partial}{\partial u} - v\frac{\partial}{\partial v}\right). \end{aligned} \tag{10.60}$$

These generators do not contain the hadronic coordinate variable $X$, as transformations of the little group do not change the hadronic momentum.

For an extended particle at rest, $J_1$, $J_2$, and $J_3$ generate the O(3)-like little group. If the particle moves along the $z$ direction with the velocity of light, the generators of the little group are $J_3$, and

$$N_1 = K_1 - J_2, \quad N_2 = K_2 + J_1, \tag{10.61}$$

as in Section 7.2. In terms of the light-cone variables,

$$N_1 = -i\sqrt{2}\left(x\frac{\partial}{\partial u} - v\frac{\partial}{\partial x}\right), \quad N_2 = -i\sqrt{2}\left(y\frac{\partial}{\partial u} - v\frac{\partial}{\partial y}\right). \tag{10.62}$$

The boost operator along the $z$ direction is

$$B(R) = \exp\left\{-\rho\left(u\frac{\partial}{\partial u} - v\frac{\partial}{\partial v}\right)\right\}. \tag{10.63}$$

We are interested in obtaining $N_1$ and $N_2$ by boosting $J_2$ and $J_1$. This boost is applied to $J_2$ and $J_1$, as in the case of Eq. 10.43,

$$W_1(R) = -i\left\{x\frac{\partial}{\partial u} - v\frac{\partial}{\partial x} - \left(\frac{1}{R}\right)^2\left(u\frac{\partial}{\partial x} - x\frac{\partial}{\partial v}\right)\right\},$$

$$W_2(R) = -i\left\{y\frac{\partial}{\partial u} - v\frac{\partial}{\partial y} - \left(\frac{1}{R}^2\right)\left(u\frac{\partial}{\partial y} - y\frac{\partial}{\partial v}\right)\right\}. \tag{10.64}$$

In the limit of large $R$, $W_1$ and $W_2$ become $F_1$ and $F_2$ respectively, and they are

$$F_1 = -i\left(x\frac{\partial}{\partial u} - v\frac{\partial}{\partial x}\right), \quad F_2 = -i\left(y\frac{\partial}{\partial u} - v\frac{\partial}{\partial y}\right). \tag{10.65}$$

As in the case of Eq. 10.40, these operators are proportional to $N_1$ and $N_2$ respectively. Since the commutation relations for the generators of the little group given in Eq. 7.8 are invariant under multiplication of $N_1$ and $N_2$ by a constant, we can use $F_1$ and $F_2$ as the generators. As a consequence, the transformation matrix is

$$D(\xi,\eta,0) = \exp\left(-i(\xi x + \eta y)\frac{\partial}{\partial u} - iv\left(\xi\frac{\partial}{\partial x} + \eta\frac{\partial}{\partial y}\right)\right), \tag{10.66}$$

which can be decomposed into

$$D(\xi,\eta,0) = \exp\left(-i(\xi x + \eta y)\frac{\partial}{\partial u}\right)\exp\left(-iv\left(\xi\frac{\partial}{\partial x} + \eta\frac{\partial}{\partial y}\right)\right)$$
$$\times \exp\left(-i\frac{v}{2}\left(\xi^2 + \eta^2\right)\frac{\partial}{\partial u}\right), \tag{10.67}$$

as in the case of Eq. 10.45.

Let us now look at the wave function to which the above operator is applicable. If the system is boosted along the $z$ axis, the relevant part of the covariant oscillator wave function is

$$\psi(z,t) = \left(\frac{1}{\pi}\right)^{1/2} \exp\left\{-\left(u^2 + v^2\right)/2\right\}, \tag{10.68}$$

from Eq. 8.72. This is the ground-state wave function in the rest frame. If we apply the $B(R)$ operator to the above wave function,

$$\psi_\beta(z,t) = \left(\frac{1}{\pi}\right)^{1/2} \exp\left\{-\left((u/R)^2 + (Rv)^2\right)/2\right\}. \tag{10.69}$$

This form is the same as the Lorentz-squeezed wave function of Eq. 8.74. The width of this function along the $u$ axis increases as $R$ becomes large, while the distribution along the $v$ axis becomes narrow. Indeed, $B(R)$ is a squeeze operator.

The width of the $v$ distribution decreases as $1/R$. When the $v$ distribution is very narrow, we can consider the transformation in the subspace where $v = 0$. Then the factors $\exp\left\{-iv\left(\xi\frac{\partial}{\partial x} + \eta\frac{\partial}{\partial y}\right)\right\}$ and $\exp\left\{-i\frac{v}{2}(\xi^2+\eta^2)\frac{\partial}{\partial u}\right\}$ in Eq. 10.66 for $D(\xi,\eta,0)$ can be dropped. As a consequence,

$$D(\xi,\eta,0) = \exp\left\{-i(\xi x + \eta y)\frac{\partial}{\partial u}\right\}. \tag{10.70}$$

This means that the terms $v\frac{\partial}{\partial x}$ and $v\frac{\partial}{\partial y}$ in Eq. 10.64 can be dropped, and $F_1$ and $F_2$ can be written

$$F_1 = -ix\frac{\partial}{\partial u}, \quad F_2 = -iy\frac{\partial}{\partial u}. \tag{10.71}$$

These operators generate translations along the $u$ axis. These operators, together with the rotation generator $J_3$ of Eq. 10.58, are the generators of the cylindrical group. The differential operators $F_1$ and $F_2$ are now the generators of gauge transformations applicable to functions with a narrow distribution in $v$.

Here again, a complete description of the little group for massive particles in the infinite-momentum limit requires both the cylindrical and Euclidean components. The Euclidean component can be deleted in the infinite-momentum limit or in the $v = 0$ subspace. As was noted in Section 10.5, this is the Lorentz condition applicable to massive particles in the infinite-momentum limit.

The same reasoning can be carried out in the momentum-energy space, and the result will be

$$F_1 = -iq_x\frac{\partial}{\partial q_v}, \quad F_2 = -iq_y\frac{\partial}{\partial q_v}. \tag{10.72}$$

As in the case of photons, these operators are generators of gauge transformations. In the oscillator regime discussed in this section, the above operators generate translations along the $q_v$ axis, where $q_v = (q_z + q_0)/\sqrt{2}$ as defined in Eq. 8.77. This variable becomes $\sqrt{2}q_z$ in the infinite-momentum limit. This means that $F_1$ and $F_2$ generate transformations which change the value of Feynman's $x$ parameter. Therefore, Feynman's $x$ is a gauge transformation parameter in the parton picture

(Kim 1989). This completes the third row of Figure 10.1 which indicates that the covariant phase space is the physical basis for a unified picture of the quark model for hadrons at rest and the parton model at the infinite-momentum limit.

## 10.7 Lorentz Transformations and Hadronic Temperature

There are no free quarks. Quarks in hadrons are permanently bound. We call them "quarks in confinement phase." Partons are almost free and independent. Since the momentum distribution is wide spread, we are tempted to call those partons "quarks in plasma phase." We are also tempted to say that the quarks are in the confinement phase at low temperature, and they are in the plasma phase at high temperature. Then how does the concept of temperature come into the picture of quarks and partons?

The answer to this question is very simple, if we compare the Lorentz-squeezed ground-state wave function of Eq. 8.74 with the two-mode squeezed-state wave function of Eq. 5.140. The Lorentz-squeezed wave function can be expanded as

$$\psi^0_\beta(z,t) = \left(1-\beta^2\right)^{1/2} \sum_n \beta^n \psi_n(z)\psi_n(t), \tag{10.73}$$

according to Eq. 3.68. The density matrix for this pure state is

$$\begin{aligned}\rho_\beta(z,t;z',t') &= \psi_\beta(z,t)\psi_\beta(z,t) \\ &= \left(\frac{1}{\pi}\right)\exp\left\{-\left(\frac{1-\beta}{1+\beta}\right)\left[\left(\frac{z+t}{2}\right)^2+\left(\frac{z'+t'}{2}\right)^2\right]\right. \\ &\quad \left. -\left(\frac{1+\beta}{1-\beta}\right)\left[\left(\frac{z-t}{2}\right)^2+\left(\frac{z'-t'}{2}\right)^2\right]\right\}.\end{aligned} \tag{10.74}$$

Within the framework of the present form of quantum measurement theory, it is not possible to measure the time-separation variable $t$. We therefore have to take the trace with respect to this variable. The resulting density matrix is

$$\rho_\beta(z,z') = \int \rho_\beta(z,t;z',t)dt. \tag{10.75}$$

We can evaluate this integral using either the series form derivable from Eq. 10.73 or the exponential form of Eq. 10.74. In the series form

$$\rho_\beta(z,z') = \left(1-\beta^2\right)\sum_n \beta^{2n}\psi_n(z)\psi_n(z'). \tag{10.76}$$

In the exponential form,

$$\rho_\beta(z,z') = \left(\frac{1-\beta^2}{\pi[1+\beta^2]}\right)^{1/2} \exp\left\{-\left(\frac{1-\beta^2}{1+\beta^2}\right)\left(\frac{z+z'}{2}\right)^2\right.$$

$$-\left(\frac{1+\beta^2}{1-\beta^2}\right)\left(\frac{z-z'}{2}\right)^2\bigg\}. \tag{10.77}$$

The diagonal element of this density matrix is

$$\rho_\beta(z) = \rho_\beta(z,z) = \left(\frac{1-\beta^2}{\pi[1+\beta^2]}\right)^{1/2} \times \exp\left\{-\left(\frac{1-\beta^2}{1+\beta^2}\right)(z)^2\right\} \tag{10.78}$$

It is straight-forward to derive the Wigner distribution function from Eq. 10.77. The result is (Kim and Li 1989)

$$W_\beta(z,p) = \left(\frac{1-\beta^2}{\pi[1+\beta^2]}\right) \times \exp\left\{-\left(\frac{1-\beta^2}{1+\beta^2}\right)\left(z^2+p^2\right)\right\}, \tag{10.79}$$

where we have used $p$ instead of $q_z$ for the momentum-energy separation variable.

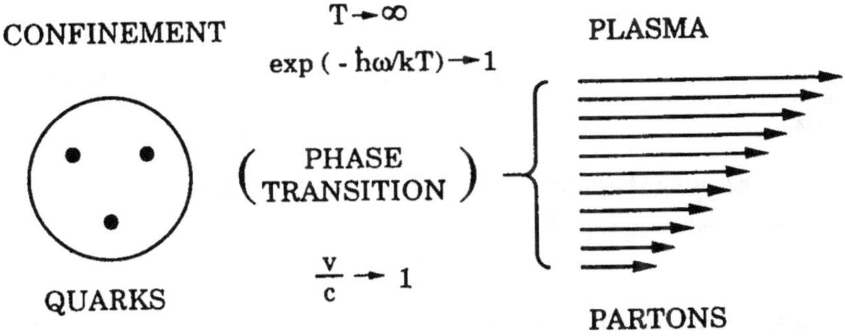


Figure 10.5: Transition in hadronic phase. If the hadron is slow or at rest, it is in the confinement phase. If its velocity is close to that of light, it is in a plasma phase consisting of incoherent partons with a wide-spread momentum distribution. Then there should be a phase transition while $\beta$ goes from 0 to 1.

Let us compare Eq. 10.76 with Eq. 2.88 or Eq. 5.137, Eq. 10.77 with Eq. 3.69, and Eq. 10.79 with Eq. 3.70 or Eq. 6.104. Then we can define the hadronic temperature through (Han *et al.* 1990a)

$$\beta^2 = \omega/kT. \tag{10.80}$$

What is $\omega$ in this hadronic system? In order to answer this question, let us go to the spring constant $\Omega$ which was used in Chapter 9. This constant comes from the

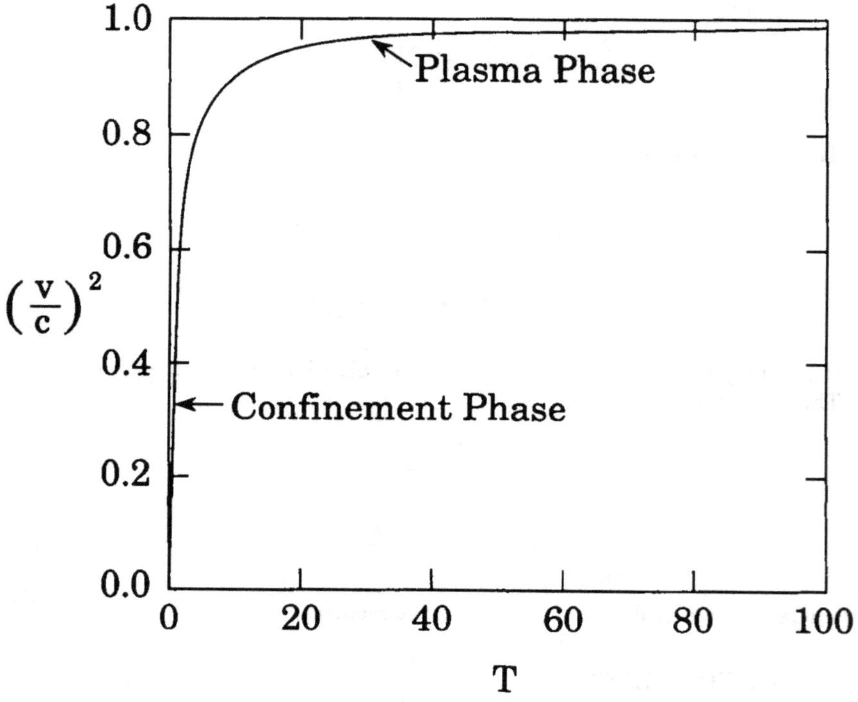


Figure 10.6: The $(v/c)^2$ factor as a function of the temperature $T$ measured in units of $\Omega/m$. There is an abrupt change in slope within the interval $5 < T < 10$. The critical temperature is within this interval.

oscillator equation (Feynman *et al.* 1971):

$$\frac{1}{2}\left\{\Omega^2 x_\mu^2 - \left(\frac{\partial}{\partial x_\mu}\right)^2\right\}\psi(x) = (\Omega\lambda)\psi(x), \tag{10.81}$$

where $\lambda$ is a dimensionless number. On the other hand, in nonrelativistic quantum mechanics, the oscillator equation takes the form

$$\frac{1}{2}\left\{(m\omega)^2 x^2 - \left(\frac{\partial}{\partial x}\right)^2\right\}\psi(x) = (m\omega)\lambda\psi(x), \tag{10.82}$$

for the $x$ component, where $m$ in this case is the reduced quark mass. Therefore we can identify $\omega$ as $\Omega/m$ and write

$$\beta^2 = \exp(-\Omega/mkT), \quad \text{or} \quad kT = (\Omega/m)/\ln\left(1 + (M/P)^2\right), \tag{10.83}$$

where $M$ and $P$ are the hadronic mass and its magnitude of momentum respectively. If the hadron is at rest with $P = 0$, $T$ vanishes.

The temperature rises as the hadronic momentum increases. As the momentum becomes very large, $kT$ increases as $(\Omega/mM^2)P^2$. As was discussed in Chapter 9, the hadron becomes a collection of partons, and behaves as though it were in a plasma state. If we add this interpretation to Figure 9.4, the result is Figure 10.5.

If the hadron goes through the transition from the confinement phase to a plasma phase, what is the critical temperature? The parton model does not specify the critical speed at which the hadron becomes a collection of plasma-like partons. This transition is known to be a gradual process. On the other hand, as is illustrated in Figure 10.6, the $\beta^2$ factor as a function of $T$ has an abrupt change in slope in the interval between $T = 5\ \Omega/m$ and $15\ \Omega/m$. The critical temperature is within this interval.

## 10.8 Decoherence and Entropy

In Section 10.7, we considered the thermal excitation of the ground-state harmonic oscillator wave function as a consequence of the Lorentz squeeze and the non-observance of the time-separation variable. It was possible to define the hadronic temperature because the resulting density matrix or Wigner function was like that for the harmonic oscillator-state in thermal equilibrium.

The question then is what interpretation can be given to other forms of wave functions, especially when there is no apparent reason to expect that the system is in thermal equilibrium? In this Section, it will be noted that it is still possible to calculate the entropy for all Lorentz-squeezed wave functions (Kim and Wigner 1990c).

The entropy is a measure of our ignorance and is computed from the density matrix (Von Neumann 1955, Wigner and Yanase 1963). In terms of the density matrix, the entropy is defined as

$$S = -Tr\,(\rho \ln(\rho)) . \tag{10.84}$$

This definition does not require that the system be in a thermal equilibrium state. Let us study the harmonic oscillator in an excited state.

In terms of the light-cone variables, the excited-state harmonic oscillator wave function takes the form

$$\psi_\beta^n(z,t) = \left(1/\sqrt{\pi}(n!)2^n\right)^{1/2} H_n\left((e^{-\lambda}u + e^{\lambda}v)/\sqrt{2}\right) \times \exp\left\{-\left(\frac{1}{2}\right)\left(e^{-2\lambda}u^2 + e^{2\lambda}v^2\right)\right\}, \tag{10.85}$$

with $e^{\lambda} = [(1+\beta)/(1-\beta)]^{1/2}$. According to Section 8.6 this wave function can be expanded as

$$\psi_\beta^n(z,t) = \left(1-\beta^2\right)^{(n+1)/2} \sum_k \left(\frac{(n+k)!}{n!k!}\right)^{1/2} \beta^k \psi_{n+k}(z)\psi_n(t). \tag{10.86}$$

From this wave function, we can construct the pure-state density matrix

$$\rho_\beta^n(z,t;z',t') = \psi_\beta^n(z,t)\psi_\beta^n(z',t'), \tag{10.87}$$

which satisfies the condition $\rho^2 = \rho$:

$$\rho_\beta^n(z,t;z',t') = \int \rho_\beta^n(z,t;z'',t'')\rho_\beta^n(z'',t'';z',t')dz''dt''. \tag{10.88}$$

However, there are at present no measurement theories which accommodate the time-separation variable $t$, as is indicated in Figure 10.7. Thus, we can take the trace of the $\rho$ matrix with respect to the $t$ variable. Then the resulting density matrix is

$$\begin{aligned}\rho_\beta^n(z,z') &= \int \psi_\beta^n(z,t)\left(\psi_\beta^n(z',t)\right) dt \\ &= \left(1-\beta^2\right)^{(n+1)} \sum_k \left(\frac{(n+k)!}{n!k!}\right) \beta^{2k}\psi_{k+n}(z)\psi_{k+n}(z').\end{aligned} \tag{10.89}$$

The trace of this density matrix is one, but the trace of $\rho^2$ is less than one, as

$$\begin{aligned}Tr\left(\rho^2\right) &= \int \rho_\beta^n(z,z')\rho_\beta^n(z',z)dz'dz \\ &= \left(1-\beta^2\right)^{2(n+1)} \sum_k \left(\frac{(n+k)!}{n!k!}\right)^2 \beta^{4k},\end{aligned} \tag{10.90}$$

which is less than one. This is due to the fact that we do not know how to deal with the time-like separation in the present formulation of quantum mechanics. Our knowledge is less than complete. The result is a non-zero entropy (Kim and Wigner (1990c).

The density matrix of Eq. 10.89 leads to

$$\begin{aligned}S = &-(n+1)\left\{\ln\left(1-\beta^2\right) + \beta^2\left(\ln\left(\beta^2\right)\right)/\left(1-\beta^2\right)\right\} \\ &- \left(1-\beta^2\right)^{(n+1)} \sum_k \left(\frac{(n+k)!}{n!k!}\right) \ln\left(\frac{(n+k)!}{n!k!}\right) \beta^{2k}.\end{aligned} \tag{10.91}$$

For the ground state, the entropy is

$$S = -\left\{\ln\left(1-\beta^2\right) + \beta^2\left(\ln\left(\beta^2\right)\right)/\left(1-\beta^2\right)\right\}. \tag{10.92}$$

This expression is consistent with the harmonic oscillator in thermal equilibrium if $\beta^2$ is identified with the Boltzmann factor $e^{-\omega/kT}$.

The integration over the time separation variable is basically a time average which was discussed in Section 2.5 for the ensemble average where the relative phases are eliminated. This process can be called the decoherence (Hartle 1990) or more specifically temporal decoherence (Kim and Noz 1991). This decoherence process

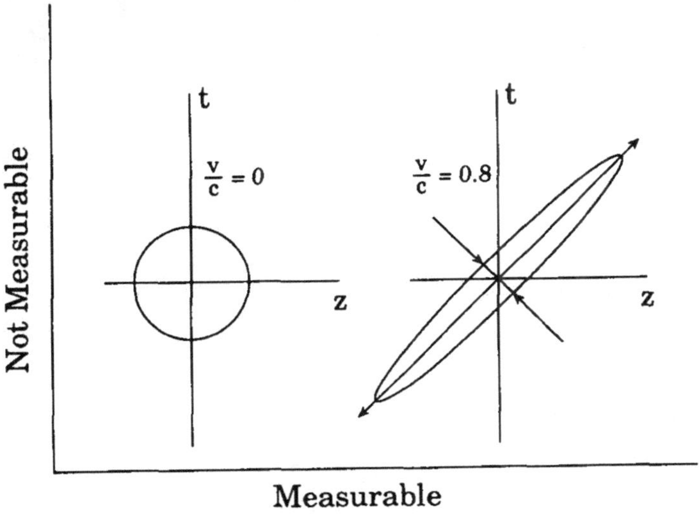


Figure 10.7: Lorentz squeeze and entropy. When the hadron is at rest, the ground-state oscillator wave function is concentrated within a circular region, and it undergoes a Lorentz squeeze as the hadronic velocity increases, as is shown in Figure 9.6. It is possible to perform a measurement along the longitudinal direction, but the time-separation variable is not measurable within the frame of the present form of quantum mechanics. This ignorance leads to an increase in entropy.

is inherent when quantum mechanics contains variable which are not measured. In his book on statistical mechanics (Feynman 1973), Feynman states this as:

*When we solve a quantum-mechanical problem, what we really do is divide the universe into two parts - the system in which we are interested and the rest of the universe. We then usually act as if the system in which we are interested comprised the entire universe. To motivate the use of density matrices, let us see what happens when we include the part of the universe outside the system.*

In this Section, we have identified Feynman's *rest of the universe* as the time-separation coordinate in a relativistic two-body problem. Our ignorance about this coordinate leads to a density matrix for a non-pure state, and consequently to an increase of entropy.

# Appendix A

# Reprinted Articles

This book serves as an introduction to the phase-space picture of quantum mechanics. While avoiding the mathematical complications which usually arise in dynamical problems, we have concentrated our efforts on the Wigner function for the ground-state harmonic oscillator and its associated symmetry. The formalism has rich applications in quantum optics and in relativistic bound-state problems.

Indeed, in this book, we deal mostly with the uncertainty principle and its symmetry properties. This is not a new subject. There are many important papers written on this subject, and there are edited volumes in the literature on this subject (e.g., Noz and Kim 1988). In this Appendix, we reprint the papers which form the foundations of the phase-space picture of quantum mechanics and its symmetry problems.

The first article is Wigner's 1932 paper on the quantum correction for thermodynamic equilibrium. In this paper, Wigner introduces the Wigner phase-space distribution function which plays the central throughout this book. The second article reprinted in this Appendix is Wigner's 1939 paper on unitary representations of the Poincaré group. While this paper deals with the fundamental space-time symmetries of relativistic particles (Kim and Noz 1986), it also teaches us how to construct representations of the Lorentz group systematically. The basic symmetry in phase space is that of the Lorentz group.

The third and fourth articles reprinted in this Appendix are Dirac's papers on harmonic oscillators. In his 1945 paper on unitary representations of the Lorentz group, Dirac suggests the use of normalizable four-dimensional oscillator wave functions. The covariant harmonic oscillator wave functions discussed in Chapter 8 were first discussed by Dirac in 1945.

In his 1963 paper on the "remarkable representation" of the (3 + 2)- dimensional de Sitter group, Dirac constructs the representation using two harmonic oscillator wave functions. In this paper, Dirac was exploiting the correspondence between the group Sp(4) and O(3,2). It is indeed remarkable that Dirac was discussing two-mode squeezed states in 1963. Dirac was fond of harmonic oscillators.

We would like thank Professor Eugene P. Wigner, Mrs. Margit Dirac, the American Physical Society, Princeton University Press, the Royal Society of London, and the American Institute of Physics for allowing us to reprint the articles in this Appendix.

# A.1 On the Quantum Correction for Thermodynamic Equilibrium

by E.P. Wigner

reprinted with permission from

Physical Review **40**, 749-759 (1932).

JUNE 1, 1932 PHYSICAL REVIEW VOLUME 40

# On the Quantum Correction For Thermodynamic Equilibrium

By E. WIGNER
Department of Physics, Princeton University
(Received March 14, 1932)

The probability of a configuration is given in classical theory by the Boltzmann formula $\exp[-V/hT]$ where $V$ is the potential energy of this configuration. For high temperatures this of course also holds in quantum theory. For lower temperatures, however, a correction term has to be introduced, which can be developed into a power series of $h$. The formula is developed for this correction by means of a probability function and the result discussed.

## 1

IN classical statistical mechanics the relative probability for the range $p_1$ to $p_1+dp_1$; $p_2$ to $p_2+dp_2$; $\cdots$; $p_n$ to $p_n+dp_n$ for the momenta and $x_1$ to $x_1+dx_1$; $x_2$ to $x_2+dx_2$; $\cdots$; $x_n$ to $x_n+dx_n$ for the coordinates is given for statistical equilibrium by the Gibbs-Boltzmann formula

$$P(x_1, \cdots, x_n; p_1, \cdots, p_n)dx_1 \cdots dx_n dp_1 \cdots dp_n = e^{-\beta\epsilon}dx_1 \cdots dx_n dp_1 \cdots dp_n \quad (1)$$

where $\epsilon$ is the sum of the kinetic and potential energy $V$

$$\epsilon = \frac{p_1^2}{2m_1} + \frac{p_2^2}{2m_2} + \cdots + \frac{p_n^2}{2m_n} + V(x_1 \cdots x_n) \quad (2)$$

and $\beta$ is the reciprocal temperature $T$ divided by the Boltzmann constant

$$\beta = 1/kT. \quad (3)$$

In quantum theory there does not exist any similar simple expression for the probability, because one cannot ask for the simultaneous probability for the coordinates and momenta. Moreover, it is not possible to derive a simple expression even for the relative probabilities of the coordinates alone—as is given in classical theory by $e^{-\beta V(x_1 \cdots x_n)}$. One sees this by considering that this expression would give at once the square of the wave function of the lowest state $|\psi_0(x_1 \cdots x_n)|^2$ when $\beta = \infty$ is inserted and on the other hand we know that it is not possible, in general, to derive a closed formula for the latter.

The thermodynamics of quantum mechanical systems is in principle, however, given by a formula of Neumann,[1] who has shown that the mean value of any physical quantity is, (apart from a normalizing constant depending only on temperature), the sum of the diagonal elements of the matrix

$$Qe^{-\beta H} \quad (4)$$

where $Q$ is the matrix (operator) of the quantity under consideration and $H$ is the Hamiltonian of the system. As the diagonal sum is an invariant under

[1] J. von Neumann, Gött. Nachr. p. 273, 1927.

transformations, one can choose any matrix or operator-representation for the $Q$ and $H$. In building the exponential of $H$ one must, of course, take into account the non-commutability of the different parts of $H$.

2

It does not seem to be easy to make explicit calculations with the form (4) of the mean value. One may resort therefore to the following method.

If a wave function $\psi(x_1 \cdots x_n)$ is given one may build the following expression[2]

$$P(x_1, \cdots, x_n; p_1, \cdots, p_n) = \left(\frac{1}{h\pi}\right)^n \int_{-\infty}^{\infty} \cdots \int dy_1 \cdots dy_n \psi(x_1 + y_1 \cdots x_n + y_n)^* \psi(x_1 - y_1 \cdots x_n - y_n) e^{2i(p_1y_1+\cdots+p_ny_n)/h} \quad (5)$$

and call it the probability-function of the simultaneous values of $x_1 \cdots x_n$ for the coordinates and $p_1 \cdots p_n$ for the momenta. In (5), as throughout this paper, $h$ is the Planck constant divided by $2\pi$ and the integration with respect to the $y$ has to be carried out from $-\infty$ to $\infty$. Expression (5) is real, but not everywhere positive. It has the property, that it gives, when integrated with respect to the $p$, the correct probabilities $|\psi(x_1 \cdots x_n)|^2$ for the different values of the coordinates and also it gives, when integrated with respect to the $x$, the correct quantum mechanical probabilities

$$\left| \int_{-\infty}^{\infty} \cdots \int \psi(x_1 \cdots x_n) e^{-i(p_1x_1+\cdots+p_nx_n)/h} dx_1 \cdots dx_n \right|^2$$

for the momenta $p_1, \cdots, p_n$. The first fact follows simply from the theorem about the Fourier integral and one gets the second by introducing $x_k+y_k = u_k; x_k-y_k=v_k$ into (5).

Hence it follows, furthermore, that one may get the correct expectation values of any function of the coordinates or the momenta for the state $\psi$ by the normal probability calculation with (5). As expectation values are additive this even holds for a sum of a function of the coordinates and a function of the momenta as, e.g., the energy $H$. In formulas, it is

$$\int_{-\infty}^{\infty} \cdots \int \int_{-\infty}^{\infty} \cdots \int dx_1 \cdots dx_n dp_1 \cdots dp_n [f(p_1 \cdots p_n) + g(x_1 \cdots x_n)] P(x_1 \cdots x_n; p_1 \cdots p_n)$$
$$= \int_{-\infty}^{\infty} \cdots \int \psi(x_1 \cdots x_n)^* \left[ f\left(\frac{h}{i}\frac{\partial}{\partial x_1}, \cdots, \frac{h}{i}\frac{\partial}{\partial x_n}\right) + g(x_1 \cdots x_n) \right] \psi(x_1 \cdots x_n) dx_1 \cdots dx_n \quad (6)$$

for any $\psi, f, g$, if $P$ is given by (5).

[2] This expression was found by L. Szilard and the present author some years ago for another purpose.

Of course $P(x_1, \cdots, x_n; p_1, \cdots, p_n)$ cannot be really interpreted as the simultaneous probability for coordinates and momenta, as is clear from the fact, that it may take negative values. But of course this must not hinder the use of it in calculations as an auxiliary function which obeys many relations we would expect from such a probability. It should be noted, furthermore, that (5) is not the only bilinear expression in $\psi$, which satisfies (6). There must be a great freedom in the expression (5), as it makes from a function $\psi$ of $n$ variables one with $2n$ variables. It may be shown, however, that there does not exist any expression $P(x_1 \cdots x_n;\ p_1 \cdots p_n)$ which is bilinear in $\psi$, satisfies (6) and is everywhere (for all values of $x_1, \cdots, x_n, p_1, \cdots, p_n$) positive, so (5) was chosen from all possible expressions, because it seems to be the simplest.

If $\psi(x_1, \cdots, x_n)$ changes according to the second Schrödinger equation

$$ih\frac{\partial\psi}{\partial t} = -\sum_{k=1}^{n}\frac{h^2}{2m_k}\frac{\partial^2\psi}{\partial x_k^2} + V(x_1, \cdots, x_n)\psi \tag{7}$$

the change of $P(x_1, \cdots, x_n; p_1, \cdots, p_n)$ is given by

$$\frac{\partial P}{\partial t} = -\sum_{k=1}^{n}\frac{p_k}{m_k}\frac{\partial P}{\partial x_k} + \sum\frac{\partial^{\lambda_1+\cdots+\lambda_n}V}{\partial x_1^{\lambda_1}\cdots\partial x_n^{\lambda_n}}\frac{(h/2i)^{\lambda_1+\cdots+\lambda_n-1}}{\lambda_1!\cdots\lambda_n!}\frac{\partial^{\lambda_1+\cdots+\lambda_n}P}{\partial p_1^{\lambda_1}\cdots\partial p_n^{\lambda_n}} \tag{8}$$

where the last summation has to be extended over all positive integer values of $\lambda_1, \cdots, \lambda_n$ for which the sum $\lambda_1+\lambda_2+\cdots+\lambda_n$ is odd. In fact we get for $\partial P/\partial t$ by (5) and (7)

$$\begin{aligned}\frac{\partial P}{\partial t} = \frac{1}{(h\pi)^n}\int\cdots\int dy_1\cdots dy_n e^{2i(p_1y_1+\cdots+p_ny_n)/h} \\ \cdot\Bigg\{\sum_k\frac{ih}{2m_k}\Bigg[-\frac{\partial^2\psi(x_1+y_1,\cdots,x_n+y_n)^*}{\partial x_k^2}\psi(x_1-y_1,\cdots,x_n-y_n) \\ +\psi(x_1+y_1,\cdots,x_n+y_n)^*\frac{\partial^2\psi(x_1-y_1,\cdots,x_n-y_n)}{\partial x_k^2} \\ +\frac{i}{h}[V(x_1+y_1,\cdots,x_n+y_n) \\ -V(x_1-y_1,\cdots,x_n-y_n)]\psi(x_1+y_1,\cdots,x_n+y_n)^*\psi(x_1-y_1,\cdots x_n-y_n)\Bigg\}.\end{aligned} \tag{9}$$

Here one can replace the differentiations with respect to $x_k$ by differentiations with respect to $y_k$ and perform in the first two terms one partial integration with respect to $y_k$. In the last term we can develop $V(x_1+y_1, \cdots, x_n+y_n)$ and $V(x_1-y_1, \cdots, x_n-y_n)$ in a Taylor series with respect to the $y$ and get

$$\begin{aligned}\frac{\partial P}{\partial t} = \frac{1}{(\pi h)^n}\int\cdots\int dy_1\cdots dy_n e^{2i(p_1y_1+\cdots+p_ny_n)/h} \\ \cdot\Bigg\{\sum_k\frac{p_k}{m_k}\Bigg[-\frac{\partial\psi(x_1+y_1,\cdots,x_n+y_n)^*}{\partial y_k}\psi(x_1-y_1,\cdots,x_n-y_n)\end{aligned}$$

$$+\psi(x_1+y_1,\cdots,x_n+y_n)^*\frac{\partial\psi(x_1-y_1,\cdots,x_n-y_n)}{\partial y_k}\Bigg] \tag{10}$$

$$+\frac{i}{h}\sum_\lambda \frac{\partial^{\lambda_1+\cdots+\lambda_n}V}{\partial x_1^{\lambda_1}\cdots\partial x_n^{\lambda_n}}\frac{y_1^{\lambda_1}\cdots y_n^{\lambda_n}}{\lambda_1!\cdots\lambda_n!}\psi(x_1+y_1,\cdots,x_n+y_n)^*$$

$$\cdot\psi(x_1-y_1,\cdots,x_n-y_n)\Bigg\},$$

which is identical with (8) if one replaces now the differentiations with respect to $y_k$ by differentiations with respect to $x_k$. Of course, (8) is legitimate only if it is possible to develop the potential energy $V$ in a Taylor series.

Eq. (8) shows the close analogy between the probability function of the classical mechanics and our $P$: indeed the equation of continuity

$$\frac{\partial P}{\partial t}=-\sum_k\frac{p_k}{m_k}\frac{\partial P}{\partial x_k}+\sum_k\frac{\partial V}{\partial x_k}\frac{\partial P}{\partial p_k}$$

differs from (8) only in terms of at least the second power of $h$ and at least the third derivative of $V$. Expression (8) is even identical with the classical when $V$ has no third and higher derivatives as, e.g., in a system of oscillators.

There is an alternative form for $\partial P/\partial t$, which however will not be used later on. It is

$$\frac{\partial}{\partial t}P(x_1,\cdots,x_n;p_1,\cdots,p_n)=-\sum_k\frac{p_k}{m_k}\frac{\partial}{\partial x_k}P(x_1,\cdots,x_n;p_1,\cdots,p_n)$$

$$+\int_{-\infty}^{\infty}\cdots\int dj_1\cdots dj_nP(x_1,\cdots,x_n;P_1+j_1,\cdots,P_n+j_n)J(x_1,\cdots,x_n;\tag{11}$$

$$j_1,\cdots,j_n)$$

where $J(x_1,\cdots,x_n;j_1,\cdots,j_n)$ can be interpreted as the probability of a jump in the momenta with the amounts $j_1,\cdots,j_n$ for the configuration $x_1,\cdots,x_n$. The probability of this jump is given by

$$J(x_1,\cdots,x_n;j_1,\cdots,j_n)$$

$$=\frac{i}{\pi^n h^{n+1}}\int_{-\infty}^{\infty}\cdots\int dy_1\cdots dy_n[V(x_1+y_1,\cdots,x_n+y_n)$$

$$-V(x_1-y_1,\cdots,x_n-y_n)]e^{-(2i/h)(y_1j_1+\cdots+y_nj_n)} \tag{11a}$$

that is, by the Fourier expansion coefficients of the potential $V(x_1,\cdots,x_n)$. This form clearly shows the quantum mechanical nature of our $P$: the momenta change discontinuously by amounts which would be half the momenta of light quanta if the potential were composed of light.[2a] To derive (11) one can insert both for $P$ and $J$ their respective values (5) and (11a) on the right hand side of (11). In the first term one can replace $p_ke^{2i(p_1y_1+\cdots+p_ny_n)/h}$ by

[2a] Cf. F. Bloch, Zeits. f. Physik **52**, 555 (1929).

$(h/2i)(\partial/\partial y_k)e^{2i(p_1y_1+\cdots+p_ny_n)/h}$ and then perform a partial integration with respect to $y_k$. Then one can replace the differentiation with respect to $y$ by differentiation with respect to $x$, upon which some terms cancel and the rest goes over to

$$\sum_k \frac{h}{2im}\int\cdots\int dy_1\cdots dy_n\left[\frac{\partial^2\psi(x_1+y_1,\cdots,x_n+y_n)^*}{\partial x_k^2}\psi(x_1-y_1,\cdots,x_n-y_n)\right.$$
$$\left.-\psi(x_1+y_1,\cdots,x_n+y_n)\frac{\partial^2\psi(x_1-y_1,\cdots,x_n-y_n)}{\partial x_k^2}\right]e^{2i(p_1y_1+\cdots+p_ny_n)/h} \quad (12)$$

which is just what we need for the left side of (11). By integrating the second term on the right side of (11)

$$\int\cdots\int dy_1\cdots dy_n\psi(x_1+y_1\cdots x_n+y_n)^*\psi(x_1-y_1\cdots x_n-y_n)$$
$$\cdot\int\cdots\int dj_1\cdots dj_n e^{(2i/h)[(p_1+j_1)y_1+\cdots+(p_n+j_n)y_n]}$$
$$\cdot\frac{i}{\pi^n h^{n+1}}\int\cdots\int dz_1\cdots dz_n[V(x_1+z_1\cdots x_n+z_n)$$
$$-V(x_1-z_1\cdots x_n-z_n)]e^{-2i(z_1j_1+\cdots+z_nj_n)/h}$$

with respect to $z$ and $j$ one gets because of the Fourier theorem[3]

$$(i/h)\int\cdots\int dy_1\cdots dy_n\psi(x_1+y_1\cdots x_n+y_n)^*\psi(x_1-y_1\cdots x_n-y_n)$$
$$e^{2i(p_1y_1+\cdots+p_ny_n)/h}\cdot[V(x_1+y_1\cdots x_n+y_n)-V(x_1-y_1\cdots x_n-y_n)] \quad (12a)$$

and this gives the second part of the left side of (11).

3

So far we have defined only a probability function for pure states, which gives us the correct expectation values for quantities $f(p_1\cdots p_n)+g(x_1\cdots x_n)$. If, however, we have a mixture,[4] e.g., the pure states $\psi_1$, $\psi_2$, $\psi_3$, $\cdots$ with the respective probabilities $w_1$, $w_2$, $w_3$, $\cdots$ (with $w_1+w_2+w_3+\cdots=1$) the normal probability calculation suggests a probability function

$$P(x_1,\cdots,x_n,p_1,\cdots,p_n)=\sum_\lambda w_\lambda P_\lambda(x_1,\cdots,x_n,\cdots,p_n) \quad (13)$$

where $P_\lambda$ is the probability function for $\psi_\lambda$. This probability function gives obviously the correct expectation values for all quantities, for which (5) gives correct expectation values and therefore will be adopted.

For a system in statistical equilibrium at the temperature $T=1/k\beta$ the relative probability of a stationary state $\psi_\lambda$ is $e^{-\beta E_\lambda}$ where $E_\lambda$ is the energy of $\psi_\lambda$. Therefore the probability function is a part from a constant

[3] Cf. e. g., R. Courant und D. Hilbert, Methoden der mathematischen Physik I. Berlin 1924. p. 62, Eq. (29).

[4] J. v. Neumann, Gött Nachr. 245, 1927. L. Landau, Zeits. f. Physik 45, 430 (1927).

$$P(x_1 \cdots x_n; p_1 \cdots p_n)$$

$$= \sum_\lambda \int \cdots \int dy_1 \cdots dy_n \psi_\lambda(x_1 + y_1 \cdots x_n + y_n)^*$$

$$e^{-\beta E_\lambda}\psi(x_1 - y_1 \cdots x_n - y_n)e^{2i(p_1 y_1 + \cdots + p_n y_n)/h}. \quad (14)$$

Now

$$\sum_\lambda \psi_\lambda (u_1 \cdots u_n)^* f(E_\lambda)\, \psi_\lambda (v_1 \cdots v_n)$$

is that matrix element of the operator $f(H)$, ($H$ is the energy operator) which is in the $u_1 \cdots u_n$ row and $v_1 \cdots v_n$ column. Therefore (14) may be written as

$$P(x_1 \cdots x_n; p_1 \cdots p_n)$$

$$= \int_{-\infty}^{\infty} \cdots \int dy_1 \cdots dy_n e^{i[(x_1+y_1)p_1 + \cdots + (x_n+y_n)p_n]/h} [e^{-\beta H}]_{x_1+y_1 \cdots x_n+y_n; x_1-y_1 \cdots x_n-y_n}$$

$$\cdot e^{-i[(x_1-y_1)p_1 + \cdots + (x_n-y_n)p_n]/h}. \quad (15)$$

so that we have under the integral sign the $x_1+y_1 \cdots x_n+y_n;\ x_1-y_1 \cdots x_n-y_n$ element of the matrix $e^{-\beta H}$ transformed by the diagonal matrix $e^{i(p_1 x_1 + \cdots + p_n x_n)/h}$. Instead of transforming $e^{-\beta H}$ we can transform $H$ first and then take the exponential with the transformed expression. By transforming $H$ we get the operator (the $p$ are numbers, not operators!)

$$H = e^{i(x_1 p_1 + \cdots + x_n p_n)/h}\left(-\sum_k \frac{h^2}{2m_k}\frac{\partial^2}{\partial x_k^2} + V(x_1 \cdots x_n)\right)e^{-i(x_1 p_1 + \cdots + x_n p_n)/h}$$

which is equal to

$$\tilde{H} = \epsilon + \sum_{k=1}^{n}\left(\frac{ihp_k}{m_k}\frac{\partial}{\partial x_k} - \frac{h^2}{2m_k}\frac{\partial^2}{\partial x_k^2}\right) \quad (16)$$

where

$$\epsilon = \sum_{k=1}^{n} \frac{p_k^2}{2m_k} + V(x_1, \cdots, x_n). \quad (17)$$

So we get for (15)

$$P(x_1, \cdots, x_n; p_1, \cdots, p_n)$$

$$= \int \cdots \int dy_1 \cdots dy_n [e^{-\beta \tilde{H}}]_{x_1+y_1 \cdots x_n+y_n; x_1-y_1 \cdots x_n-y_n}. \quad (18)$$

By calculating the mean value of a quantity $Q = f(p_1, \cdots, p_n) + g(x_1, \cdots, x_n)$ by (18) one has to obtain the same result as by using the original expression (4) of Neumann.

If we are dealing with a system, the behavior of which in statistical equilibrium is nearly correctly given by the classical theory, we can expand (18) into a power of $h$ and keep the first few terms only. The term with the zero power of $h$ is $\sum_r (-\beta)^r \epsilon^r / r!$ Now $\epsilon^r$ is the operator of multiplication with the $r$

power of (17). Its $x_1+y_1, \cdots, x_n+y_n; x_1-y_1, \cdots, x_n-y_n$ element is consequently

$$\epsilon(x_1+y_1, \cdots, x_n+y_n)^r \delta(x_1+y_1, x_1-y_1) \cdots \delta(x_n+y_n, x_n-y_n).$$

As $\delta$ (also $\delta', \delta'', \cdots$) only depends on the difference of its two arguments, one can write $\delta(-2y_1) \cdots \delta(-2y_n)$ for the last factors and perform the integration by introducing $-2y_1, \cdots, -2y_n$ as new variables. The terms with the zero power of $h$, arising from the first part of (16) only, give thus

$$(1/2^n) \sum_r (-\beta)^r \epsilon(x_1, \cdots, x_n)^r / r! = e^{-\beta\epsilon}/2^n \tag{19}$$

which is just the classical expression.

The higher approximations of the probability function can be calculated in a very similar way. The terms of $e^{-\beta\bar{H}}$, involving the first power of the second part of $\bar{H}$ only, are

$$\sum_{r=0}^{\infty} \frac{(-\beta)^r}{r!} \sum_{\rho=1}^{r} \epsilon^{\rho-1} \sum_k \left( \frac{ihp_k}{m_k} \frac{\partial}{\partial x_k} - \frac{h^2}{2m_k} \frac{\partial^2}{\partial x_k^2} \right) \epsilon^{r-\rho} \tag{20}$$

By replacing all operators by symbolic integral-kernels one gets for the $x_1+y_1, \cdots, x_n+y_n; x_1-y_1, \cdots, x_n-y_n$ element of the operator (20)

$$\sum_r \frac{(-\beta)^r}{r!} \sum_{\rho=1}^{r} \epsilon(x_1+y_1, \cdots, x_n+y_n)^{\rho-1}$$

$$\cdot \sum_k \left[ \frac{ihp_k}{m_k} \delta(-2y_1) \cdots \delta'(-2y_k) \cdots \delta(-2y_n) \right.$$

$$\left. - \frac{h^2}{2m_k} \delta(-2y_1) \cdots \delta''(-2y_k) \cdots \delta(-2y_n) \right] \epsilon(x_1-y_1, \cdots, x_n-y_n)^{r-\rho}.$$

Now

$$\sum_{\rho=1}^{r} \epsilon_+^{\rho-1} \epsilon_-^{r-\rho} = \sum_{\rho=1}^{r} \epsilon_-^{r-1} \left( \frac{\epsilon_+}{\epsilon_-} \right)^{\rho-1} = \frac{\epsilon_+^r - \epsilon_-^r}{\epsilon_+ - \epsilon_-}$$

so that the summation over $\rho$ and $r$ can be performed in (21). By introducing again new variables $w_1, \cdots, w_n$ for $-2y_1, \cdots, -2y_n$ and performing the integration one has

$$\frac{1}{2^n} \sum_k \left[ \frac{ihp_k}{m_k} \frac{\partial}{\partial w_k} - \frac{h^2}{2m_k} \frac{\partial^2}{\partial w_k^2} \right]$$

$$\frac{e^{-\beta\epsilon(x_1, \cdots, x_k - w_k/2, \cdots, x_n)} - e^{-\beta\epsilon(x_k, \cdots, x_k + w_k/2, \cdots, x_n)}}{\epsilon(x_1, \cdots, x_k - \frac{1}{2}w_k, \cdots, x_n) - \epsilon(x_1, \cdots, x_k + \frac{1}{2}w_k, \cdots, x_n)}$$

where $w_k=0$ must be inserted after differentiation. The first differential quotient vanishes at $w_k=0$, as the expression to be differentiated is an even function of $w_k$. The second part gives

$$\frac{e^{-\beta\epsilon}}{2^n}\sum_k \frac{h^2}{m_k}\left(-\frac{\beta^2}{8}\frac{\partial^2\epsilon}{\partial x_k^2}+\frac{\beta^3}{24}\left(\frac{\partial\epsilon}{\partial x_k}\right)^2\right). \tag{21}$$

In principle it is possible to calculate in the same way the terms involving the higher powers of the second part of $\tilde{H}$ also, the summation over $r$ and the quantities corresponding to our $\rho$ can always be performed in a very similar way. In practice, however, the computation becomes too laborious. Still it is clear, that if we develop our probability function for thermal equilibrium in a power series of $h$

$$P(x_1,\cdots,x_n;p_1,\cdots,p_n)=e^{-\beta\epsilon}+hf_1+h^2f_2+\cdots \tag{22}$$

(we can omit the factor $1/2^n$ before $e^{-\beta\epsilon}$, as we are dealing with relative probabilities anyway) all terms will be quite definite functions of the $p$, $V$ and the different partial derivatives of the latter. Furthermore it is easy to see, that $f_k$ will not involve higher derivatives of $V$ than the $k$-th nor higher powers of $p$ than the $k$-th. These facts enable us to calculate the higher terms of (22) in a somewhat simpler way, than the direct expansion of (18) would be.

The state (22) is certainly stationary, so that it would give identically $\partial P/\partial t=0$ when inserted into (8). By equating the coefficients of the different powers of $h$ in $\partial P/\partial t$ to zero one gets the following equations:

$$\sum_k -\frac{p_k}{m_k}\frac{\partial e^{-\beta\epsilon}}{\partial x_k}+\sum_k\frac{\partial V}{\partial x_k}\frac{\partial e^{-\beta\epsilon}}{\partial p_k}=0 \tag{23, 0}$$

$$\sum_k -\frac{p_k}{m_k}\frac{\partial f_1}{\partial x_k}+\sum_k\frac{\partial V}{\partial x_k}\frac{\partial f_1}{\partial p_k}=0 \tag{23, 1}$$

$$\begin{aligned}\sum_k -\frac{p_k}{m_k}\frac{\partial f_2}{\partial x_k}+\sum_k\frac{\partial V}{\partial x_k}\frac{\partial f_2}{\partial p_k}-\sum_k\frac{\partial^3V}{\partial x_k^3}\frac{h^2}{24}\frac{\partial^3e^{-\beta\epsilon}}{\partial p_k^3}&\\ -\sum_{k\neq l}\frac{\partial^3 v}{\partial x_k^2\partial x_l}\frac{h^2}{8}\frac{\partial^3 e^{-\beta\epsilon}}{\partial p_k^2\partial p_l}=0&\end{aligned} \tag{23, 2}$$

and so on. The first of these equations is an identity because of (17), as it must be; (23, a), (23, 2), $\cdots$ will determine $f_1, f_2, \cdots$ respectively. All Eqs. (23, a) are linear inhomogeneous partial differential equations for the unknown $f$. From one solution $f_a$ of (23, a) one obtains the general solution by adding to it the general solution $F$ of the homogeneous part of (23, a), which is always

$$\sum_k -\frac{p_k}{m_k}\frac{\partial F}{\partial x_k}+\sum_k\frac{\partial V}{\partial x_k}\frac{\partial F}{\partial p_k}=0.$$

This equation in turn is the classical equation for the stationary character of the probability distribution $F(x_1,\cdots,x_n;p_1,\cdots,p_n)$. It has in general only one solution which contains only a finite number of derivatives of $V$, namely

$$F(x_1,\cdots,x_n;p_1,\cdots,p_n)=F\left(\sum_k\frac{p_k^2}{2m_k}+v(x_1\cdots x_n)\right)=F(\epsilon).$$

In fact, if it had other integrals, like

$$F(p_1, \cdots, p_n;\, V, \partial V/\partial x_1, \partial V/\partial x_2, \cdots) \tag{24}$$

then all mechanical problems would have in addition to the energy-integral further integrals of the form (24) which, of course, is not true.

One solution of (23, 1) is $f_1 = 0$ and the most general we have to consider is therefore $f_1 = F(\epsilon)$. We have to take however $F(\epsilon) = 0$ as $f_1$ has to vanish for a constant $V$. So we get $f_1 = 0$, as we know it already from the direct expansion of (18). The same holds consequently for $f_3, f_5, \cdots$, as the inhomogeneous part of the equation for $f_3$ only contains $f_1$, the inhomogeneous part of the equation for $f_5$ only $f_1$ and $f_3$, and so on.

For $f_2$ one easily gets

$$f_2 = e^{-\beta\epsilon}\left[\sum_k\left(-\frac{\beta^2}{8m_k}\frac{\partial^2 V}{\partial x_k^2} + \frac{\beta^3}{24m_k}\left(\frac{\partial V}{\partial x_k}\right)^2\right) + \sum_{k,l}\frac{\beta^3 p_k p_l}{24m_k m_l}\frac{\partial^2 V}{\partial x_k \partial x_l}\right] \tag{25}$$

as a solution of (23, 2) and it is also clear, that this is the solution we need. The first two terms of $f_2$ we have already directly computed (21), the third arises from terms with the second power of the second part of $\tilde{H}$. Similarly $f_4$ is for one degree of freedom ($n = 1$)

$$\begin{aligned} 64m^2\beta^{-2}e^{\beta\epsilon}f_4 = {} & H_4(q)\,[\beta^2 V''^2/72 - \beta V''''/120] \\ & + H_2(q)\,[\beta^3 V'^2 V''/18 - 2\beta^2 V''^2/15 - \beta^2 V' V'''/15 + \beta V''''/15] \\ & + H_0(q)\,[\beta^4 V'^4/18 - 22\beta^3 V'^2 V''/45 + 2\beta^2 V''^2/5 + 8\beta^2 V' V'''/15 \\ & - 4\beta V''''/15] \end{aligned} \tag{26}$$

where $H_r$ is the $r$-th Hermitean polynomial and $q = \beta^{1/2} p/(2m)^{1/2}$.

It does not seem to be easy to get a simple closed expression for $f_k$, but it is quite possible to calculate all of them successively. A discussion of Eqs. (23) shows, that the $g$ in

$$P(x_1, \cdots, x_n;\, p_1, \cdots, p_n) = e^{-\beta\epsilon}(1 + h^2 g_2 + h^4 g_4 + \cdots) \tag{27}$$

are rational expressions in the derivatives of $V$ only (do not contain $V$ itself) and all terms of $g_k$ contain $k$ differentiations and as functions of the $p$ are polynomials of not higher than the $k$-th degree. The first term in (27) with the zero power of $h$ is the only one, which occurs in classical theory. There is no term with the first power, so that if one can develop a property in a power series with respect to $h$, the deviation from the classical theory goes at least with the second power of $h$ in thermal equilibrium. One familiar example for this is the inner energy of the oscillator, where the term with the first power of $h$ vanishes just in consequence of the zero point energy. The second term can be interpreted as meaning that a quick variation of the probability function with the coordinates is unlikely, as it would mean a quick variation, a short wave-length, in the wave functions. This however would have the consequence of a high kinetic energy. The quantum mechanical probability is therefore something like the integral of the classical expression $e^{-\beta\epsilon}$ over a finite range of coordinates of the magnitude $\sim h/\bar{p}$ where $\bar{p}$ is the mean momentum $\sim (kTm)^{1/2}$. The correction terms of (27) have, among other effects, the consequence that the probability for a particle being in a narrow hole is smaller than would be in classical statistics. From now on we will keep only the first two terms of (27).

4

From (25) one easily calculates the relative probabilities of the different configurations by integration with respect to the $p$:

$$\int \cdots \int dp_1 \cdots dp_n P(x_1 \cdots x_n; p_1 \cdots p_n)$$

$$= e^{-\beta V}\left[1 - \frac{h^2\beta^2}{12} \sum_k \frac{1}{m_k} \frac{\partial^2 V}{\partial x_k^2} + \frac{h^2\beta^3}{24} \sum_k \frac{1}{m_k}\left(\frac{\partial V}{\partial x_k}\right)^2\right]. \quad (28)$$

Hence the mean potential energy is

$$\bar{V} = \frac{\int V e^{-\beta V} dx}{\int e^{-\beta V} dx} + \frac{h^2\beta^2}{24} \frac{\int \sum_k \frac{1}{m_k} \frac{\partial^2 V}{\partial x_k^2} e^{-\beta V} dx \int V e^{-\beta V} dx}{\left(\int e^{-\beta V} dx\right)^2}$$
$$+ \frac{h^2\beta}{24} \frac{\int \sum_k \frac{1}{m_k} \frac{\partial^2 V}{\partial x_k^2}(1 - \beta V) e^{-\beta V} dx}{\int e^{-\beta V} dx} \quad (29)$$

where $dx$ is written for $dx_1 \cdots dx_n$ and the higher power terms of $h$ are omitted. Similarly the mean value of the kinetic energy is

$$\sum_k \frac{\overline{p_k^2}}{2m_k} = \frac{n}{2\beta} + \frac{h^2\beta}{24} \frac{\int \sum_k \frac{1}{m_k} \frac{\partial^2 V}{\partial x_k^2} e^{-\beta V} dx}{\int e^{-\beta V} dx}. \quad (30)$$

This formula also is correct only within the second power of $h$; in order to derive it one has to perform again some partial integrations with respect to the $x$. Eqs. (28), (29), (30) have a strict quantum mechanical meaning and it should be possible to derive them also from (4). One sees that the kinetic energy is in all cases larger than the classical expression $\frac{1}{2}nkT$.

5

One fact still needs to be mentioned. We assumed that the probability of a state with the energy $E$ is given by $e^{-\beta E}$. This is not true in general, since the Pauli principle forbids some states altogether. The corrections thus introduced by the Bose or Fermi statistics even give terms with the first power of $h$, so that it seems, that as long as one cannot take the Bose of Fermi statistics into account, Eq. (25) cannot be applied to an assembly of identical particles, as, e.g., a gas. There is reason to believe however, that because of the large radii of the atoms this is not true and the corrections due to Fermi and Bose statistics may be neglected for moderately low temperatures.

The second virial coefficient was first calculated in quantum mechanics by F. London on the basis of his theory of inneratomic forces.[5] He also pointed out that quantum effects should be taken into account at lower temperatures. Slater and Kirkwood[6] gave a more exact expression for the inneratomic potential of He and Kirkwood and Keyes[7] calculated on this basis the classical part of the second virial coefficient of He. H. Margenau[8] and Kirkwood[9] performed the calculations for the quantum-correction. The present author also tried to calculate it by the method just outlined. He got results, which differ from those of Margenau and Kirkwood in some cases by more than 100 percent.[10] It does not seem however to be easy to compare these results with experiment, as the classical part of the second virial coefficient is at low temperatures so sensitive to small variations of the parameters occurring in the expression of the interatomic potential, that it changes by more than 20 percent if the parameter in the exponential (2.43) is changed by $\frac{1}{2}$ percent and it does not seem to be possible to determine the latter within this accuracy.

[5] F. London, Zeits. f. Physik **63**, 245 (1930).

[6] J. C. Slater and J. G. Kirkwood, Phys. Rev. **37**, 682 (1931).

[7] J. G. Kirkwood and F. G. Keyes, Phys. Rev. **38**, 516 (1931).

[8] H. Margenau, Proc. Nat. Acad. **18**, 56, 230 (1932). Cf. also J. C. Slater, Phys. Rev. **38**, 237 (1931).

[9] J. G. Kirkwood, Phys. Zeits. **33**, 39 (1932).

[10] I am very much indebted to V. Rojansky for his kind assistance with these calculations. The reason for the disagreement between our results and those of Margenau and Kirkwood may be the fact that they did not apply any corrections for the continuous part of the spectrum.

In a paper which appeared recently in the Zeits. f. Physik (**74**, 295 (1932)) F. Bloch gets results which are somewhat similar to those of the present paper. (*Note added at proof.*)

# A.2 On Unitary Representations of the Inhomogeneous Lorentz Group

by E.P. Wigner

reprinted with permission from

Annals of Mathematics **40**, 149-204 (1939).

# ON UNITARY REPRESENTATIONS OF THE INHOMOGENEOUS LORENTZ GROUP*

E. Wigner

(Received December 22, 1937)

## 1. Origin and Characterization of the Problem

It is perhaps the most fundamental principle of Quantum Mechanics that the system of states forms a *linear manifold*,[1] in which a *unitary scalar* product is defined.[2] The states are generally represented by wave functions [3] in such a way that $\varphi$ and constant multiples of $\varphi$ represent the same physical state. It is possible, therefore, to normalize the wave function, i.e., to multiply it by a constant factor such that its scalar product with itself becomes 1. Then, only a constant factor of modulus 1, the so-called phase, will be left undetermined in the wave function. The linear character of the wave function is called the superposition principle. The square of the modulus of the unitary scalar product $(\psi, \varphi)$ of two normalized wave functions $\psi$ and $\varphi$ is called the transition probability from the state $\psi$ into $\varphi$, or conversely. This is supposed to give the probability that an experiment performed on a system in the state $\varphi$, to see whether or not the state is $\psi$, gives the result that it is $\psi$. If there are two or more different experiments to decide this (e.g., essentially the same experiment, performed at different times) they are all supposed to give the same result, i.e., the transition probability has an invariant physical sense.

---

*Parts of the present paper were presented at the Pittsburgh Symposium on Group Theory and Quantum Mechanics. Cf. Bull. Amer. Math. Soc., 41, p.306, 1935.

[1]The possibility of a future non linear character of the quantum mechanics must be admitted, of course. An indication in this direction is given by the theory of the positron, as developed by P.A.M. Dirac (Proc. Camb. Phil Soc. *30*, 150, 1934, cf. also W. Heisenberg, Zeits. f. Phys. *90*, 209, 1934; *92*, 623, 1934; W. Heisenberg and H. Euler, ibid. *98*, 714, 1936 and R. Serber, Phys. Rev. *48*, 49, 1935; *49*, 545, 1936) which does not use wave functions and is a non linear theory.

[2]Cf. P.A.M. Dirac, *The Principles of Quantum Mechanics*, Oxford 1935, Chapters I and II; J. v. Neumann, *Mathematische Grundlagen der Quantenmechanik*, Berlin 1932, pages 19-24.

[3]The wave functions represent throughout this paper states in the sense of the "Heisenberg picture," i.e. a single wave function represents the state for all past and future. On the other hand, the operator which refers to a measurement at a certain time $t$ contains this $t$ as a parameter. (Cf. e.g. Dirac, l.c. ref. 2, pages 115-123). One obtains the wave function $\varphi_s(t)$ of the Schrödinger picture from the wave function $\varphi_H$ of the Heisenberg picture by $\varphi_s(t) = \exp(-iHt/\hbar)\varphi_H$. The operator of the Heisenberg picture is $Q(t) = \exp(iHt/\hbar)Q\exp(-iHt/\hbar)$, where $Q$ is the operator in the Schrödinger picture which does not depend on time. Cf also E. Schroedinger, Sitz. d. Koen. Preuss. Akad. p. 418, 1930.

The wave functions are complex quantities and the undetermined factors in them are complex also. Recently attempts have been made toward a theory with real wave functions. Cf. E. Majorana, Nuovo Cim. *14*, 171, 1937 and P. A. M. Dirac, in print.

The wave functions form a description of the physical state, not an invariant however, since the same state will be described in different coordinate systems by different wave functions. In order to put this into evidence, we shall affix an index to our wave functions, denoting the Lorentz frame of reference for which the wave function is given. Thus $\varphi_l$ and $\varphi_{l'}$ represent the same state, but they are different functions. The first is the wave function of the state in the coordinate system $l$, the second in the coordinate system $l'$. If $\varphi_l = \psi_{l'}$ the state $\varphi$ behaves in the coordinate system $l$ exactly as $\psi$ behaves in the coordinate system $l'$. If $\varphi_l$ is given, all $\varphi_{l'}$ are determined up to a constant factor. Because of the invariance of the transition probability we have

$$|(\varphi_l, \psi_l)|^2 = |(\varphi_{l'}, \psi_{l'})|^2 \tag{1}$$

and it can be shown [4] that the aforementioned constants in the $\varphi_{l'}$ can be chosen in such a way that the $\varphi_{l'}$ are obtained from the $\varphi_l$ by a linear unitary operation, depending, of course, on $l$ and $l'$.

$$\varphi_{l'} = D(l', l)\varphi_l. \tag{2}$$

The unitary operators $D$ are determined by the physical content of the theory up to a constant factor again, which can depend on $l$ and $l'$. Apart from this constant however, the operations $D(l', l)$ and $D(l'_1, l_1)$ must be identical if $l'$ arises from $l$ by the same Lorentz transformation, by which $l'_1$ arises from $l_1$. If this were not true, there would be a real difference between the frames of reference $l$ and $l_1$. Thus the unitary operator $D(l', l) = D(L)$ is in every Lorentz invariant quantum mechanical theory (apart from the constant factor which has no physical significance) completely determined by the Lorentz transformation $L$ which carries $l$ into $l' = Ll$. One can write, instead of (2)

$$\varphi_{Ll} = D(L)\varphi_l. \tag{2a}$$

By going over from a first system of reference $l$ to a second $l' = L_1 l$ and then to a third $l'' = L_2 L_1 l$ or directly to the third $l'' = (L_2 L_1)l$, one must obtain—apart from the above mentioned constant—the same set of wave functions. Hence from

$$\varphi_{l''} = D(l'', l')D(l', l)\varphi_l$$
$$\varphi_{l''} = D(l'', l)\varphi_l$$

it follows

$$D(l'', l')D(l', l) = \omega D(l'', l) \tag{3}$$

or

$$D(L_2)D(L_1) = \omega D(L_2 L_1), \tag{3a}$$

[4] E. Wigner, *Gruppentheorie und ihre Anwendungen auf die Quantenmechanik det Atoms-pektren.* Braunschweig 1931, pages 251-254.

where $\omega$ is a number of modulus 1 and can depend on $L_2$ and $L_1$. Thus the $D(L)$ form, up to a factor, a representation of the inhomogeneous Lorentz group by linear, unitary operators.

We see thus[5] that there corresponds to every invariant quantum mechanical system of equations such a representation of the inhomogeneous Lorentz group. This representation, on the other hand, though not sufficient to replace the quantum mechanical equations entirely, can replace them to a large extent. If we knew, e.g., the operator $K$ corresponding to the measurement of a physical quantity at the time $t = 0$, we could follow up the change of this quantity throughout time. In order to obtain its value for the time $t = t_1$, we could transform the original wave function $\varphi_l$ by $D(l', l)$ to a coordinate system $l'$ the time scale of which begins a time $t_1$ later. The measurement of the quantity in question in this coordinate system for the time 0 is given—as in the original one—by the operator $K$. This measurement is identical, however, with the measurement of the quantity at time $t_1$ in the original system. One can say that the representation can replace the equation of motion, it cannot replace, however, connections holding between operators at one instant of time.

It may be mentioned, finally, that these developments apply not only in quantum mechanics, but also to all linear theories, e.g., the Maxwell equations in empty space. The only difference is that there is no arbitrary factor in the description and the $\omega$ can be omitted in (3a) and one is led to real representations instead of representations up to a factor. On the other hand, the unitary character of the representation is not a consequence of the basic assumptions.

The increase in generality, obtained by the present calculus, as compared with the usual tensor theory, consists in that no assumptions regarding the field nature of the underlying equations are necessary. Thus more general equations, as far as they exist (e.g., in which the coordinate is quantized, etc.) are also included in the present treatment. It must be realized, however, that some assumptions concerning the continuity of space have been made by assuming Lorentz frames of reference in the classical sense. We should like to mention, on the other hand, that the previous remarks concerning the time-parameter in the observables, have only an explanatory character and we do not make assumptions of the kind that measurements can be performed instantaneously.

We shall endeavor, in the ensuing sections, to determine all the continuous[6] unitary representations up to a factor of the inhomogeneous Lorentz group, i.e., all continuous systems of linear, unitary operators satisfying (3a).

## 2. Comparison With Previous Treatments and Some Immediate Simplifications

[5]E. Wigner, l.c. Chapter XX.

[6]The exact definition of the continuous character of a representation up to a factor will be given in Section 5A. The definition of the inhomogeneous Lorentz group is contained in Section 4A.

## A. Previous treatments

The representations of the Lorentz group have been investigated repeatedly. The first investigation is due to Majorana,[7] who in fact found all representations of the class to be dealt with in the present work excepting two sets of representations. Dirac[8] and Proca[8] gave more elegant derivations of Majorana's results and brought them into a form which can be handled more easily. Klein's work[9] does not endeavor to derive irreducible representations and seems to be in a less close connection with the present work.

The difference between the present paper and that of Majorana and Dirac lies—apart from the finding of new representations—mainly in its greater mathematical rigor. Majorana and Dirac freely use the notion of infinitesimal operators and a set of functions to all members of which every infinitesimal operator can be applied. This procedure cannot be mathematically justified at present, and no such assumption will be used in the present paper. Also the conditions of reducibility and irreducibility could be, in general, somewhat more complicated than assumed by Majorana and Dirac. Finally, the previous treatments assume from the outset that the space and time coordinates will be continuous variables of the wave function in the usual way. This will not be done, of course, in the present work.

## B. Some immediate simplifications

Two representations are physically equivalent if there is a one to one correspondence between the states of both which is 1. invariant under Lorentz transformations and 2. of such a character that the transition probabilities between corresponding states are the same.

It follows from the second condition[5] that there either exists a unitary operator $S$ by which the wave functions $\Phi^{(2)}$ of the second representation can be obtained from the corresponding wave functions $\Phi^{(1)}$ of the first representation

$$\Phi^{(2)} = S\Phi^{(1)} \tag{4}$$

or that this is true for the conjugate imaginary of $\Phi^{(2)}$. Although, in the latter case, the two representations are still equivalent physically, we shall, in keeping with the mathematical convention, not call them equivalent.

The first condition now means that if the states $\Phi^{(1)}$, $\Phi^{(2)} = S\Phi^{(1)}$ correspond to each other in one coordinate system, the states $D^{(1)}(L)\Phi^{(1)}$ and $D^{(2)}(L)\Phi^{(2)}$ correspond to each other also. We have then

$$D^{(2)}(L)\Phi^{(2)} = SD^{(1)}(L)\Phi^{(1)} = SD^{(1)}(L)S^{-1}\Phi^{(2)}. \tag{4a}$$

[7] E. Majorana, Nuovo Cim. *9*, 335, 1932.
[8] P. A. M. Dirac, Proc. Roy. Soc. A. *155*, 447, 1936; Al. Proca, J. de Phys. Rad. *7*, 347, 1936.
[9] Klein, Arkiv f. Matem. Astr. och Fysik, *25A*, No. 15, 1936. I am indebted to Mr. Darling for an interesting conversation on this paper.

As this shall hold for every $\Phi^{(2)}$, the existence of a unitary $S$ which transforms $D^{(1)}$ into $D^{(2)}$ is the condition for the equivalence of these two representations. Equivalent representations are not considered to be really different and it will be sufficient to find one sample from every infinite class of equivalent representations.

If there is a closed linear manifold of states which is invariant under all Lorentz transformations, i.e., which contains $D(L)\psi$ if it contains $\psi$, the linear manifold perpendicular to this one will be invariant also. In fact, if $\varphi$ belongs to the second manifold, $D(L)\varphi$ will be, on account of the unitary character of $D(L)$, perpendicular to $D(L)\psi'$ if $\psi'$ belongs to the first manifold. However, $D(L^{-1})\psi$ belongs to the first manifold if $\psi$ does and thus $D(L)\varphi$ will be orthogonal to $D(L)D(L^{-1})\psi = \omega\psi$ i.e. to all members of the first manifold and belong itself to the second manifold also. The original representation then "decomposes" into two representations, corresponding to the two linear manifolds. It is clear that, conversely, one can form a representation, by simply "adding" several other representations together, i.e. by considering as states linear combinations of the states of several representations and assume that the states which originate from different representations are perpendicular to each other.

Representations which are equivalent to sums of already known representations are not really new and, in order to master all representations, it will be sufficient to determine those, out of which all others can be obtained by "adding" a finite or infinite number of them together.

Two simple theorems shall be mentioned here which will be proved later (Sections 7A and 8C respectively). The first one refers to unitary representations of any closed group, the second to irreducible unitary representations of any (closed or open) group.

The representations of a closed group by unitary *operators* can be transformed into the sum of unitary representations with matrices of finite dimensions.

Given two non equivalent irreducible unitary representations of an arbitrary group. If the scalar product between the wave functions is invariant under the operations of the group, the wave functions belonging[28] to the first representation are orthogonal to all wave functions belonging to the second representation.

## C. Classification of unitary representations according to von Neumann and Murray[10]

Given the operators $D(L)$ of a unitary representation, or a representation up to a factor, one can consider the algebra of these operators, i.e. all linear combinations

$$a_1 D(L_1) + a_2 D(L_2) + a_3 D(L_3) + \cdots$$

of the $D(L)$ and all limits of such linear combinations which are bounded operators. According to the properties of this representation algebra, three classes of unitary representations can be distinguished.

[10] F. J. Murray and J. v. Neumann, Ann. of Math. *37*, 116, 1936; J. v. Neumann, to be published soon.

The first class of *irreducible* representations has a representation algebra which contains all bounded operators, i.e. if $\psi$ and $\varphi$ are two arbitrary states, there is an operator $A$ of the representation algebra for which $A\psi = \varphi$ and $A\psi' = 0$ if $\psi'$ is orthogonal to $\psi$. It is clear that the center of the algebra contains only the unit operator and multiply thereof. In fact, if $C$ is in the center one can decompose $C\psi = \alpha\psi + \psi'$ so that $\psi'$ shall be orthogonal to $\psi$. However, $\psi'$ must vanish since otherwise $C$ would not commute with the operator which leaves $\psi$ invariant and transforms every function orthogonal to it into 0. For similar reasons, $\alpha$ must be the same for all $\psi$. For irreducible representations there is no closed linear manifold of states, (excepting the manifold of all states) which is invariant under all Lorentz transformations. In fact, according to the above definition, a $\varphi'$ arbitrarily close to any $\varphi$ can be represented by a finite linear combination

$$a_1 D(L_1)\psi + a_2 D(L_2)\psi + \cdots + a_n D(L_n)\psi.$$

Hence, a closed linear invariant manifold contains every state if it contains one. This is, in fact, the more customary definition for irreducible representations and the one which will be used subsequently. It is well known that all finite dimensional representations are sums of irreducible representations. This is not true, [10] in general, in an infinite number of dimensions.

The second class of representations will be called factorial. For these, the center of the representation algebra still contains only multiples of the unit operator. Clearly, the irreducible representations are all factorial, but not conversely. For finite dimensions, the factorial representations may contain one irreducible representation several times. This is also possible in an infinite number of dimensions, but in addition to this, there are the "continuous" representations of Murray and von Neumann.[10] These are not irreducible as there are invariant linear manifolds of states. On the other hand, it is impossible to carry the decomposition so far as to obtain as parts only irreducible representations. In all the examples known so far, the representations into which these continuous representations can be decomposed, are equivalent to the original representation.

The third class contains all possible unitary representations. In a finite number of dimensions, these can be decomposed first into factorial representations, and these, in turn, in irreducible ones. Von Neumann[10] has shown that the first step still is possible in infinite dimensions. We can assume, therefore, from the outset that we are dealing with factorial representations.

In the theory of representations of finite dimensions, it is sufficient to determine only the irreducible ones, all others are equivalent to sums of these. Here, it will be necessary to determine all factorial representations. Having done that, we shall know from the above theorem of von Neumann, that all representations are equivalent to finite or infinite sums of factorial representations.

It will be one of the results of the detailed investigation that the inhomogeneous Lorentz group has no "continuous" representations, all representations can be decomposed into irreducible ones. Thus the work of Majorana and Dirac appears to be justified from this point of view a posteriori.

### D. Classification of unitary representations from the point of view of infinitesimal operators

The existence of an infinitesimal operator of a continuous one parametric (cyclic, abelian) unitary group has been shown by Stone.[11] He proved that the operators of such a group can be written as $\exp(iHt)$ where $H$ is a (bounded or unbounded) hermitian operator and $t$ is the group parameter. However, the Lorentz group has many one parametric subgroups, and the corresponding infinitesimal operators $H_1, H_2, \cdots$ are all unbounded. For every $H$ an everywhere dense set of functions $\varphi$ can be found such that $H_i\varphi$ can be defined. It is not clear, however, that an everywhere dense set can be found to all members of which every H can be applied. In fact, it is not clear that one such $\varphi$ can be found.

Indeed, it may be interesting to remark that for an irreducible representation the existence of one function $\varphi$ to which all infinitesimal operators can be applied, entails the existence of an everywhere dense set of such functions. This again has the consequence that one can operate with infinitesimal operators to a large extent in the usual way.

**Proof**: Let $Q(t)$ be a one parametric subgroup such that $Q(t)Q(t') = Q(t+t')$. If the infinitesimal operator of all subgroups can be applied to $\varphi$, the

$$\lim_{t=0} t^{-1}(Q(t)-1)\varphi \tag{5}$$

exists. It follows, then, that the infinitesimal operators can be applied to $R\varphi$ also where $R$ is an arbitrary operator of the representation: Since $R^{-1}Q(t)R$ is also a one parametric subgroup

$$\lim_{t=0} t^{-1}(R^{-1}Q(t)R-1)\varphi = \lim_{t=0} R^{-1}\cdot t^{-1}(Q(t)-1)R\varphi$$

also exists and hence also ($R$ is unitary)

$$\lim_{t=0} t^{-1}(Q(t)-1)R\varphi.$$

Every infinitesimal operator can be applied to $R\varphi$ if they all can be applied to $\varphi$, and the same holds for sums of the kind

$$a_1R_1\varphi + a_2R_2\varphi + \cdots + a_nR_n\varphi. \tag{6}$$

These form, however, an everywhere dense set of functions if the representation is irreducible.

[11] M. H. Stone, Proc. Nat. Acad. *16*, 173, 1930, Ann of Math. *33*, 643, 1932, also J. v. Neumann, ibid, *33*, 567, 1932.

If the representation is not irreducible, one can consider the set $N_0$ of such wave functions to which every infinitesimal operator can be applied. This set is clearly linear and, according to the previous paragraph, invariant under the operations of the group (i.e. contains every $R\varphi$ if it contains $\varphi$). The same holds for the closed set $N$ generated by $N_0$ and also of the set $P$ of functions which are perpendicular to all functions of $N$. In fact, if $\varphi_l$, is perpendicular to all $\varphi_n$ of $N$, it is perpendicular also to all $R^{-1}\varphi_n$ and, for the unitary character of $R$, the $R\varphi_p$ is perpendicular to all $\varphi_n$, i.e. is also contained in the set $P$.

We can decompose thus, by a unitary transformation, every unitary representation into a "normal" and a "pathological" part. For the former, there is an everywhere dense set of functions, to which all infinitesimal operators can be applied. There is no single wave functions to which all infinitesimal operators of a "pathological" representation could be applied.

According to Murray and von Neumann, if the original representation was factorial, all representations into which it can be decomposed will be factorial also. Thus every representation is equivalent to a sum of factorial representations, part of which is "normal," the other part "pathological."

It will turn out again that the inhomogeneous Lorentz group has no pathological representations. Thus this assumption of Majorana and Dirac also will be justified a posteriori. Every unitary representation of the inhomogeneous Lorentz group can be decomposed into normal irreducible representations. It should be stated, however, that the representations in which the unit operator corresponds to every translation have not been determined to date (cf. also section 3, end). Hence, the above statements are not proved for these representations, which are, however, more truly representations of the homogeneous Lorentz group, than of the inhomogeneous group.

While all these points may be of interest to the mathematician only, the new representation of the Lorentz group which will be described in section 7 may interest the physicist also. It describes a particle with a continuous spin.

*Acknowledgment.* The subject of this paper was suggested to me as early as 1928 by P. M. Dirac who realized even at that date the connection of representations with quantum mechanical equations. I am greatly indebted to him also for many fruitful conversations about this subject, especially during the years 1934/35, the outgrowth of which the present paper is.

I am indebted also to J. v. Neumann for his help and friendly advice.

## 3. Summary of Ensuing Sections

Section 4 will be devoted to the definition of the inhomogeneous Lorentz group and the theory of characteristic values and characteristic vectors of a homogeneous (ordinary) Lorentz transformation. The discussion will follow very closely the corresponding, well-known theory of the group of motions in ordinary space and the theory of characteristic values of orthogonal transformations.[12] It will contain only a straightforward generalization of the methods usually applied in those discussions.

In section 5, it will be proved that one can determine the physically meaningless constants in the $D(L)$ in such a way that instead of (3a) the more special equation

$$D(L_1)D(L_2) = \pm D(L_1L_2) \tag{7}$$

will be valid. This means that instead of a representation up to a factor, we can consider representations up to the sign. For the case that either $L_1$ or $L_2$ is a pure translation, Dirac[13] has given a proof of (7) using infinitesimal operators. A consideration very similar to his can be carried out, however, also using only finite transformations.

For representations with a finite number of dimensions (corresponding to an only finite number of linearly independent states), (7) could be proved also if both $L_1$ and $L_2$ are homogeneous Lorentz transformations, by a straightforward application of the method of Weyl and Schreier.[14] However, the Lorentz group has no finite dimensional representation (apart from the trivial one in which the unit operation corresponds to every L). Thus the method of Weyl and Schreier cannot be applied. Its first step is to normalize the indeterminate constants in every matrix $D(L)$ in such a way that the determinant of $D(L)$ becomes 1. No determinant can be defined for general unitary operators.

The method to be employed here will be to decompose every $L$ into a product of two involutions $L = MN$ with $M^2 = N^2 = 1$. Then $D(M)$ and $D(N)$ will be normalized so that their squares become unity and $D(L) = D(M)D(N)$ set. It will be possible, then, to prove (7) without going back to the topology of the group.

Sections 6, 7, and 8 will contain the determination of the representations. The pure translations form an invariant subgroup of the whole inhomogeneous Lorentz group and Frobenius' method[15] will be applied in Section 6 to build up the representations of the whole group out of representations of the subgroup, by means of a "little group." In Section 6, it will be shown on the basis of an as yet unpublished work[24] of J. v. Neumann that there is a characteristic (invariant) set of "momentum vectors" for every irreducible representation. The irreducible representations of the Lorentz group will be divided into four classes. The momentum vectors of the

*1st class* are time-like,
*2nd class* are null-vectors, but not all their components will be zero,
*3rd class* vanish (i.e., all their components will be zero),
*4th class* are space-like.

[12] Cf. e.g. E. Wigner, l.c.. Chapter III. O. Veblen and J.W. Young, *Projective Geometry*, Boston 1917. Vol 2, especially Chapter VII.

[13] P. A. M. Dirac, mimeographed notes of lectures delivered at Princeton University, 1934/35, page 5a.

[14] H. Weyl, Mathem. Zeits. *23*, 271; *24*, 328, 377, 789, 1925; O. Schreier, Abhandl. Mathem. Seminar Hamburg, *4*, 15, 1926; *5*, 233, 1927.

[15] G. Frobenius, Sitz. d. Kön. Preuss. Akad. p. 501, 1898, I. Schur, ibid, p. 164, 1906; F. Seitz, Ann. of Math. *37*, 17, 1936.

Only the first two cases will be considered in Section 7, although the last case may be the most interesting from the mathematical point of view. I hope to return to it in another paper. I did not succeed so far in giving a complete discussion of the 3rd class. (All these restrictions appear in the previous treatments also.)

In Section 7, we shall find again all known representations of the inhomogeneous Lorentz group (i.e., all known Lorentz invariant equations) and two new sets.

Sections 5, 6, 7 will deal with the "restricted Lorentz group" only, i.e. Lorentz transformations with determinant 1 which do not reverse the direction of the time axis. In section 8, the representations of the extended Lorentz group will be considered, the transformations of which are not subject to these conditions.

## 4. Description of the Inhomogeneous Lorentz Group

### A.

An inhomogeneous Lorentz transformation $L = (a, \Lambda)$ is the product of a translation by a real vector $a$

$$x_i' = x_i + a_i \qquad (i = 1, 2, 3, 4) \tag{8}$$

and a homogeneous Lorentz transformation $\Lambda$ with real coefficients

$$x_i' = \sum_{k=1}^{4} \Lambda_{ik} x_k. \tag{9}$$

The translation shall be performed after the homogeneous transformation. The coefficients of the homogeneous transformation satisfy three conditions: (1) They are real and $\Lambda$ leaves the indefinite quadratic form $-x_1^2 - x_2^2 - x_3^2 + x_4^2$ invariant:

$$\Lambda F \Lambda' = F \tag{10}$$

where the prime denotes the interchange of rows and columns and $F$ is the diagonal matrix with the diagonal elements $-1, -1, -1, +1$.—(2) The determinant $|\Lambda_{ik}| = 1$ and—(3) $\Lambda_{44} > 0$.

We shall denote the Lorentz-hermitian product of two vectors $x$ and $y$ by

$$\{x, y\} = -x_1^* y_1 - x_2^* y_2 - x_3^* y_3 - x_4^* y_4. \tag{11}$$

(The star denotes the conjugate imaginary.) If $\{x, x\} < 0$ the vector $x$ is called space-like, if $\{x, x\} = 0$, it is a null vector, if $\{x, x\} > 0$, it is called time-like. A real time-like vector lies in the positive light cone if $x_4 > 0$; it lies in the negative light cone if $x_4 < 0$. Two vectors $x$ and $y$ are called orthogonal if $\{x, y\} = 0$.

On account of its linear character a homogeneous Lorentz transformation is completely defined if $\Lambda v$ is given for four linearly independent vectors $v^{(1)}$, $v^{(2)}$, $v^{(3)}$, $v^{(4)}$.

From (11) and (10) it follows that $\{v, w\} = \{\Lambda v, \Lambda w\}$ for every pair of vectors $v, w$. This will be satisfied for every pair if it is satisfied for all pairs $v^{(i)}$, $v^{(k)}$ of four linearly independent vectors. The reality condition is satisfied if $(\Lambda v^{(i)})^* = \Lambda(v^{(i)*})$ holds for four such vectors.

The scalar product of two vectors $x$ and $y$ is positive if both lie in the positive light cone or both in the negative light cone. It is negative if one lies in the positive, the other in the negative light cone. Since both $x$ and $y$ are time-like $|x_4|^2 > |x_1|^2 + |x_2|^2 + |x_3|^2$; $|y_4|^2 > |y_1|^2 + |y_2|^2 + |y_3|^2$. Hence, by Schwarz's inequality $|x_4^* y_4| > |x_1^* y_1 + x_2^* y_2 + x_3^* y_3|$ and the sign of the scalar product of two real time-like vectors is determined by the product of their time components.

A time-like vector is transformed by a Lorentz transformation into a time-like vector. Furthermore, on account of the condition $\Lambda_{44} > 0$, the vector $v^{(0)}$ with the components $0, 0, 0, 1$ remains in the positive light cone, since the fourth component of $\Lambda v^{(0)}$ is $\Lambda_{44}$. If $v^{(1)}$ is another vector[16] in the positive light cone $\left\{v^{(1)}, v^{(0)}\right\} > 0$ and hence also $\left\{\Lambda v^{(1)}, \Lambda v^{(0)}\right\} > 0$ and $\Lambda v^{(1)}$ is in the positive light cone also. The third condition for a Lorentz transformation can be formulated also as the requirement that every vector in (or on) the positive light cone shall remain in (or, respectively, on) the positive light cone.

This formulation of the third condition shows that the third condition holds for the product of two homogeneous Lorentz transformations if it holds for both factors. The same is evident for the first two conditions.

From $\Lambda F \Lambda' = F$ one obtains by multiplying with $\Lambda^{-1}$ from the left and $\Lambda'^{-1} = (\Lambda^{-1})'$ from the right $F = \Lambda^{-1} F (\Lambda^{-1})'$ so that the reciprocal of a homogeneous Lorentz transformation is again such a transformation. The homogeneous Lorentz transformations form a group, therefore.

One easily calculates that the product of two inhomogeneous Lorentz transformations $(b, M)$ and $(c, N)$ is again an inhomogeneous Lorentz transformation $(a, \Lambda)$

$$(b, M)(c, N) = (a, \Lambda) \tag{12}$$

where

$$\Lambda_{ik} = \textstyle\sum_j M_{ij} N_{jk}; \qquad a_i = b_i + \sum_j M_{ij} c_j \tag{12a}$$

or, somewhat shorter

$$\Lambda = MN; \qquad a = b + Mc. \tag{12b}$$

## B. Theory of characteristic values and characteristic vectors

[16]Wherever a confusion between vectors and vector components appears to be possible, upper indices will be used for distinguishing different vectors and lower indices for denoting the components of a vector.

**of a homogeneous Lorentz transformation**

Linear homogeneous transformations are most simply described by their characteristic values and vectors. Before doing this for the homogeneous Lorentz group, however, we shall need two rules about orthogonal vectors.

[1] *If* $\{v,w\}=0$ *and* $\{v,v\}>0$, *then* $\{w,w\}>0$; *if* $\{v,w\}=0, \{v,v\}=0$, *then* $w$ *is either space-like, or parallel to* $v$ (*either* $\{w,w\}>0$, *or* $w=cv$).

**Proof**:

$$v_4^* w_4 = v_1^* w_1 + v_2^* w_2 + v_3^* w_3. \tag{13}$$

By Schwarz's inequality, then

$$|v_4|^2|w_4|^2 \le (|v_1|^2+|v_2|^2+|v_3|^2)(|w_1|^2+|w_2|^2+|w_3|^2). \tag{14}$$

For $|v_4|^2 > |v_1|^2+|v_2|^2+|v_3|^2$ it follows that $|w_4|^2 < |w_1|^2+|w_2|^2+|w_3|^2$. If $|v_4|^2 = |v_1|^2+|v_2|^2+|v_3|^2$ the second inequality still follows if the inequality sign holds in (14). The equality sign can hold only, however, if the first three components of the vectors $v$ and $w$ are proportional. Then, on account of (13) and both being null vectors, the fourth components are in the same ratio also.

[2] *If four vectors* $v^{(1)}, v^{(2)}, v^{(3)}, v^{(4)}$ *are mutually orthogonal and linearly independent, one of them is time-like, three are space-like.*

**Proof**: It follows from the previous paragraph that only one of four mutually orthogonal, linearly independent vectors can be time-like or a null vector. It remains to be shown therefore only that one of them is time-like. Since they are linearly independent, it is possible to express by them any time-like vector

$$v^{(t)} = \sum_{k=1}^{4} \alpha_k v^{(k)}.$$

The scalar product of the left side of this equation with itself is positive and therefore

$$\left\{\sum_k \alpha_k v^{(k)}, \sum_k \alpha_k v^{(k)}\right\} > 0$$

or

$$\sum_k |\alpha_k|^2 \left\{v^{(k)}, v^{(k)}\right\} > 0 \tag{15}$$

and one $\left\{v^{(k)}, v^{(k)}\right\}$ must be positive. Four mutually orthogonal vectors are not necessarily linearly independent, because a null vector is perpendicular to itself. The linear independence follows, however, if none of the four is a null vector.

We go over now to the characteristic values $\lambda$ of $\Lambda$. These make the determinant $|\Lambda-\lambda 1|$ of the matrix $\Lambda-\lambda 1$ vanish.

[3] *If* $\lambda$ *is a characteristic value,* $\lambda^*$, $\lambda^{-1}$ *and* $\lambda^{*-1}$ *are characteristic values also.*

**Proof**: For $\lambda^*$ this follows from the fact that $\Lambda$ is real. Furthermore, from $|\Lambda-\lambda 1|=0$ also $|\Lambda'-\lambda 1|=0$ follows, and this multiplied by the determinants of $\Lambda F$ and $F^{-1}$ gives

$$|\Lambda F|\cdot|\Lambda'-\lambda 1|\cdot|F|^{-1} = |\Lambda F \Lambda' F^{-1} - \lambda\Lambda| = |1-\lambda\Lambda| = 0,$$

so that $\lambda^{-1}$ is a characteristic value also.

[4] *The characteristic vectors $v_1$ and $v_2$ belonging to two characteristic values $\lambda_1$ and $\lambda_2$ are orthogonal if $\lambda_1^*\lambda_2 \neq 1$.*

**Proof:**

$$\{v_1, v_2\} = \{\Lambda v_1, \Lambda v_2\} = \{\lambda_1 v_1, \lambda_2 v_2\} = \lambda_1^* \lambda_2 \{v_1, v_2\}.$$

Thus if $\{v_1, v_2\} \neq 0$, $\lambda_1^*\lambda_2 = 1$.

[5] *If the modulus of a characteristic value $\lambda$ is $|\lambda| \neq 1$, the corresponding characteristic vector $v$ is a null vector and $\lambda$ itself real and positive.*

From $\{v, v\} = \{\Lambda v, \Lambda v\} = |\lambda|^2 \{v, v\}$ the $\{v, v\} = 0$ follows immediately for $|\lambda| \neq 1$. If $\lambda$ were complex, $\lambda^*$ would be a characteristic value also. The characteristic vectors of $\lambda$ and $\lambda^*$ would be two different null vectors and, because of [4], orthogonal to each other. This is impossible on account of [1]. Thus $\lambda$ is real and $v$ a real null vector. Then, on account of the third condition for a homogeneous Lorentz transformation, $\lambda$ must be positive.

[6] *The characteristic value $\lambda$ of a characteristic vector $v$ of length null is real and positive.*

If $\lambda$ were not real, $\lambda^*$ would be a characteristic value also. The corresponding characteristic vector $v^*$ would be different from $v$, a null vector also, and perpendicular to $v$ on account of [4]. This is impossible because of [1].

[7] *The characteristic vector $v$ of a complex characteristic value $\lambda$ (the modulus of which is 1 on account of [5]) is space-like: $\{v, v\} < 0$.*

**Proof:** $\lambda^*$ is a characteristic value also, the corresponding characteristic vector is $v^*$. Since $(\lambda^*)^*\lambda = \lambda^2 \neq 1$, $\{v^*, v\} = 0$. Since they are different, at least one is space-like. On account of $\{v, v\} = \{v^*, v^*\}$ both are space-like. If all four characteristic values were complex and the corresponding characteristic vectors linearly independent (which is true except if $\Lambda$ has elementary divisors) we should have four space-like, mutually orthogonal vectors. This is impossible, on account of [2]. Hence

[8] *There is not more than one pair of conjugate complex characteristic values, if $\Lambda$ has no elementary divisors. Similarly, under the same condition, there is not more than one pair $\lambda$, $\lambda^{-1}$ of characteristic values whose modulus is different from* 1. Otherwise their characteristic vectors would be orthogonal, which they cannot be, being null vectors.

For homogeneous Lorentz transformations which do not have elementary divisors, the following possibilities remain:

(a) There is a pair of complex characteristic values, their modulus is 1, on account of [5]

$$\lambda_1 = \lambda_2^* = \lambda_2^{-1}; \quad |\lambda_1| = |\lambda_2| = 1, \tag{16}$$

and also a pair of characteristic values $\lambda_3$, $\lambda_4$, the modulus of which is not 1. These must be real and positive:

$$\lambda_4 = \lambda_3^{-1}; \quad \lambda_3 = \lambda_3^* > 0. \tag{16a}$$

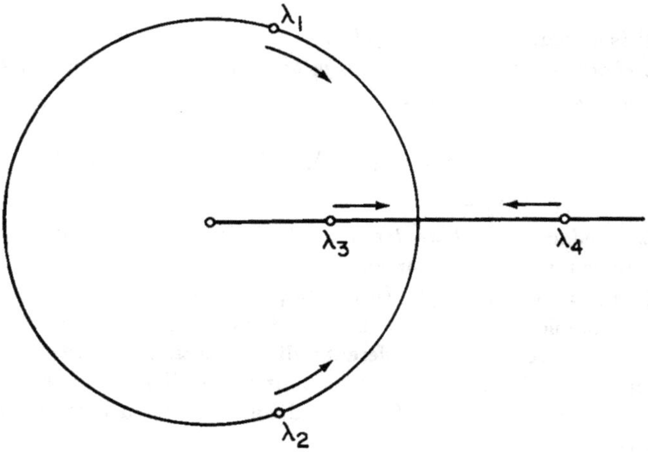


Figure 1: Position of the characteristic values for the general case a) in the complex plane. In case b), $\lambda_3$ and $\lambda_4$ coincide and are equal 1; in case c), $\lambda_1$ and $\lambda_2$ coincide and are either +1 or −1. In case d) both pairs $\lambda_3 = \lambda_4 = 1$ and $\lambda_1 = \lambda_2 = \pm 1$ coincide.

The characteristic vectors of the conjugate complex characteristic values are conjugate complex, perpendicular to each other and space-like so that they can be normalized to −1

$$v_1 = v_2^*; \quad \{v_1, v_2\} = \{v_1, v_1^*\} = 0$$
$$\{v_1, v_1\} = \{v_2, v_2\} = -1 \tag{17}$$

those of the real characteristic values are real null vectors, their scalar product can be normalized to 1

$$v_3 = v_3^* \quad v_4 = v_4^* \quad \{v_3, v_4\} = 1$$
$$\{v_3, v_3\} = \{v_4, v_4\} = 0. \tag{17a}$$

Finally, the former pair of characteristic vectors is perpendicular to the latter kind

$$\{v_1, v_3\} = \{v_1, v_4\} = \{v_2, v_3\} = \{v_2, v_4\} = 0. \tag{17b}$$

It will turn out that all the other cases in which $\Lambda$ has no elementary divisor are special cases of (a).

(b) There is a pair of complex characteristic values $\lambda_1$, $\lambda_2 = \lambda_1^{-1} = \lambda_1^*$, $\lambda_1 \neq \lambda_1^*$, $|\lambda_1| = |\lambda_2| = 1$. No pair with $|\lambda_3| \neq 1$, however. Then on account of [8], still $\lambda_3 = \lambda_3^*$ which gives with $|\lambda_3| = 1$, $\lambda_3 = \pm 1$. Since the product $\lambda_1\lambda_2\lambda_3\lambda_4 = 1$, on account of the second condition for homogeneous Lorentz transformations, also $\lambda_4 = \lambda_3 = \pm 1$. The double characteristic value $\pm 1$ has two linearly independent characteristic vectors $v_3$ and $v_4$ which can be assumed to be perpendicular to each other, $\{v_3, v_4\} = 0$. According to [2], one of the four characteristic vectors must be time-like and since those of $\lambda_1$ and $\lambda_2$ are space-like, the time-like one must belong to $\pm 1$. This must be positive, therefore $\lambda_3 = \lambda_4 = 1$. Out of the time-like and space-like vectors $\{v_3, v_3\} = -1$ and $\{v_4, v_4\} = 1$, one can build two null vectors $v_4 + v_3$ and $v_4 - v_3$. Doing this, case (b) becomes the special case of (a) in which the real positive characteristic values become equal $\lambda_3 = \lambda_4^{-1} = 1$.

(c) All characteristic values are real; there is however one pair $\lambda_3 = \lambda_3^*$, $\lambda_4 = \lambda_3^{-1}$, the modulus of which is not unity. Then $\{v_3, v_3\} = \{v_4, v_4\} = 0$ and $\lambda_3 > 0$ and one can conclude for $\lambda_1$ and $\lambda_2$, as before for $\lambda_3$ and $\lambda_4$ that $\lambda_1 = \lambda_2 = \pm 1$. This again is a special case of (a); here the two characteristic values of modulus 1 become equal.

(d) All characteristic values are real and of modulus 1. If all of them are $+1$, we have the unit matrix which clearly can be considered as a special case of (a). The other case is $\lambda_1 = \lambda_2 = -1$, $\lambda_3 = \lambda_4 = +1$. The characteristic vectors of $\lambda_1$ and $\lambda_2$ must be space-like, on account of the third condition for a homogeneous Lorentz transformation; they can be assumed to be orthogonal and normalized to $-1$. This is then a special case of (b) and hence of (a) also. The cases (a), (b), (c), (d), are illustrated in Fig. 1.

The cases remain to be considered in which $\Lambda$ has an elementary divisor. We set therefore

$$\Lambda_e v_e = \lambda_e v_e; \quad \Lambda_e w_e = \lambda_e w_e + v_e. \tag{18}$$

It follows from [5] that either $|\lambda_e| = 1$, or $\{v_e, v_e\} = 0$. We have $\{v_e, w_e\} = \{\Lambda_e v_e, \Lambda_e w_e\} = |\lambda_e|^2 \{v_e, w_e\} + \{v_e, v_e\}$. From this equation

$$\{v_e, v_e\} = 0 \tag{19}$$

follows for $|\lambda_e| = 1$, so that (19) holds in any case. It follows then from [6] that $\lambda_e$ is real, positive and $v_e$, $w_e$ can be assumed to be real also. The last equation now becomes $\{v_e, w_e\} = \lambda_e^2 \{v_e, w_e\}$ so that either $\lambda_e = 1$ or $\{v_e, w_e\} = 0$. Finally, we have

$$\{w_e, w_e\} = \{\Lambda_e w_e, \Lambda_e w_e\} = \lambda_e^2 \{w_e, w_e\} + 2\lambda_e \{w_e, v_e\} + \{v_e, v_e\}.$$

This equation now shows that

$$\{w_e, v_e\} = 0 \tag{19a}$$

even if $\lambda_e = 1$. From (19), (19a) it follows that $w_e$ is space-like and can be normalized to

$$\{w_e, w_e\} = -1. \tag{19b}$$

Inserting (19a) into the preceding equation we finally obtain

$$\lambda_e = 1. \tag{19c}$$

[9] *If $\Lambda_e$ has an elementary divisor, all its characteristic roots are* 1.

From (19c) we see that the root of the elementary divisor is 1 and this is at least a double root. If $\Lambda$ had a pair of characteristic values $\lambda_1 \neq 1$, $\lambda_2 = \lambda_1^{-1}$, the corresponding characteristic vectors $v_1$ and $v_2$ would be orthogonal to $v_e$ and therefore space-like. On account of [5], then $|\lambda_1| = |\lambda_2| = 1$ and $\{v_1, v_2\} = 0$. Furthermore, from $\{w_e, v_1\} = \{\Lambda_e w_e, \Lambda_e v_1\} = \lambda_1 \{w_e, v_1\} + \lambda_1 \{v_e, v_1\}$ and from $\{v_e, v_1\} = 0$ also $\{w_e, v_1\} = 0$ follows. Thus all the four vectors $v_1$, $v_2$, $v_e$, $w_e$ would be mutually orthogonal. This is excluded by [2] and (19).

Two cases are conceivable now. Either the fourfold characteristic root has only one characteristic vector, or there is in addition to $v_e$ (at least) another characteristic vector $v_1$. In the former case four linearly independent vectors $v_e$, $w_e$, $z_e$, $x_e$ could be found such that

$$\Lambda_e v_e = v_e \qquad \Lambda_e w_e = w_e + v_e$$
$$\Lambda_e z_e = z_e + w_e \qquad \Lambda_e x_e = x_e + z_e$$

However $\{v_e, x_e\} = \{\Lambda_e v_e, \Lambda_e x_e\} = \{v_e, x_e\} + \{v_e, z_e\}$ from which $\{v_e, z_e\} = 0$ follows. On the other hand

$$\{w_e, z_e\} = \{\Lambda_e w_e, \Lambda_e z_e\} = \{w_e, z_e\} + \{w_e, w_e\} + \{v_e, z_e\} + \{v_e, w_e\}.$$

This gives with (19a) and (19b) $\{v_e, z_e\} = 1$ so that this case must be excluded.

(e) There is thus a vector $v_1$ so that in addition to (18)

$$\Lambda_e v_1 = v_1 \tag{18a}$$

holds. From $\{w_e, v_1\} = \{\Lambda_e w_e, \Lambda_e v_1\} = \{w_e, v_1\} + \{v_e, v_1\}$ follows

$$\{v_e, v_1\} = 0. \tag{19d}$$

The equations (18), (18a) will remain unchanged if we add to $w_e$ and $v_1$ a multiple of $v_e$. We can achieve in this way that the fourth components of both $w_e$ and $v_1$ vanish. Furthermore, $v_1$ can be normalized to $-1$ and added to $w_e$ also with an arbitrary coefficient, to make it orthogonal to $v_1$. Hence, we can assume that

$$v_{14} = w_{e4} = 0; \quad \{v_1, v_1\} = -1; \quad \{w_e, v_1\} = 0. \tag{19e}$$

We can finally define the null vector $z_e$ to be orthogonal to $w_e$ and $v_1$ and have a scalar product 1 with $v_e$

$$\{z_e, z_e\} = \{z_e, w_e\} = \{z_e, v_1\} = 0; \quad \{z_e, v_e\} = 1. \tag{19f}$$

Then the null vectors $v_e$ and $z_e$ represent the momenta of two light beams in opposite directions. If we set $\Lambda_e z_e = av_e + bw_e + cz_e + dv_1$ the conditions $\{z_e, v\} = \{\Lambda_e z_e, \Lambda_e v\}$ give, if we set for $v$ the vectors $v_e$, $w_e$, $z_e$, $v_1$, the conditions $c = 1$; $b = c$; $2ac - b^2 - d^2 = 0$; $d = 0$. Hence

$$\begin{aligned} \Lambda_e v_e &= v_e \quad & \Lambda_e w_e &= w_e + v_e \\ \Lambda_e v_1 &= v_1 \quad & \Lambda_e z_e &= z_e + w_e + 1/2 v_e. \end{aligned} \tag{20}$$

A Lorentz transformation with an elementary divisor can be best characterized by the null vector $v_e$ which is invariant under it and the space part of which forms with the two other vectors $w_e$ and $v_1$ three mutually orthogonal vectors in ordinary space. The two vectors $w_e$ and $v_1$ are normalized, $v_1$ is invariant under $\Lambda_e$ while the vector $v_e$ is added to $w_e$ upon application of $\Lambda_e$. The result of the application of $\Lambda_e$ to a vector which is linearly independent of $v_e$, $w_e$ and $v_1$ is, as we saw, already determined by the expressions for $\Lambda_e v_e$, $\Lambda_e w_e$ and $\Lambda_e v_1$.

The $\Lambda_e(\gamma)$ which have the invariant null vector $v_e$ and also $w_e$ (and hence also $v_1$) in common and differ only by adding to $w_e$ different multiples $\gamma v_e$ of $v_e$, form a cyclic group with $\gamma = 0$, the unit transformation as unity:

$$\Lambda_e(\gamma)\Lambda_e(\gamma') = \Lambda_e(\gamma + \gamma').$$

The Lorentz transformation $M(\alpha)$ which leaves $v_1$ and $w_e$ invariant but replaces $v_e$ by $\alpha v_e$ (and $z_e$ by $\alpha^{-1} z_e$) has the property of transforming $\Lambda_e(\gamma)$ into

$$M(\alpha)\Lambda_e(\gamma)M(\alpha)^{-1} = \Lambda_e(\alpha\gamma). \tag{+}$$

An example of $\Lambda_e(\gamma)$ and $M(\alpha)$ is

$$\Lambda_e(\gamma) = \begin{pmatrix} 1 & 0 & 0 & 0 \\ 0 & 1 & \gamma & \gamma \\ 0 & -\gamma & 1 - \frac{1}{2}\gamma^2 & -\frac{1}{2}\gamma^2 \\ 0 & \gamma & \frac{1}{2}\gamma^2 & 1 + \frac{1}{2}\gamma^2 \end{pmatrix};$$

$$M(\alpha) = \begin{pmatrix} 1 & 0 & 0 & 0 \\ 0 & 1 & 0 & 0 \\ 0 & 0 & \frac{1}{2}(\alpha + \alpha^{-1}) & \frac{1}{2}(\alpha - \alpha^{-1}) \\ 0 & 0 & \frac{1}{2}(\alpha - \alpha^{-1}) & \frac{1}{2}(\alpha + \alpha^{-1}) \end{pmatrix}.$$

These Lorentz transformations play an important role in the representations with space like momentum vectors.

A behavior like (+) is impossible for finite unitary matrices because the characteristic values of $M(\alpha)^{-1}\Lambda_e(\gamma)M(\alpha)$ and $\Lambda_e(\gamma)$ are the same—those of $\Lambda_e(\gamma\alpha) = \Lambda_e(\gamma)^\alpha$ the $\alpha^{\text{th}}$ powers of those of $\Lambda_e(\gamma)$. This shows very simply that the Lorentz group has no true unitary representation in a finite number of dimensions.

## C. Decomposition of a homogeneous Lorentz transformation into rotations and an acceleration in a given direction

The homogeneous Lorentz transformation is, from the point of view of the physicist, a transformation to a uniformly moving coordinate system, the origin of which coincided at $t = 0$ with the origin of the first coordinate system. One can, therefore, first perform a rotation which brings the direction of motion of the second system into a given direction—say the direction of the third axis—and impart it a velocity in this direction, which will bring it to rest. After this, the two coordinate systems can differ only in a rotation. This means that every homogeneous Lorentz transformation can be decomposed in the following way[17]

$$\Lambda = RZS \tag{21}$$

where $R$ and $S$ are pure rotations, (i.e. $R_{i4} = R_{4i} = S_{i4} = S_{4i} = 0$ for $i \neq 4$ and $R_{44} = S_{44} = 1$, also $R' = R^{-1}$, $S' = S^{-1}$) and $Z$ is an acceleration in the direction of the third axis, i.e.

$$Z = \begin{pmatrix} 1 & 0 & 0 & 0 \\ 0 & 1 & 0 & 0 \\ 0 & 0 & a & b \\ 0 & 0 & b & a \end{pmatrix}$$

with $a^2 - b^2 = 1$, $a > b > 0$. The decomposition (21) is clearly not unique. It will be shown, however, the $Z$ is uniquely determined, i.e. the same in every decomposition of the form (21).

In order to prove this mathematically, we chose $R$ so that in $R^{-1}\Lambda = I$ the first two components in the fourth column $I_{14} = I_{24} = 0$ become zero: $R^{-1}$ shall bring the vector with the components $\Lambda_{14}$, $\Lambda_{24}$, $\Lambda_{34}$ into the third axis. Then we take $I_{34} = (\Lambda_{14}^2 + \Lambda_{24}^2 + \Lambda_{34}^2)^{\frac{1}{2}}$ and $I_{44} = \Lambda_{44}$ for $b$ and $a$ to form $Z$; they satisfy the equation $I_{44}^2 - I_{34}^2 = 1$. Hence, the first three components of the fourth column of $J = Z^{-1}I = Z^{-1}R^{-1}\Lambda$ will become zero and $J_{44} = 1$, because of $J_{44}^2 - J_{14}^2 - J_{24}^2 - J_{34}^2 = 1$. Furthermore, the first three components of the fourth row of $J$ will vanish also, on account of $J_{44}^2 - J_{41}^2 - J_{42}^2 - J_{43}^2 = 1$, i.e. $J = S = Z^{-1}R^{-1}\Lambda$ is a pure rotation. This proves the possibility of the decomposition (21).

The trace of $\Lambda\Lambda' = RZ^2R^{-1}$ is equal to the trace of $Z^2$, i.e. equal to $2a^2+2b^2+2 = 4a^2 = 4b^2 + 4$ which shows that the $a$ and $b$ of $Z$ are uniquely determined. In particular $a = 1$, $b = 0$ and $Z$ the unit matrix if $\Lambda\Lambda' = 1$, i.e. $\Lambda$ a pure rotation.

It is easy to show now that the group space of the homogeneous Lorentz transformations is only doubly connected. If a continuous series $\Lambda(t)$ of homogeneous Lorentz transformations is given, which is unity both for $t = 0$ and $t = 1$, we can decompose it according to (21)

$$\Lambda(t) = R(t)Z(t)S(t). \tag{21a}$$

[17]Cf. e.g. L. Silberstein, *The Theory of Relativity*, London 1924, p. 142.

It is also clear from the foregoing, that $R(t)$ can be assumed to be continuous in $t$, except for values of $t$, for which $\Lambda_{14} = \Lambda_{24} = \Lambda_{34} = 0$, i.e. for which $\Lambda$ is a pure rotation. Similarly, $Z(t)$ will be continuous in $t$ and this will hold even where $\Lambda(t)$ is a pure rotation. Finally, $S = Z^{-1}R^{-1}\Lambda$ will be continuous also, except where $\Lambda(t)$ is a pure rotation.

Let us consider now the series of Lorentz transformations

$$\Lambda_s(t) = R(t)Z(t)^s S(t) \tag{21b}$$

where the $b$ of $Z(t)^s$ is $s$ times the $b$ of $Z(t)$. By decreasing $s$ from 1 to 0 we continuously deform the set $\Lambda_1(t) = \Lambda(t)$ of Lorentz transformations into a set of rotations $\Lambda_0(t) = R(t)S(t)$. Both the beginning $\Lambda_0(0) = 1$ and the end $\Lambda_s(1) = 1$ of the set remain the unit matrix and the sets $\Lambda_s(t)$ remain continuous in $t$ for all values of $s$. This last fact is evident for such $t$ for which $\Lambda(t)$ is not a rotation: for such $t$ all factors of (21b) are continuous. But it is true also for $t_0$ for which $\Lambda(t_0)$ is a rotation, and for which, hence $Z(t_0) = 1$ and $\Lambda_s(t_0) = \Lambda_1(t_0) = \Lambda(t_0)$. As $Z(t)$ is everywhere continuous, there will be a neighborhood of $t_0$ in which $Z(t)$ and hence also $Z(t)^s$ is arbitrarily close to the unit matrix. In this neighborhood $\Lambda_s(t) = \Lambda(t)\cdot S(t)^{-1}Z(t)^{-1}Z(t)^s S(t)$ is arbitrarily close to $\Lambda(t)$; and, if the neighborhood is small enough, this is arbitrarily close to $\Lambda(t_0) = \Lambda_s(t_0)$.

Thus (21b) replaces the continuous set $\Lambda(t)$ of Lorentz transformations by a continuous set of rotations. Since these form an only doubly connected manifold, the manifold of Lorentz transformations can not be more than doubly connected. The existence of a two valued representation[18] shows that it is actually doubly and not simply connected.

We can form a new group[14] from the Lorentz group, the elements of which are the elements of the Lorentz group, together with a way $\Lambda(t)$, connecting $\Lambda(1) = \Lambda$ with the unity $\Lambda(0) = E$. However, two ways which can be continuously deformed into each other are not considered different. The product of the element "$\Lambda$ with the way $\Lambda(t)$" with the element "$I$ with the way $I(t)$" is the element $\Lambda I$ with the way which goes from $E$ along $\Lambda(t)$ to $\Lambda$ and hence along $\Lambda I(t)$ to $\Lambda I$. Clearly, the Lorentz group is isomorphic with this group and two elements (corresponding to the two essentially different ways to $\Lambda$) of this group correspond to one element of the Lorentz group. It is well known,[18] that this group is holomorphic with the group of unimodular complex two dimensional transformations.

Every continuous representation of the Lorentz group "up to the sign" is a singlevalued, continuous representation of this group. The transformation which corresponds to "$\Lambda$ with the way $\Lambda(t)$" is that $d(\Lambda)$ which is obtained by going over

[18] Cf. H. Weyl, *Gruppentheorie und Quantenmechanik*, 1st. ed. Leipzig 1928, pages 110-114, 2nd ed. Leipzig 1931, pages 130-133. It may be interesting to remark that essentially the same isomorphism has been recognized already by L. Silberstein, l.c. pages 148-157.

from $d(E) = d(\Lambda(0)) = 1$ continuously along $d(\Lambda(t))$ to $d(\Lambda(1)) = d(\Lambda)$.

### D. The homogeneous Lorentz group is simple

It will be shown, first, that an invariant subgroup of the homogeneous Lorentz group contains a rotation (i.e. a transformation which leaves $x_4$ invariant).—We can write an arbitrary element of the invariant subgroup in the form $RZS$ of (21). From its presence in the invariant subgroup follows that of $S \cdot RZS \cdot S^{-1} = SRZ = TZ$. If $X_\pi$ is the rotation by $\pi$ about the first axis, $X_\pi Z X_\pi = Z^{-1}$ and $X_\pi TZX_\pi^{-1} = X_\pi TX_\pi X_\pi ZX_\pi = X_\pi TX_\pi Z^{-1}$ is contained in the invariant subgroup also and thus the transform of this with $Z$, i.e. $Z^{-1}X_\pi TX_\pi$ also. The product of this with $TZ$ is $TX_\pi TX_\pi$ which leaves $x_4$ invariant. If $TX_\pi TX_\pi = 1$ we can take $TY_\pi TY_\pi$. If this is the unity also, $TX_\pi TX_\pi = TY_\pi TY_\pi$ and $T$ commutes with $X_\pi Y_\pi$, i.e. is a rotation about the third axis. In this case the space like (complex) characteristic vectors of $TZ$ lie in the plane of the first two coordinate axes. Transforming $TZ$ by an acceleration in the direction of the first coordinate axis we obtain a new element of the invariant subgroup for which the space like characteristic vector will have a not vanishing fourth component. Taking this for $RZS$ we can transform it with $S$ again to obtain a new $SRZ = TZ$. However, since $S$ leaves $x_4$ invariant, the fourth component of the space like characteristic vectors of this $TZ$ will not vanish and we can obtain from it by the procedure just described a rotation which must be contained in the invariant subgroup.

It remains to be shown that an invariant subgroup which contains a rotation, contains the whole homogeneous Lorentz group. Since the three-dimensional rotation group is simple, all rotations must be contained in the invariant subgroup. Thus the rotation by $\pi$ around the first axis $X_\pi$ and also its transform with $Z$ and also

$$ZX_\pi Z^{-1} \cdot X_\pi = Z \cdot X_\pi Z^{-1} X_\pi = Z^2$$

is contained in the invariant subgroup. However, the general acceleration in the direction of the third axis can be written in this form. As all rotations are contained in the invariant subgroup also, (21) shows that this holds for all elements of the homogeneous Lorentz group.

It follows from this that the homogeneous Lorentz group has apart from the representation with unit matrices only true representations. It follows then from the remark at the end of part B, that these have all infinite dimensions. This holds even for the two-valued representations to which we shall be led in Section 5 equ. (52D), as the group elements to which the positive or negative unit matrix corresponds must form an invariant subgroup also, and because the argument at the end of part B holds for two-valued representations also. One easily sees furthermore from the equations (52B), (52C) that it holds for the inhomogeneous Lorentz group equally well.

## 5. Reduction of Representations Up to a Factor to Two-Valued Representations

The reduction will be effected by giving each unitary transformation, which is defined by the physical content of the theory and the consideration of reference only up to a factor of modulus unity, a "phase," which will leave only the sign of the representation operators undetermined. The unitary operator corresponding to the translation $a$ will be denoted by $T(a)$, that to the homogeneous Lorentz transformation $\Lambda$ by $d(\Lambda)$. To the general inhomogeneous Lorentz transformation then $D(a,\Lambda) = T(a)d(\Lambda)$ will correspond. Instead of the relations (12), we shall use the following ones.

$$T(a)T(b) = \omega(a,b)T(a+b) \tag{22B}$$
$$d(\Lambda)T(a) = \omega(\Lambda,a)T(\Lambda a)d(\Lambda) \tag{22C}$$
$$d(\Lambda)d(I) = \omega(\Lambda,I)d(\Lambda I). \tag{22D}$$

The $\omega$ are numbers of modulus 1. They enter because the multiplication rules (12) hold for the representative only up to a factor. Otherwise, the relations (22) are consequences of (12) and can in their return replace (12). We shall replace the $T(a)$, $d(\Lambda)$ by $\Omega(a)T(a)$ and $\Omega(\Lambda)d(\Lambda)$ respectively, for which equations similar to (22) hold, however with

$$\omega(a,b) = 1; \quad \omega(\Lambda,a) = 1; \quad \omega(\Lambda,I) = \pm 1. \tag{22'}$$

## A.

It is necessary, first, to show that the undetermined factors in the representation $D(L)$ can be assumed in such a way that the $\omega(a,b)$, $\omega(\Lambda,a)$, $\omega(\Lambda,I)$ become—apart from regions of lower dimensionality—continuous functions of their arguments. This is a consequence of the continuous character of the representation and shall be discussed first.

(a) From the point of view of the physicist, the natural definition of the continuity of a representation up to a factor is as follows. The neighborhood $\delta$ of a Lorentz transformation $L_0 = (b, I)$ shall contain all the transformations $L = (a,\Lambda)$ for which $|a_k - b_k| < \delta$ and $|\Lambda_{ik} - I_{ik}| < \delta$. The representation up to a factor $D(L)$ is continuous if there is to every positive number $\epsilon$, every normalized wave function $\varphi$ and every Lorentz transformation $L_0$ such a neighborhood $\delta$ of $L_0$ that for every $L$ of this neighborhood one can find an $\Omega$ of modulus 1 (the $\Omega$ depending on $L$ and $\varphi$) such that $(u_\varphi, u_\varphi) < \epsilon$ where

$$u_\varphi = (D(L_0) - \Omega D(L)\varphi). \tag{23}$$

Let us now take a point $L_0$ in the group space and find a normalized wave function $\varphi$ for which $|(\varphi, DL_0\varphi)| > 1/6$. There always exists a $\varphi$ with this property, if $|(\varphi, D(L_0)\varphi)| < 1/6$ then $\psi = \alpha\varphi + \beta D(L_0)\varphi$ with suitably chosen $\alpha$ and $\beta$ will be normalized and $|(\psi, D(L_0)\psi)| > 1/6$. We consider then such a neighborhood $\aleph$ of $L_0$ for all $L$ of which $|(\varphi, D(L)\varphi)| > 1/12$. It is well known[19] that the whole group space can be covered with such neighborhoods. We want to show now that the $D(L)\varphi$ can be multiplied with such phase factors (depending on $L$) of modulus unity that it becomes strongly continuous in the region $\aleph$.

We shall chose that phase factor so that $(\varphi, D(L)\varphi)$ becomes real and positive. Denoting then

$$(D(L_1) - D(L))\varphi = U_\varphi, \tag{23'}$$

the $(U_\varphi, U_\varphi)$ can be made arbitrarily small by letting $L$ approach sufficiently near to $L_1$, if $L_1$ is in $\aleph$. Indeed, on account of the continuity, as defined above, there is an $\Omega = e^{ik}$ such that $(u, u) < \epsilon$ if $L$ is sufficiently near to $L_1$ where

$$u = (D(L_1) - e^{ik}D(L))\varphi.$$

Taking the absolute value of the scalar product of $u$ with $\varphi$ one obtains

$$|(\varphi, D(L_1)\varphi) - \cos k(\varphi, D(L)\varphi) - i \sin k(\varphi, D(L)\varphi)| = |(\varphi, u)| \leq \sqrt{\epsilon},$$

because of Schwartz's inequality. If only $\sqrt{\epsilon} < 1/12$, the $k$ must be smaller than $\pi/2$ because the absolute value is certainly greater than the real part and both $(\varphi, D(L_1)\varphi)$ and $(\varphi, D(L)\varphi)$ are real and greater than 1/12.

As the absolute value is also greater than the imaginary part, we

$$\sin k < 12\sqrt{\epsilon}$$

On the other hand,

$$U_\varphi = u + (e^{ik} - 1)D(L)\varphi,$$

and thus

$$(U_\varphi, U_\varphi)^{\frac{1}{2}} \leq (u, u)^{\frac{1}{2}} + |e^{ik} - 1| \leq \sqrt{\epsilon} + 2 \sin k/2$$

$$(U_\varphi, U_\varphi) \leq 625\epsilon.$$

(b) It shall be shown next that if $D(L)\varphi$ is strongly continuous in a region and $D(L)$ is continuous in the sense defined at the beginning of this section, then $D(L)\psi$ with an arbitrary $\psi$ is (strongly) continuous in that region also. We shall see, hence, that the $D(L)$, with any normalization which makes a $D(L)\varphi$ strongly continuous, is continuous in the ordinary sense: There is to every $L_1$, $\epsilon$ and *every* $\psi$ a $\delta$ so that $(U_\psi, U_\psi) < \epsilon$ where

$$U_\psi = (D(L_1) - D(L))\psi$$

if $L$ is in the neighborhood $\delta$ of $L_1$.

[19] This condition is the "separability" of the group. Cf. e.g. A. Haar, Ann. of Math., *34*, 147, 1933.

It is sufficient to show the continuity of $D(L)\psi$ where $\psi$ is orthogonal to $\varphi$. Indeed, every $\psi'$ can be decomposed into two terms, $\psi' = \alpha\varphi + \beta\psi$ the one of which is parallel, the other perpendicular to $\varphi$. Since $D(L)\varphi$ is continuous, according to supposition, $D(L)\psi' = \alpha D(L)\varphi + \beta D(L)\psi$ will be continuous also if $D(L)\psi$ is continuous.

The continuity of the representation up to a factor requires that it is possible to achieve that $(u_\psi, u_\psi) < \epsilon$ and $(u_{\psi+\varphi}, u_{\psi+\varphi}) < \epsilon$ where

$$u_\psi = (D(L_1) - \Omega_\psi D(L))\psi, \tag{23a}$$

$$u_{\psi+\varphi} = (D(L_1) - \Omega_{\psi+\varphi}D(L))(\psi + \varphi), \tag{23b}$$

with suitably chosen $\Omega$'s. According to the foregoing, it also is possible to choose $L$ and $L_1$ so close that $(U_\varphi, U_\varphi) < \epsilon$.

Subtracting (23′) and (23a) from (23b) and applying $D(L)^{-1}$ on both sides gives

$$(\Omega_\psi - \Omega_{\psi+\varphi})\psi + (1 - \Omega_{\psi+\varphi})\varphi = D(L)^{-1}(u_{\psi+\varphi} - u_\psi - U_\varphi)$$

The scalar product of the right side with itself is less than $9\epsilon$. Hence both $|\Omega_\psi - \Omega_{\psi+\varphi}| < 3\epsilon^{\frac{1}{2}}$ and $|1 - \Omega_{\psi+\varphi}| < 3\epsilon^{\frac{1}{2}}$ or $|1 - \Omega_\psi| < 6\epsilon^{\frac{1}{2}}$. Because of $U_\psi = u_\psi - (1 - \Omega_\psi)D(L)\psi$, the $(U_\psi, U_\psi)^{\frac{1}{2}} < (u_\psi, u_\psi)^{\frac{1}{2}} + |1 - \Omega_\psi|$ and thus $(U_\psi, U_\psi) < 49\epsilon$.

This completes the proof of the theorem stated under (b). It also shows that not only the continuity of $D(L)\psi$ has been achieved in the neighborhood of $L_0$ by the normalization used in (a) but also that of $D(L)\psi$ with every $\psi$, i.e., the continuity of $D(L)$.

It is clear also that every finite part of the group space can be covered by a finite number of neighborhoods in which $D(L)$ can be made continuous. It is easy to see that the $\omega$ of (22) will be also continuous in these neighborhoods so that it is possible to make them continuous, apart from regions of lower dimensionality than their variables have. In the following only the fact will be used that they can be made continuous in the neighborhood of any $a$, $b$, and $\Lambda$.

## B.

(a) We want to show next that all $T(a)$ commute. From (22B) we have

$$T(a)T(b)T(a)^{-1} = c(a,b)T(b) \tag{24}$$

where $c(a,b) = \omega(a,b)/\omega(b,a)$ and hence

$$c(a,b) = c(b,a)^{-1}. \tag{24a}$$

Transforming (24) with $T(a')$ one obtains

$$T(a')T(a)T(b)T(a)^{-1}T(a')^{-1} = c(a,b)T(a')T(b)T(a')^{-1}$$

or
$$\omega(a',a)T(a'+a)T(b)\omega(a',a)^{-1}T(a'+a)^{-1} = c(a,b)c(a',b)T(b)$$

or

$$c(a,b)c(a',b) = c(a+a',b). \tag{25}$$

It follows[20] from (25) and the partial continuity of $c(a,b)$ that

$$c(a,b) = \exp\left(2\pi i \sum_{k=1}^{4} a_k f_k(b)\right) \tag{26}$$

and, since this is equal to $c(b,a)^{-1} = \exp(-2\pi i \sum b_k f_k(a))$

$$\sum_{k=1}^{4}(a_k f_k(b) + b_k f_k(a)) = n(a,b) \tag{27}$$

where $n(a,b)$ is an integer. Setting in (27) for $b$ the vector $e^{(\lambda)}$ the $\lambda$ component of which is 1, all the others zero and for $f_k(e^{(\lambda)}) = -f_{k\lambda}$

$$f_\lambda(a) = n(a, e^{(\lambda)}) + \sum_k a_k f_{k\lambda},$$

and putting this back into (27) we obtain

$$\sum_{k,\lambda=1}^{4} f_{k\lambda}(a_\lambda b_k + b_\lambda a_k) + \sum_{k=1}^{4} a_k n(b, e^{(k)}) + b_k n(a, e^{(k)}) = n(a,b). \tag{28}$$

Assuming for the components of $a$ and $b$ such values which are transcendental both with respect to each other and the $f_{k\lambda}$ (which are fixed numbers), one sees that (28) cannot hold except if the coefficient of every one vanishes

$$f_{k\lambda} + f_{\lambda k} = 0; \quad n(b, e^{(k)}) = 0, \tag{29}$$

so that (26) becomes

$$c(a,b) = \exp\left(2\pi i \sum_{k,\lambda=1}^{4} f_{k\lambda}, a_\lambda b_k\right). \tag{30}$$

It is necessary now to consider the existence of an operator $d(\Lambda)$ satisfying (22C). Transforming this equation with the similar equation containing $b$ instead of $a$

$$\begin{aligned} & d(\Lambda)T(b)d(\Lambda)^{-1}d(\Lambda)T(a)d(\Lambda)^{-1}d(\Lambda)T(b)^{-1}d(\Lambda)^{-1} \\ &= \omega(\Lambda,b)T(\Lambda b)\omega(\Lambda,a)T(\Lambda a)\omega(\Lambda,b)^{-1}T(\Lambda b)^{-1} \\ &= \omega(\Lambda,a)c(\Lambda b,\Lambda a)T(\Lambda a), \end{aligned}$$

[20]G. Hamel, Math. Ann. *60*, 460, 1905, quoted from H. Hahn, *Theorie der reellen Funktionen*. Berlin 1921, pages 581-583.

while the first line is clearly $d(\Lambda)c(b,a)T(a)d(\Lambda)^{-1} = \omega(\Lambda,a)c(b,a)T(\Lambda a)$ whence

$$c(b,a) = c(\Lambda b, \Lambda a) \tag{31}$$

holds for every Lorentz transformation $\Lambda$. Combined with (30) this gives

$$\sum_{k\lambda}\left(f_{k\lambda}a_k b_\lambda - \sum_{\nu\mu} f_{\nu\mu}\Lambda_{\nu k}\Lambda_{\mu\lambda}a_k b_\lambda\right) = n'(a,b),$$

where $n'(a,b)$ is again an integer. As this equation holds for every $a$, $b$

$$f_{k\lambda} = \sum_{\nu\mu} f_{\nu\mu}\Lambda_{\nu k}\Lambda_{\mu\lambda}; \qquad f = \Lambda' f \Lambda$$

must hold also, for every Lorentz transformation. However, the only form invariant under all Lorentz transformations are multiples of the $F$ of (10). Actually, because of (29), $f$ must vanish and $c(a,b) = 1$, all the operators corresponding to translations commute

$$T(a)T(b) = T(b)T(a). \tag{32}$$

It is well to remember that it was necessary for obtaining this result to use the existence of $d(\Lambda)$ satisfying (22C).

(b) Equation (32) is clearly independent of the normalization of the $T(a)$. If we could fix the translation operators in four linearly independent directions $e^{(1)}$, $e^{(2)}$, $e^{(3)}$, $e^{(4)}$ so that for each of these directions

$$T(ae^{(k)})T(be^{(k)}) = T((a+b)e^{(k)}) \tag{33}$$

be valid for every pair of numbers $a$, $b$, then the normalization

$$T(a_1e^{(1)} + a_2e^{(2)} + a_3e^{(3)} + a_4e^{(4)})$$

$$= T(a_1e^{(1)})T(a_2e^{(2)})T(a_3e^{(3)})T(a_4e^{(4)}) \tag{33a}$$

and (32) would ensure the general validity of

$$T(a)T(b) = T(a+b). \tag{34}$$

As the four linearly independent directions $e^{(1)}, \cdots, e^{(4)}$ we shall take four null vectors. If $e$ is a null vector, there is, according to section 3, a homogeneous Lorentz transformation[21] $\Lambda_e$ such that $\Lambda_e e = 2e$.

We normalize $T(e)$ so that

$$d(\Lambda_e)T(e)d(\Lambda_e)^{-1} = T(e)^2. \tag{35}$$

[21]The index $e$ denotes here the vector $e$ for which $\Lambda_e e = 2e$; this $\Lambda_e$ has no elementary divisor.

This is clearly independent of the normalization of $d(\Lambda_e)$. We further normalize for all (positive and negative) integers $n$

$$d(\Lambda_e)^n T(e) d(\Lambda_e)^{-n} = T(2^n e). \tag{35a}$$

It follows from this equation also that

$$\begin{aligned} T(2^n e)^2 &= d(\Lambda_e)^n T(e)^2 d(\Lambda_e)^{-n} \\ &= d(\Lambda_e)^n d(\Lambda_e) T(e) d(\Lambda_e)^{-1} d(\Lambda_e)^{-n} = T(2^{n+1} e). \end{aligned} \tag{36}$$

This allows us to normalize for every positive integer $k$

$$T(k \cdot 2^{-n} e) = T(2^{-n} e)^k \tag{35b}$$

in such a way that the normalizaton remains the same if we replace $k$ by $2^m k$ and $n$ by $n+m$. This ensures, together with (36), the validity of

$$\begin{aligned} T(\nu e) T(\mu e) &= T((\nu + \mu) e) \\ d(\Lambda_e) T(\nu e) d(\Lambda_e)^{-1} &= T(2 \nu e) \end{aligned} \tag{36a}$$

for all dyadic fractions $\nu$ and $\mu$.

It must be shown that if $\nu_1, \nu_2, \nu_3, \cdots$ is a sequence of dyadic fractions, converging to 0, $\lim T(\nu_i e) = 1$. From $T(a) \cdot T(0) = \omega(a,0) T(a)$ it follows that $T(0)$ is a constant. According to the theorem of part (A)(b), the $T(\nu e)$, if multiplied by proper constants $\Omega_\nu$ will converge to 1, i.e., by choosing an arbitrary $\varphi$, it is possible to make both $(1 - \Omega_\nu T(\nu e))\varphi = u$ and $(1 - \Omega_\nu T(\nu e)) \cdot d(\Lambda_e)^{-1} \varphi = u'$ arbitrarily small, by making $\nu$ small. Applying $d(\Lambda_e)$ to the second expression, one obtains, for (36a), that $(1 - \Omega_\nu T(2\nu e))\varphi = d(\Lambda_e) u'$ is also small. On the other hand, applying $T(\nu e)$ to the first expression one sees that $(T(\nu e) - \Omega_\nu T(2 \nu e))\varphi = T(\nu e) u$ approaches zero also. Hence, the difference of these two quantities $(1 - T(\nu e))\varphi$ goes to zero, i.e. $T(\nu_i e)\varphi$ converges to $\varphi$ if $\nu_1, \nu_2, \nu_3, \cdots$ is a sequence of dyadic fractions approaching 0.

Now $\nu_1, \nu_2, \nu_3, \cdots$ be a sequence of dyadic fractions coverging to an arbitary number $a$. It will be shown then that $T(\nu_i e)$ converges to a multiple of $T(ae)$ and this multiple of $T(ae)$ will be the normalized $T(ae)$. Again, it follows from the continuity that there are such $\Omega_i$ that $\Omega_i T(\nu_i e)\varphi$ converges to $T(ae)\varphi$. The $\Omega_j^{-1} T(\nu_j e)^{-1} \Omega_i T(\nu_i e)\varphi$ will converge to $\varphi$, therefore, as both $i$ and $j$ tend to infinity. However, according to the previous paragraph, $T((\nu_i - \nu_j)e)\varphi$ tends to $\varphi$ and thus $\Omega_j^{-1}\Omega_i$ tends to 1. It follows that $\Omega_i^{-1}$ converges to a definite number $\Omega$. Hence $\Omega_i^{-1} \cdot \Omega_i T(\nu_i e)\varphi$ converges to $\Omega T(ae)\varphi$ which will be denoted, henceforth, by $T(ae)$. For the $T(ae)$, normalized in this way, (33) will hold, since if $\mu_1, \mu_2, \mu_3, \cdots$ are dyadic fractions converging to $b$, we obtain, with the help (36a)

$$T(ae)T(be)\varphi = \lim_{i,j=\infty} T((v_i + \mu_i)e)\varphi = T((a+b)e)\varphi.$$

This argument not only shows that it is possible to normalize the $T(ae^{(k)})$ and hence by (33a) the $T(a)$ so that (34) holds for them but, in addition to this, that these $T(a)$ will be continuous in the ordinary sense.

### C.

It is clear that (34) will remain valid if one replaces $T(a)$ by $\exp(2\pi i\,\{a,c\})T(a)$ where $c$ is an arbitrary vector. This remaining freedom in the normalization of $T(a)$ will be used to eliminate the $\omega(\Lambda,a)$ from (22C).

Transforming (22C) $d(\Lambda)T(a)d(\Lambda)^{-1} = \omega(\Lambda,a)T(\Lambda a)$ with $d(M)$ one obtains on the left side $\omega(M,\Lambda)d(M\Lambda)T(a)\omega(M,\Lambda)^{-1}d(M\Lambda)^{-1} = \omega(M\Lambda,a)T(M\Lambda a)$ while the right side becomes $\omega(\Lambda,a)\omega(M,\Lambda a)T(M\Lambda a)$. Hence

$$\omega(M\Lambda,a) = \omega(M,\Lambda a)\omega(\Lambda,a). \tag{37}$$

On the other hand, the product of two equations (22C) with the same $\Lambda$ but with $a$ and $b$ respectively, instead of $a$ yields with the help of (34)

$$\omega(\Lambda,a)\omega(\Lambda,b) = \omega(\Lambda(a+b)).$$

Hence

$$\omega(\Lambda,a) = \exp(2\pi i\,\{a,f(\Lambda)\}),$$

where $f(\Lambda)$ is a vector which can depend on $\Lambda$. Inserting this back into (37) one obtains

$$\{a,f(M\Lambda)\} = \{\Lambda a, f(M)\} + \{a,f(\Lambda)\} + n,$$
$$\left\{a, f(M\Lambda) - \Lambda^{-1}f(M) - f(\Lambda)\right\} = n,$$

where $n$ is an integer which must vanish since it is a linear function of $a$. Hence

$$f(M\Lambda) = \Lambda^{-1}f(M) + f(\Lambda). \tag{38}$$

If we can show that the most general solution of the equation is

$$f(\Lambda) = (\Lambda^{-1} - 1)v_0, \tag{39}$$

where $v_0$ is a vector independent of $\Lambda$, the $\omega(\Lambda,\ a)$ will become $\omega(\Lambda,\ a) = \exp(2\pi i\,\{(\Lambda-1)a, v_0\})$. Then $\omega(\Lambda,a)$ in (22C) will disappear if we replace $T(a)$ by $\exp(2\pi i\,\{a,v_0\})T(a)$.

The proof that (39) is a consequence of (38) is somewhat laborious. One can first consider the following homogeneous Lorentz transformations.

$$X(\alpha_1,\gamma_1) = \begin{pmatrix} C_1 & 0 & 0 & S_1 \\ 0 & c_1 & s_1 & 0 \\ 0 & -s_1 & c_1 & 0 \\ S_1 & 0 & 0 & C_1 \end{pmatrix}; \quad Y(\alpha_2,\gamma_2) = \begin{pmatrix} c_2 & 0 & -s_2 & 0 \\ 0 & C_2 & 0 & S_2 \\ s_2 & 0 & c_2 & 0 \\ 0 & S_2 & 0 & C_2 \end{pmatrix}$$

$$Z(\alpha_3,\gamma_3) = \begin{pmatrix} c_3 & s_3 & 0 & 0 \\ -s_3 & c_3 & 0 & 0 \\ 0 & 0 & C_3 & S_3 \\ 0 & 0 & S_3 & C_3 \end{pmatrix} \tag{40}$$

where $c_i = \cos\alpha_i$; $s_i = \sin\alpha_i$; $C_i = Ch\gamma_i$; $S_i = Sh\gamma_i$. All the $X(\alpha,\gamma)$ commute. Let us choose, therefore, two angles $\alpha_1$, $\gamma_1$ for which $1 - X(\alpha_1,\gamma_1)^{-1}$ has a reciprocal. It follows then from (38)

$$X(\alpha,\gamma)^{-1}f(X(\alpha_1,\gamma_1)) + f(X(\alpha,\gamma))X(\alpha_1,\gamma_1)^{-1}f(X(\alpha,\gamma)) + f(X(\alpha_1,\gamma_1))$$
$$\text{or } f(X(\alpha,\gamma)) = [1 - X(\alpha_1,\gamma_1)^{-1}]^{-1}[1 - X(\alpha,\gamma)^{-1}]f(X(\alpha_1,\gamma_1))$$
$$f(X(\alpha,\gamma)) = (1 - X(\alpha,\gamma)^{-1})v_x, \tag{41}$$

where $v_x$ is independent of $\alpha$, $\gamma$. Similar equations hold for the $f(Y(\alpha,\gamma))$ and $f(Z(\alpha\gamma))$. Let us denote now $X(\pi,0) = X$; $Y(\pi,0) = Y$; $Z(\pi,0) = Z$. These anticommute in the following sense with the transformations (40):

$$YX(\alpha,\gamma)Y = ZX(\alpha,\gamma)Z = X(\alpha,\gamma)^{-1}. \tag{42}$$

From (38) one easily calculates

$$f(YX(\alpha,\gamma)Y) = (YX(\alpha,\gamma)^{-1} + 1)f(Y) + Yf(X(\alpha,\gamma)),$$

or, because of (41) and (42), after some trivial transformations

$$(1 - X(\alpha,\gamma))(1 - Y)(v_X - v_Y) = 0. \tag{43}$$

As $\alpha$, $\gamma$ can be taken arbitrarily, the first factor can be dropped. This leaves $(1 - Y)(v_X - v_Y) = 0$, or that the first and third components of $v_X$ and $v_Y$ are equal. One similarly concludes, however, that $(1 - X)(v_Y - v_X) = 0$ and thus that the first three components of $v_X$, $v_Y$ and also of $v_Z$ are equal.

For $\gamma_1 = \gamma_2 = \gamma_3 = 0$ the transformations (40) are the generators of all rotations, i.e. all Lorentz transformations $R$ not affecting the fourth coordinate. As the 4-4 matrix element of these transformations is 1, the expression $(1 - R^{-1})v$ is independent of the fourth component of $v$ and $(1 - R^{-1})v_X = (1 - R^{-1})v_Y = (1 - R^{-1})v_Z$. It follows from (38) that if $f(R) = (1 - R^{-1})v_X$ and $f(S) = (1 - S^{-1})v_X$, then $f(SR) = (1 - R^{-1}S^{-1})v_X$. Thus $f(R) = (1 - R^{-1})v_X$ is valid with the same $v_X$ for all rotations.

Now

$$f(X(\alpha,\gamma)R) = R^{-1}(1 - X(\alpha,\gamma)^{-1})v_X + (1 - R^{-1})v_X$$
$$= (1 - (X(\alpha,\gamma)R)^{-1})v_X.$$

One easily concludes from (38) that the $f(E)$ corresponding to the unit operation vanishes and $f(\Lambda^{-1}) = -\Lambda f(\Lambda)$. Hence $f(R^{-1}X(\alpha,\gamma)^{-1}) = (1 - X(\alpha,\gamma R)v_X$; and one concludes further that for all Lorentz transformations $\Lambda = RX(\alpha,\gamma)S$, (39) holds with $v_0 = -v_X$ if $R$ and $S$ are rotations. However, every homogeneous Lorentz transformation can be brought into this form (Section 4C). This completes the proof of (39) and thus of $\omega(\Lambda,\alpha) = 1$.

## D.

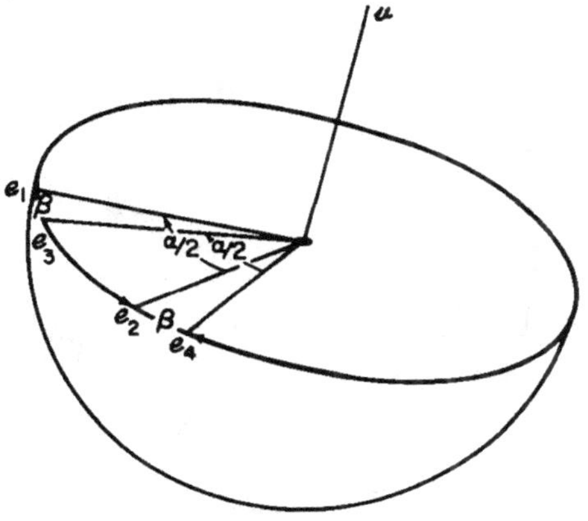


Figure 2.

The quantities $\omega(a,b)$ and $\omega(\Lambda,a)$ for which it has just been shown that they can be assumed to be 1, are independent from the normalization of $d(\Lambda)$. We can affix therefore an arbitrary factor of modulus 1 to all the $d(\Lambda)$, without interfering with the normalizations so far accomplished. In consequence hereof, the ensuing discussion will be simply a discussion of the normalization of the operators for the homogeneous Lorentz group and the result to be obtained will be valid for the group also.

Partly because the representations up to a factor of the three dimensional rotation group may be interesting in themselves, but more particularly because the procedure to be followed for the Lorentz group can be especially simply demonstrated for this group, the three dimensional rotation group shall be taken up first.

It is well known that the nomalization cannot be carried so far that $\omega(\Lambda, I) = 1$ in (22D) and there are well known representations for which $\omega(\Lambda, I) = \pm 1$. We shall allow this ambiguity therefore from the outset.

One can observe, first, that the operator corresponding to the unity of the group is a constant. This follows simply from $d(\Lambda)d(E) = \omega(\Lambda,\ E)d(\Lambda)$. The square of an operator corresponding to an involution is a constant, therefore.

The operator corresponding to the rotation about the axis $e$ by the angle $\pi$; normalized so that its square be actually 1, will be denoted by $\tilde{e}$; $\tilde{e}^2 = 1$. The $\tilde{e}$ are—apart from the sign—uniquely defined.

A rotation $R$ about $v$ by the angle $\alpha$ is the product of two rotations by $\pi$ about $e_1$ and $e_2$ where $e_1$ and $e_2$ are perpendicular to $v$ and $e_2$ arises from $e_1$ by rotation about $v$ and with $\alpha/2$. Choosing for every $v$ an arbitrary $e_1$ perpendicular to $v$, we can normalize, therefore

$$d(R) = \pm\tilde{e}_1\tilde{e}_2. \tag{44}$$

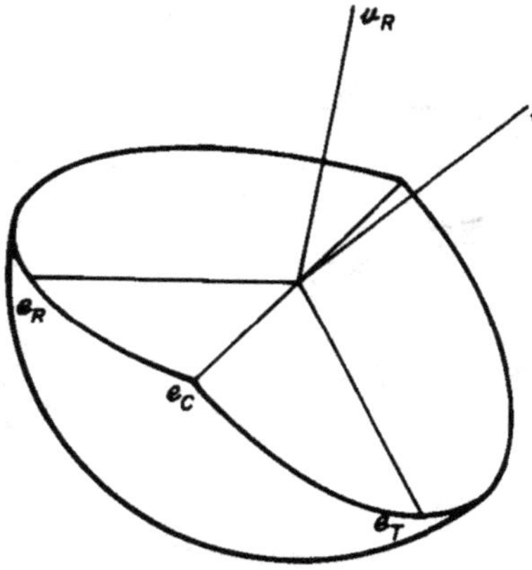


Figure 3.

Now $d(R)$ commutes with every $d(S)$ if $S$ is also a rotation about $v$. This is proved in equations (24)-(30). The $f_{11}$ in (30) must vanish on account of (29). (Also, both $R$ and $S$ can be arbitrarily accurately represented as powers of a very small rotation about $v$). Hence, transforming (44) by $d(S)$ one obtains

$$d(R) = \pm d(S)\tilde{e}_1 d(S)^{-1} \cdot d(S)\tilde{e}_2 d(S)^{-1}. \tag{44a}$$

Now $d(S)\tilde{e}_1 d(S)^{-1}$ corresponds to a rotation by $\pi$ about an axis, perpendicular to $v$ and enclosing an angle $\beta$ with $e_1$, where $\beta$ is the angle of rotation of $S$. Since the square of $d(S)\tilde{e}_1 d(S)^{-1}$ is also 1, (44a) is simply another way of writing $d(R) = \tilde{e}_3\tilde{e}_4$ as a product of two $\tilde{e}$ and we see that the normalization (44) is independent of the choice of the axis $e_1$ (Cf. Fig. 2).

For computing $d(R)d(T)$ we can draw the planes perpendicular to the axes of rotation of $R$ and $T$ and use for $d(R) = \tilde{e}_R\tilde{e}_C$ such a development that the axis $e_C$ of the second involution coincide with the intersection line of the above-mentioned planes, while for $d(T) = \tilde{e}_C\tilde{e}_T$ we choose the first involution to be a rotation about this intersection line (Fig. 3). Then, the product

$$d(R)d(T) = \pm\tilde{e}_R\tilde{e}_C\tilde{e}_C\tilde{e}_T = \pm\tilde{e}_R\tilde{e}_T \tag{45}$$

will automatically have the normalization corresponding to (44). This shows that the operators normalized in (44) give a representation up to the sign.

For the Lorentz group, the proof can be performed along the same line, only the underlying geometrical facts are less obvious. Let $\Lambda$ be a Lorentz transformation without elementary divisors with the characteristic values $e^{2i\gamma}$, $e^{-2i\gamma}$, $e^{2\chi}$, $e^{-2\chi}$ and the characteristic vectors $v_1$, $v_2 = v_1^*$, $v_3$, $v_4$, as described in section 4B.

We want to make $\Lambda = MN$ with $M^2 = N^2 = 1$. For $\Lambda N = M$, we have $\Lambda N \Lambda N = 1$ and thus $\Lambda N \Lambda = N$. Setting $Nv_i = \sum_k \alpha_{ik} v_k$, we obtain $\Lambda N \Lambda v_i = \sum \lambda_k \alpha_{ik} \lambda_i v_i = \sum \alpha_{ik} v_k$. Because of the linear independence of the $v_k$ this amounts to $\lambda_i \lambda_k \alpha_{ik} = \alpha_{ik}$: all $\alpha_{ik}$ are zero, except those for which $\lambda_i \lambda_k = 1$. As in none of the cases (a), (b), (c), (d) of section 4B is $\lambda_1$ or $\lambda_2$ reciprocal to one of the last two $\lambda$, the vectors $v_1$ and $v_2$ will be transformed by $N$ into a linear combination of $v_1$ and $v_2$ again, and the same holds for $v_3$ and $v_4$. This means the $N$ can be considered as the product of two transformations $N = N_s N_t$, the first in the $v_1 v_2$ plane, the second in the $v_3 v_4$ plane. (Instead of $v_1 v_2$ plane one really should say $v_1 + v_2, iv_1 - iv_2$ plane, as $v_1$ and $v_2$ are complex themselves. This will be meant always by $v_1 v_2$ plane, etc.). The same holds for $M$ also.

Both $N_s$ and $N_t$ must satisfy the first and third condition for Lorentz transformations (cf. 4A) and both determinants must be either 1, or $-1$. Furthermore, the square of both of them must be unity.

If both determinants were $+1$, the $N_t$ had to be unity itself, while $N_s$ could be the unity or a rotation by $\pi$ in the $v_1 v_2$ plane. Thus $v_1$, $v_2$, $v_3$, $v_4$ would be characteristic vectors of $N$ itself.

If both determinants are $-1$ (this will turn out to be the case), $N_s$ is a reflection on a line in the $v_1 v_2$ plane and $N_t$ a reflection in the $v_3 v_4$ plane, interchanging $v_3$ and $v_4$. In this case $v_1$, $v_2$, $v_3$, $v_4$ would not all be characteristic vectors of $N$.

If $v_1$, $v_2$, $v_3$, $v_4$ are characteristic vectors of $N$, they are characteristic vectors of $M = \Lambda N$ also. Then both $M$ and $N$ would be either unity, or a rotation by $\pi$ in the $v_1 v_2$ plane. If both of them were rotations in the $v_1 v_2$ plane, their product $\Lambda$ would be the unity which we want to exclude for the present. We can exclude the remaining cases in which the determinants of $N_s$ and $N_t$ are $+1$ by further stipulating that neither $M$ nor $N$ shall be the unity in the decomposition $\Lambda = MN$.

Hence $N$ is the product of a reflection in the $v_1 v_2$ plane

$$N s'_\nu = s'_\nu; \quad N s_\nu = s_\nu \tag{46a}$$

where $s_\nu$ and $s'_\nu$ are two perpendicular real vectors in the $v_1 v_2$ plane

$$s'_\nu = e^{i\nu} v_1 + e^{-i\nu} v_2; \quad s_\nu = i(e^{i\nu} v_1 - e^{-i\nu} v_2), \tag{46b}$$

and of a reflection in the $v_3 v_4$ plane

$$N t'_\mu = t'_\mu; \quad N t_\mu = -t_\mu, \tag{46c}$$

where again $t_\mu$, $t'_\mu$ are real vectors in the $v_3 v_4$ plane, perpendicular to each other, $t_\mu$ being space-like, $t'_\mu$ time-like:

$$t'_\mu = e^{\mu} v_3 + e^{-\mu} v_4; \quad t_\mu = e^{\mu} v_3 - e^{-\mu} v_4. \tag{46d}$$

Thus $N$ becomes a rotation by $\pi$ in the purely space like $s_\nu t_\mu$ plane. The $M$ can be calculated from $M = \Lambda N$

$$\begin{aligned}
Ms'_\nu &= \Lambda N s'_\nu = \Lambda s'_\nu = e^{i\nu+2i\gamma}v_1 + e^{-i\nu-2i\gamma}v_2 \\
&= \tfrac{1}{2}e^{2i\gamma}(s'_\nu - is_\nu) + \tfrac{1}{2}e^{-2i\gamma}(s'_\nu + is_\nu) = \cos 2\gamma \cdot s'_\nu + \sin 2\gamma \cdot s_\nu \\
Ms_\nu &= \sin 2\gamma \cdot s'_\nu - \cos 2\gamma \cdot s_\nu \\
Mt'_\mu &= \Lambda N t'_\mu = \Lambda t'_\mu = e^{\mu+2\chi}v_3 + e^{-\mu-2\chi}v_4 \\
&= \tfrac{1}{2}e^{2\chi}(t'_\mu + t_\mu) + \tfrac{1}{2}e^{2\chi}(t'_\mu - t_\mu) = \mathrm{Ch}2\chi \cdot t'_\mu + \mathrm{Sh}2\chi \cdot t_\mu \\
Mt_\mu &= -\mathrm{Sh}2\chi \cdot t'_\mu - \mathrm{Ch}2\chi \cdot t_\mu. \qquad (46e)
\end{aligned}$$

Thus $M$ also becomes a product of two reflections, one in the $v_1 v_2 = s'_\nu s_\nu$ the other in the $v_3 v_4 = t'_\mu t_\mu$ plane. This completes the decomposition of $\Lambda$ into two involutions. One of the involutions can be taken to be a rotation by $\pi$ in an arbitrary space like plane, intersecting both the $v_1 v_2$ and the $v_3 v_4$ planes, as the freedom in choosing $\nu$ and $\mu$ allows us to fix the lines $s_\nu$, and $t_\mu$ arbitrarily in those planes. The involution characterized by (46) will be called $N_{\nu\mu}$ henceforth. The other involution $M$ is then a similar rotation, in a plane, however, which is completely determined once the $s_\nu t_\mu$ plane is fixed. It will be denoted by $M_{\nu\mu}$ (it is, in fact $M_{\nu\mu} = N_{\nu+\gamma\ \mu+\chi}$). One sees the complete analogy to the three dimensional case if one remembers that $\gamma$ and $\chi$ are the half angles of rotation.

The $d(M)$ and $d(N)$ so normalized that their squares be 1 shall be denoted by $d_1(M_{\nu\mu})$ and $d_1(N_{\nu\mu})$. We must show that the normalization for

$$d(\Lambda) = \pm d_1(M_{nu\mu})d_1(N_{\nu\mu})1 \qquad (47)$$

is independent of $\nu$ and $\mu$. For this purpose, we transform

$$d(\Lambda) = \pm d_1(M_{00})d_1(N_{00}) \qquad (47a)$$

with $d(\Lambda_1)$ where $\Lambda_1$ has the same characteristic vectors as $\Lambda$ but different characteristic values, namely $e^{i\nu}$, $e^{-i\nu}$, $e^{\mu}$ and $e^{-\mu}$. Since $\Lambda_1 M_{00}\Lambda_1^{-1} = M_{\nu\mu}$ and $\Lambda_1 N_{00}\Lambda_1^{-1} = N_{\nu\mu}$ we have $d(\Lambda_1)d_1(M_{00})d(\Lambda_1)^{-1} = \omega d_1(M_{\nu\mu})$ where $\omega = \pm 1$, as the squares of both sides are 1. Hence, (47a) becomes if transformed with $d(\Lambda_1)$ just

$$d(\Lambda_1)d(\Lambda)d(\Lambda_1)^{-1} = \pm d_1(M_{\nu\mu})d_1(N_{\nu\mu}). \qquad (47b)$$

The normalization (47) would be clearly independent of $\nu$ and $\mu$ if $d(\Lambda_1)$ commuted with $d(\Lambda)$.

Again, the argument contained in equations (24)-(30) can be applied and shows that

$$d(\Lambda_1)d(\Lambda)d(\Lambda_1)^{-1} = \exp(2\pi i f(2\gamma\mu - 2\chi\nu))d(\Lambda) \qquad (48)$$

holds for every $\gamma$, $\chi$, $\nu$, $\mu$. However, the exponential in (48) must be 1 if $\gamma = 0$; $\nu = 2\pi/n$; $\chi = \frac{1}{2}n\mu$ since in this case $\Lambda = \Lambda_1^n$. Thus $\exp(-4\pi^2 i f\mu) = 1$ for every $\mu$ and $f = 0$ and the left side of (47b) can be replaced by $d(\Lambda)$; the normalization in (47) is independent of $\nu$ and $\mu$.

In order to have the analogue of (45), we must show that, having two Lorentz transformations $\Lambda = M_{\nu\mu}N_{\nu\mu}$ and $I = P_{\alpha\beta}Q_{\alpha\beta}$ we can choose $\nu$, $\mu$ and $\alpha$, $\beta$ so that $N_{\nu\mu} = P_{\alpha\beta}$ i.e. that the plane of rotation $s_\nu t_\mu$ of $N_{\nu\mu}$ coincide with the plane of rotation of $P_{\alpha\beta}$. As the latter plane can be made to an arbitrary space like plane intersecting both the $w_1w_2$ and the $w_3w_4$ planes (where $w_1$, $w_2$, $w_3$, $w_4$ are the characteristic vectors of $I$), we must show the existence of a space like plane, intersecting all four planes $v_1v_2$, $v_3v_4$, $w_1w_2$, $w_3w_4$. Both the first and the second pair of planes are orthogonal.

One can show[22] that if $\Lambda$ and $I$ have no common null vector as characteristic

[22]We first suppose the existence of a real plane $p$ intersecting all four planes $v_1v_2$, $v_3v_4$, $w_1w_2$, $w_3w_4$. If $p$ intersects $v_1v_2$ the plane $q$ perpendicular to $p$ will intersect the plane $v_3v_4$ perpendicular to $v_1v_2$. Indeed, the line which is perpendicular to both $p$ and $v_1v_2$ (there is such a line as $p$ and $v_1v_2$ intersect) is contained in both $q$ and $v_3v_4$. This shows that if there is a plane intersecting all four planes, the plane perpendicular to this will have this property also.

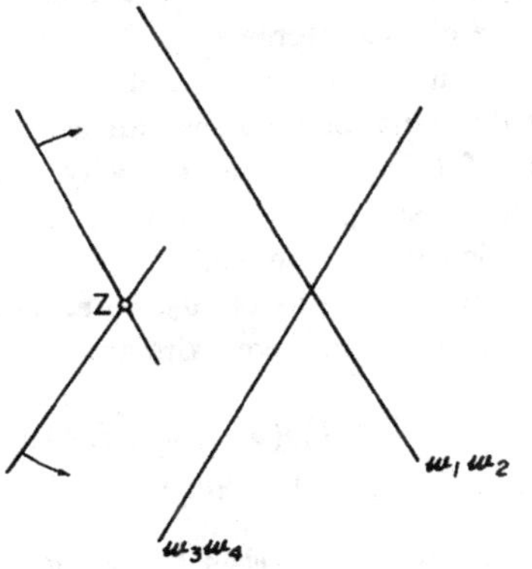


Figure 4 gives a projection of all lines into the $x_1x_2$ plane. One sees that there are, in general, two intersecting planes, only in exceptional cases is there only one.

If the plane $p$—the existence of which we suppose for the time being—contains a time-like vector, $q$ will be space-like (Section 4B, [1]). Both in this case and if $p$ contains only space-like vectors, the theorem in the text is valid. There is a last possibility, that $p$ is tangent to the light cone, i.e. contains only space like vectors and a null vector $\nu$. The space-like vectors of $p$ are all orthogonal to $v$, otherwise $p$ would contain time-like vectors also. In this case the plane $q$, perpendicular to $p$ will contain $v$ also. The line in which $v_1v_2$ intersects $p$ is space-like and orthogonal to $v$, otherwise $p$ would contain time-like vectors also. In this case the plane $q$, perpendicular to $p$ will contain $v$ also. The line in which $v_1v_2$ intersects $p$ is space-like and orthogonal to the vector in which $v_3v_4$ intersects $p$. The latter intersection must coincide with $v$, therefore, as no other vector of $p$ is orthogonal to any space-like vector in it. Hence, $v$ is the intersection of $p$ and $v_3v_4$ and is either $v_3$ or $v_4$. One can conclude in the same way that $v$ coincides with either $w_3$, or $w_4$ also and we see that if $p$ is tangent to the light cone the two transformations $\Lambda$ and $I$ have a common null vector as characteristic vector.

vector, there are always two planes, perpendicular to each other which intersect four such planes. One of these is always space like. It is possible to assume, therefore, that both $N_{\nu\mu}$ and $P_{\alpha\beta}$ are the rotation by $\pi$ in this plane. Thus

$$\begin{aligned} d(\Lambda)d(I) &= \pm d_1(M_{\nu\mu})d_1(N_{\nu\mu})d_1(P_{\alpha\beta})d_1(Q_{\alpha\beta}) \\ &= \pm d_1(M_{\nu\mu})d_1(Q_{\alpha\beta}), \end{aligned} \tag{49}$$

and $d(\Lambda)d(I)$ has the normalization corresponding to the product of two involutions, neither of which is unity. This is, however, also the normalization adopted for $d(\Lambda I)$. Hence

$$d(\Lambda)d(I) = \pm d(\Lambda I) \tag{49a}$$

holds if $\Lambda$, $I$ and $\Lambda I$ are Lorentz transformations corresponding to one of the cases (a), (b), (c) or (d) of section 4B and if $\Lambda$ and $I$ have no common characteristic null vector. In addition to this (49a) holds also, assuming $d(E) = \pm 1$, if any of the transformations $\Lambda$, $I$, $\Lambda I$ is unity, or if both characteristic null vectors of $\Lambda$ and $I$ are equal, as in this case the planes $v_3v_4$ and $w_3w_4$ and also $v_1v_2$ and $w_1w_2$ conincide and there are many space like planes intersecting all.

If $\Lambda$ and $I$ have one common characteristic null vector, $v_3 = w_3$, the others, $v_4$ and $w_4$ respectively, being different, one can use an artifice to prove (49a) which will be used in later parts of this section extensively. One can find a Lorentz transformation $J$ so that none of the pairs $I - J$; $\Lambda - IJ$; $\Lambda IJ - J^{-1}$ has a common characteristic null vector. This will be true, e.g. if the characteristic null vectors of $J$ are $v_4$ and another null vector, different from $v_3$, $w_4$ and the characteristic vectors of $\Lambda I$. Then (49a) will hold for all the above pairs and

$$\begin{aligned} d(\Lambda)d(I) &= \pm d(\Lambda)d(I)d(J)d(J^{-1}) = \pm d(\Lambda)d(IJ)d(J^{-1}) \\ &= \pm d(\Lambda IJ)d(J^{-1}) = \pm d(\Lambda I). \end{aligned}$$

---

Thus the theorem in the text is correct if we can show the existence of an arbitrary real plane $p$ intersecting all four planes $v_1v_2$, $v_3v_4$, $w_1w_2$, $w_3w_4$.

Let us draw a coordinate system in our four dimensional space, the $x_1x_2$ plane of which is the $v_1v_2$ plane, the $x_3$ and $x_4$ axes having the directions of the vectors $v_3 - v_4$ and $v_3 + v_4$, respectively. The three dimensional manifold $M$ characterized by $x_4 = 1$ intersects all planes in a line, the $v_1v_2$ plane in the line at infinity of the $x_1x_2$ plane, the $v_3v_4$ plane in the $x_3$ axis. The intersection of $M$ with the $w_1w_2$ and $w_3w_4$ planes will be lines in $M$ with directions perpendicular to each other. They will have a common normal through the origin of $M$, intersecting it at reciprocal distances. This follows from their orthogonality in the four dimensional space.

A plane intersecting $v_1v_2$ and $v_3v_4$ will be a line parallel to $x_1x_2$ through the $x_3$ axis. If we draw such lines through all points of the line corresponding to $w_1w_2$, the direction of this line will turn by $\pi$ if we go from one end of this line to the other. Similarly, the lines going through the line corresponding to $w_3w_4$ will turn by $\pi$ in the *opposite* direction. Thus the first set of lines will have at least one line in common with the second set and this line will correspond to a real plane intersecting all four planes $v_1v_2$, $v_3v_4$, $w_1w_2$, $w_3w_4$. This completes the proof of the theorem referred to in the text.

This completes the proof of (49a) for all cases in which $\Lambda$, $I$ and $\Lambda I$ have no elementary divisors. It is evident also that we can replace in the normalization (47) the $d_1$ by $d$. One also concludes easily that $d(M)^2$ is in the same representation either $+1$ for all involutions $M$, or $-1$ for every involution. The former ones will give real representations, the latter ones representations up to the sign.

If $\Lambda$ has an elementary divisor, it can be expressed in the $v_e$, $w_e$, $z_e$, $v_1$ scheme as the matrix (Cf. equ. (20))

$$\Lambda_e = \begin{pmatrix} 1 & 1 & \frac{1}{2} & 0 \\ 0 & 1 & 1 & 0 \\ 0 & 0 & 1 & 0 \\ 0 & 0 & 0 & 1 \end{pmatrix}$$

and can be written, in the same scheme, as the product of two Lorentz transformations with the square 1

$$\Lambda_e = M_0 N_0 = \begin{pmatrix} 1 & -1 & \frac{1}{2} & 0 \\ 0 & -1 & 1 & 0 \\ 0 & 0 & 1 & 0 \\ 0 & 0 & 0 & -1 \end{pmatrix} \cdot \begin{pmatrix} 1 & 0 & 0 & 0 \\ 0 & -1 & 0 & 0 \\ 0 & 0 & 1 & 0 \\ 0 & 0 & 0 & -1 \end{pmatrix}.$$

We can normalize therefore $d(\Lambda_e) = \pm d(M_0)d(N_0)$. If $\Lambda$ can be written as the product of two other involutions also $\Lambda_e = M_1 N_1$ the corresponding normalization will be identical with the original one. In order to prove this, let us consider a Lorentz transformation $J$ such that neither of the Lorentz transformations $J$, $N_0 J$, $N_1 J$, $\Lambda_e J = M_0 N_0 J = M_1 N_1 J$ have an elementary divisor. Since the number of free parameters is only 4 in case (e), while 6 for case (a), this is always possible. Then, for (45a)

$$\begin{aligned} d(M_0)d(N_0)d(J) &= \pm d(M_0)d(N_0 J) = \pm d(M_0 N_0 J) \\ &= \pm d(M_1 N_1 J) = \pm d(M_1)d(N_1 J) \\ &= \pm d(M_1)d(N_1)d(J) \end{aligned}$$

and thus $d(M_0)d(N_0) = \pm d(M_1)d(N_1)$. This shows also that even if $\Lambda I$ is in case (e), $\omega(\Lambda, I) = \pm 1$, since (49) leads to the correct normalization.

If $\Lambda = MN$ has an elementary divisor, $I$ not, $d(\Lambda)d(I)$ still will have the normalization corresponding to the product of two involutions. One can find again a $J$ such that neither of the transformations $J$, $J^{-1}$, $IJ$, $NIJ$, $MNIJ$, have an elementary divisor. Then

$$\begin{aligned} d(M)d(N)d(I) &= \pm d(M)d(N)d(I)d(J)d(J)^{-1} \\ &= \pm d(M)d(N)d(IJ)d(J)^{-1} = \pm d(M)d(NIJ)d(J)^{-1} \\ &= d(\Lambda IJ)d(J^{-1}). \end{aligned}$$

The last product has, however, the normalization corresponding to two involutions, as was shown in (49a), since neither $\Lambda IJ$, nor $J^{-1}$ is in case (e).

Lastly, we must consider the case when both $\Lambda$ and $I$ may have an elementary divisor. In this case, we need a $J$ such that neither $J$, $J^{-1}$, $IJ$ have one. Then, because of the generalization of (49a) just proved, in which the first factor is in case (e)

$$\begin{aligned} d(\Lambda)d(I) &= \pm d(\Lambda)d(I)d(J)d(J^{-1}) = \pm d(\Lambda)d(IJ)d(J^{-1}) \\ &= \pm d(\Lambda IJ)d(J^{-1}) \end{aligned}$$

which has the right normalization.

This completes the proof of

$$\omega(\Lambda, I) = \pm 1 \tag{50}$$

for all possible cases, and the normalization of all $D(L)$ of a representation of the inhomogeneous Lorentz group up to a factor, is carried out in such a way that the normalized operators give a representation up to the sign. It is even carried so far that in the first two of equations (22) $\omega = 1$ can be set. We shall consider henceforth systems of operators satisfying (7), or more specifically, (22B) and (22C) with $\omega(a, b) = \omega(\Lambda, a) = 1$ and (22D) with $\omega(\Lambda, I) = \pm 1$.

### E.

Lastly, it shall be shown that the renormalization not only did not spoil the partly continuous character of the representation, attained at the first normalization in part (A) of this section, but that the same holds now *everywhere*, in the ordinary sense for $T(a)$ and, apart from the ambiguity of sign, also for $d(\Lambda)$. For $T(a)$ this was proved in part (B)(b) of this section, for $d(\Lambda)$ it means that to every $\Lambda_1$, $\epsilon$ and $\varphi$ there is such a $\delta$ that *one* of the two quantities

$$((d(\Lambda_1) \mp d(\Lambda))\varphi, (d(\Lambda_1) \mp d(\Lambda))\varphi) < \epsilon \tag{51}$$

if $\Lambda$ is in the neighborhood $\delta$ of $\Lambda_1$. The inequality (51) is equivalent to

$$((1 \mp d(\Lambda_0))\varphi, (1 \mp d(\Lambda_0))\varphi) < \epsilon, \tag{51a}$$

where $\Lambda_0 = \Lambda_1^{-1}\Lambda$ now can be assumed to be in the neighborhood of the unity. Thus, the continuity of $d(\Lambda)$ at $\Lambda = E$ entails the continuity everywhere.[23] In fact, it would be sufficient to show that the $d(X)$, $d(Y)$ and $d(Z)$ corresponding to the transformations (40) converge to $\pm 1$, as $\alpha$, $\gamma$ approach 0, since one can write every transformation in the neighborhood of the unit element as a product $\Lambda = Z(0,\gamma_3)Y(0,\gamma_2)X(0,\gamma_1)X(\alpha_1,0)Y(\alpha_2,0)Z(\alpha_3,0)$ and the parameters $\alpha_1, \cdots, \gamma_3$ will converge to 0 as $\Lambda$ converges to 1. However, we shall carry out the proof for an arbitrary $\Lambda$ without an elementary divisor.

[23] J. von Neumann, Sitz. d. Kön. Preuss. Akad. p. 76, 1927.

For $d(\Lambda)$, equations (46) show that as $\Lambda$ approaches $E$ (i.e., as $\varphi$ and $\chi$ approach zero) both $M_{00}$ and $N_{00}$ approach the same involution, which we shall call $K$. Let us now consider a wave function $\psi = \varphi + d_1(K)\varphi$ or, if this vanishes $\psi = \varphi - d_1(K)\varphi$. We have $d_1(K)\psi = \pm\psi$. If $\Lambda$ is sufficiently near to unity, $d_1(N_{00})\psi$ will be sufficiently near to $\Omega d_1(K)\psi = \pm\Omega\psi$ and all we have to show is that $\Omega$ approaches $\pm 1$. The same thing will hold for $d_1(M_{00})$. Indeed from $d_1(N_{00})\psi - \Omega\psi = u$ it follows by applying $d_1(N_{00})$ on both sides $\psi - \Omega^2\psi = (d_1(N_{00}) + \Omega)u$. As $(u, u)$ goes to zero, $\Omega$ must go to $\pm 1$, and consequently, also $d_1(N_{00})\psi$ goes to $\psi$ or to $-\psi$. Applying $d_1(M_{00})$ to this, one sees that $d_1(M_{00})d_1(N_{00})\psi = d(\Lambda)\psi$ goes to $\pm\psi$ as $\Lambda$ goes to unity. The argument given in (A)(b) shows that this holds not only for $\psi$ but for every other function also, i.e. $d\Lambda$ converges to $\pm 1 = d(E)$ as $\Lambda$ approaches $E$. Thus $d(\Lambda)$ is continuous in the neighborhood of $E$ and hence everywhere.

According to the last remark in part 4, the operators $\pm d(\Lambda)$ form a single valued representation of the group of complex unimodular two dimensional matrices $C$. Let us denote the homogeneous Lorentz transformation which corresponds in the isomorphism to $C$ by $\tilde{C}$. Our task of solving the equs. (22) has been reduced to finding all single valued unitary representation of the group with the elements $[a, C] = [a, 1][0, C]$, the multiplication rule of which is $[a, C_1][b, C_2] = [a + \tilde{C}_1 b, C_1 C_2]$. For the representations of this group $D[a, C] = T(a)d[C]$ we had

$$\begin{aligned} T(a)T(b) &= T(a+b) \\ d[C]T(a) &= T(\tilde{C}a)d[C] \\ d[C_1]d[C_2] &= d[C_1 C_2] \end{aligned} \tag{52a}$$

It would be more natural, perhaps, from the mathematical point of view, to use henceforth this new notation for the representations and let the $d$ depend on the $C$ rather than on the $\tilde{C}$ or $\Lambda$. However, in order to be reminded on the geometrical significance of the group elements, it appeared to me to be better to keep the old notion. Instead of the equations (22B), (22C), (22D) we have, then

$$T(a)T(b) = T(a+b) \tag{52B}$$

$$d(\Lambda)T(a) = T(\Lambda a)d(\Lambda) \tag{52C}$$

$$d(\Lambda)d(I) = \pm d(\Lambda I). \tag{52D}$$

## 6. Reduction of the Representations of the Inhomogeneous Lorentz Group to Representations of a "Little Group"

This section, unlike the other ones, will often make use of methods, which though commonly accepted in physics, must be further justified from a rigorous mathematical point of view. This has been done, in the meanwhile, by J. von Neumann in an as yet unpublished article and I am much indebted to him for his cooperation in this respect and for his readiness in communicating his results to me. A reference to this paper[24] will be made whenever his work is necessary for making inexact considerations of this section rigorous.

**A.**

Since the translation operators all commute, it is possible[24] to introduce such a coordinate system in Hilbert space that the wave functions $\varphi(p,\zeta)$ contain momentum variables $p_1$, $p_2$, $p_3$, $p_4$ and a discrete variable $\zeta$ so that

$$T(a)\varphi(p,\zeta) = e^{i\{p,a\}}\varphi(p,\zeta). \tag{53}$$

$p$ will stand for the four variables $p_1$, $p_2$, $p_3$, $p_4$.

Of course, the fact that the Lorentzian scalar product enters in the exponent, rather than the ordinary, is entirely arbitrary and could be changed by changing the signs of $p_1$, $p_2$, $p_3$.

The unitary scalar product of two wave functions is not yet completely defined by the requirements so far made on the coordinate system. It can be a summation over $\zeta$ and an arbitrary Stieltjes integral over the components of $p$:

$$(\psi,\varphi) = \sum_\zeta \int \psi(p,\zeta)^*\varphi(p,\zeta)df(p,\zeta). \tag{54}$$

The importance of introducing a weight factor, depending on $p$, for the scalar product lies not so much in the possibility of giving finite but different weights to different regions in $p$ space. Such a weight distribution $g(p,\zeta)$ always could be absorbed into the wave functions, replacing all $\varphi(p,\zeta)$ by $\sqrt{g(p,\zeta)}\cdot\varphi(p,\zeta)$. The necessity of introducing the $f(p,\zeta)$ lies rather in the possibility of some regions of $p$ having zero weight while, on the other hand, at other places points may have finite weights. On account of the definite metric in Hilbert space, the integral $\int df(p,\zeta)$ over any region $r$, for any $\zeta$, is either positive, or zero, since it is the scalar product of that function with itself, which is 1 in the region $r$ of $p$ and the value $\zeta$ of the discrete variable, zero otherwise.

Let us now define the operators

$$P(\Lambda)\varphi(p,\zeta) = \varphi(\Lambda^{-1}p,\zeta). \tag{55}$$

This equation defines the function $P(\Lambda)\varphi$, which is, at the point $p$, $\zeta$, as great as the function $\varphi$ at the point $\Lambda^{-1}p,\zeta$. The operator $P(\Lambda)$ is not necessarily unitary, on account of the weight factor in (54). We can easily calculate

$$\begin{aligned} P(\Lambda)T(a)\varphi(p,\zeta) &= T(a)\varphi(\Lambda^{-1}p,\zeta) = e^{i\{\Lambda^{-1}p\}}\varphi(\Lambda^{-1}p,\zeta), \\ T(\Lambda a)P(\Lambda)\varphi(p,\zeta) &= e^{i\{p,\Lambda a\}}P(\Lambda)\varphi(p,\zeta) = e^{i\{p,\Lambda a\}}\varphi(\Lambda^{-1}p,\zeta), \end{aligned}$$

[24] J. von Neumann, Ann. of Math. to appear shortly.

so that, for $\{\Lambda^{-1}p, a\} = \{p, \Lambda a\}$, we have

$$P(\Lambda)T(a) = T(\Lambda a)P(\Lambda). \tag{56}$$

This, together with (52C), shows that $d(\Lambda)P(\Lambda)^{-1} = Q(\Lambda)$ commutes with all $T(a)$ and, therefore, with the multiplication with every function of $p$, since the exponentials form a complete set of functions of $p_1$, $p_2$, $p_3$, $p_4$. Thus

$$d(\Lambda) = Q(\Lambda)P(\Lambda), \tag{57}$$

where $Q(\Lambda)$ is an operator in the space of the $\zeta$ alone[24] which can depend, however, on the particular value of $p$ in the underlying space:

$$Q(\Lambda)\varphi(p,\zeta) = \textstyle\sum_\eta Q(P,\Lambda)_{\zeta\eta}\varphi(p,\eta). \tag{57a}$$

Here, $Q(p,\Lambda)_{\zeta\eta}$ are the components of an ordinary (finite or infinite) matrix, depending on $p$ and $\Lambda$. From (57), we obtain

$$\begin{aligned} d(\Lambda)\varphi(p,\zeta) &= \textstyle\sum_\eta Q(p,\Lambda)_{\zeta\eta}P(\Lambda)\varphi(p,\eta) \\ &= \textstyle\sum_\eta Q(p,\Lambda)_{\zeta\eta}\varphi(\Lambda^{-1}p,\eta). \end{aligned} \tag{57b}$$

As the exponentials form a complete set of functions, we can approximate the operation of multiplication with any function of $p_1$, $p_2$, $p_3$, $p_4$ by a linear combination

$$f(p)\varphi = \sum_n c_n T(a_n)\varphi. \tag{58}$$

If we choose $f(p)$ to be such a function that

$$f(p) = f(\Lambda p) \tag{58a}$$

the operation of multiplication with $f(p)$ will commute with all operations of the group. It commutes evidently with the $T(a)$ and the $Q(p,\Lambda)$, and on account of (56) and (58), (58a) also with $P(\Lambda)$. Thus the operation of (58) belongs to the centrum of the algebra of our representation. Since, however, we assume that the representation is factorial (cf. 2), the centrum contains only multiples of the unity and

$$f(p)\varphi(p,\zeta) = c\varphi(p,\zeta). \tag{58b}$$

This can be true only if $\varphi$ is different from zero only for such momenta $p$ which can be obtained from each other by homogeneous Lorentz transformations, because $f(p)$ needs to be equal to $f(p')$ only if there is a $\Lambda$ which brings them into each other.

It will be sufficient, henceforth, to consider only such representations, the wave functions of which vanish except for such momenta which can be obtained from one by homogeneous Lorentz transformations. One can restrict, then, the definition domain of the $\varphi$ to these momenta.

These representations can now naturally be divided into the four classes enumerated in section 3, and two classes contain two subclasses. There will be representations, the wave functions of which are defined for $p$ such that

$$(1)\ \{p,p\} = P > 0 \qquad (3)\ p = 0$$
$$(2)\ \{p,p\} = P = 0;\ \ p \neq 0 \qquad (4)\ \{p,p\} = P < 0.$$

The classes 1 and 2 contain two sub-classes each. In the positive subclasses $P_+$ and $0_+$ the time components of all momenta are $p_4 > 0$, in the negative subclasses $P_-$ and $0_-$ the fourth components of the momenta are negative. Class 3 will be denoted by $0_0$. If $P$ is negative, it has no index.

From the condition that $d(\Lambda)$ shall be a unitary operator, it is possible to infer [24] that one can introduce a coordinate system in Hilbert space in such a way that

$$\int_r df(p,\zeta) = \int_{\Lambda r} df(p,\eta) \tag{59}$$

if $Q(p,\Lambda)_{\zeta\eta} \neq 0$ for the $p$ of the domain $r$. Otherwise, $r$ is an arbitrary domain in the space of $p_1, p_2, p_3, p_4$, and $\Lambda r$ is the domain which contains $\Lambda p$ if $r$ contains $p$. Equation (59) holds for all $\zeta, \eta$, except for such pairs for which $Q(p,\Lambda)_{\zeta\eta} = 0$. It is possible, hence, to decompose the original representation in such a way that (59) holds within every reduced part. Neither $T(a)$ nor $d(\Lambda)$ can have matrix elements between such $\eta$ and $\zeta$ for which (59) does not hold.

In the third class of representations, the variable $p$ can be dropped entirely, and $T(a)\varphi(\zeta) = \varphi(\zeta)$, i.e., all wave functions are invariant under the operations of the invariant subgroup, formed by the translations. The equation $T(a)\varphi(\zeta) = \varphi(\zeta)$ is an invariant characterization of the representations of the third class, i.e., a characterization which is not affected by a similarity transformation. Hence, the reduced parts of a representation of class 3 also belong to this class.

Since no wave function of the other classes can remain invariant under all translations, no representation of the third class can be contained in any representation of one of the other classes. In the other classes, the variability domain of $p$ remains three dimensional. It is possible, therefore, to introduce instead of $p_1, p_2, p_3, p_4$ three independent variables. In the cases 1 and 2 with which we shall be concerned most, $p_1, p_2, p_3$ can be kept for these three variables. On account of (59), the Stieltjes integral can be replaced by an ordinary integral[24] over these variables, the weight factor being $|p_4|^{-1} = (P + p_1^2 + p_2^2 + p_3^2)^{-\frac{1}{2}}$

$$\{\psi,\varphi\} = \textstyle\sum_\zeta \int\int\int_{-\infty}^{\infty} \psi(p,\zeta)^* \varphi(p,\zeta) |p_4|^{-1} dp_1 dp_2 dp_3. \tag{59a}$$

In fact, with the weight factor $|p_4|^{-1}$ the weight of the domain $r$ i.e., $W_r$ =

$\int\int\int_r |p_4|^{-1} dp_1 dp_2 dp_3$ is equal to the weight of the domain $W_{\Lambda r}$ as required[25] by (59). Having the scalar product fixed in this way, $P(\Lambda)$ becomes a unitary operator and, hence, $Q(\Lambda)$ will be unitary also.

We want to give next a characterization of the representations with a given $P$, which is independent of the coordinate system in Hilbert space. It follows from (53), that in a representation with a given $P$ the wave functions $\psi_1, \psi_2, \cdots$ which are different from zero only in a finite domain of $p$, form an everywhere dense set, to all elements of which the infinitesimal operators of translation can be applied arbitrarily often

$$\begin{aligned}\lim_{h=0} h^{-n}(T(he)-1)^n\psi &= \lim_{h=0} h^{-n}(e^{ih\{p,e\}}-1)^n\psi \\ &= i^n \{p,e\}^n \psi,\end{aligned} \tag{60}$$

where $e$ will be a unit vector in the direction of a coordinate axis or oppositely directed to it. Hence for all members $\psi$ of this everywhere dense set

$$\lim_{h=0}\sum_k \pm h^{-2}(T(2he_k)-2T(he_k)+1)\psi = (p_1^2+p_2^2+p_3^2-p_4^2)\psi = -P\psi, \tag{61}$$

where $e_k$ is a unit vector in (or opposite) the $k^{\text{th}}$ coordinate axis and the $\pm$ is $+$ for $k=4$, and $-$ for $k=1,2,3$.

On the other hand, there is no $\varphi$ for which

$$\lim_{h=0}\textstyle\sum_k \pm h^{-2}(T(2he_k)-2T(he_k)+1)\varphi \tag{61a}$$

if it exists, would be different from $-P\varphi$. Suppose the limit in (61a) exists and is $-P\varphi+\varphi'$. Let us choose then a normalized $\psi$, from the above set, such that $(\psi,\varphi')=\delta$ with $\delta>0$ and an $h$ so that the expression after the lim sign in (61a) assumes the value $-P\varphi+\varphi'+u$ with $(u,u)<\delta/3$ and also the expression after the lim sign in (61), with oppositely directed $e_k$ becomes $-P\psi+u'$ with $(u',u')<\delta/3$. Then, on account of the unitary character of $T(a)$ and because of $T(-a)=T(a)^{-1}$

$$\begin{aligned}&\left(\psi,\sum_k \pm h^{-2}(T(2he_k)-2T(he_k)+1)\varphi\right), \\ &\quad=\left(\sum_k \pm h^{-2}(T(-2he_k)-2T(-he_k)+1)\psi,\varphi\right),\end{aligned}$$

or

$$-P(\psi,\varphi)+(\psi,\varphi')+(\psi,u) = -P(\psi,\varphi)+(u',\varphi),$$

[25]The invariance of integrals of the character of (59a) is frequently made use of in relativity theory. One can prove it by calculating the Jacobian of the transformation

$$p_i' = \Lambda_{i1}p_1+\Lambda_{i2}p_2+\Lambda_{i3}p_3+(P+p_1^2+p_2^2+p_3^2)^{\frac{1}{2}} \quad (i=1,2,3)$$

which comes out to be $(P+p_1^2+p_2^2+p_3^2)^{\frac{1}{2}}(P+p_1'^2+p_2'^2+p_3'^2)^{-\frac{1}{2}}$. Equ. (59a) will not be used in later parts of this paper.

which is clearly impossible.

Thus if the lim in (61a) exists, it is $-P\varphi$ and this constitutes a characterization of the representation which is independent of similarity transformations. Since, according to the foregoing, it is always possible to find wave functions for a representation, to which (61a) can be applied, every reduced part of a representation with a given $P$ must have this same $P$ and no representation with one $P$ can be contained in a representation with an other $P$. The same argument can be applied evidently to the positive and negative sub-classes of class 1 and 2.

## B.

Every automorphism $L \rightarrow L^0$ of the group allows us to construct from one representation $D(L)$ another representation

$$D^0(L) = D(L^0). \tag{62}$$

This principle will allow us to restrict ourselves, for representations with finite, positive or negative $P$, to one value of $P$ which can be taken respectively, to be $+1_+$ and $-1$. It will also allow in cases 1 and 2 to construct the representations of the negative sub-classes out of representations of the positive sub-classes.

The first automorphism is $a^0 = \alpha a$, $\Lambda^0 = \Lambda$. Evidently Equs. (12) are invariant under this transformation. If we set, however,

$$T^0(a)\varphi = T(\alpha a)\varphi; \quad d^0(\Lambda)\varphi = d(\Lambda)\varphi,$$

then the occurring $p$

$$T^0(a)\varphi = T(\alpha a)\varphi = e^{i\{p,\alpha a\}}\varphi = e^{i\{\alpha p,a\}}\varphi,$$

will be the $p$ occurring for the unprimed representation, multiplied by $\alpha$. This allows, with a real positive $\alpha$, to construct all representations with all possible numerical values of $P$, from all representation with one numerical value of $P$. If we take $\alpha$ negative, the representations of the negative sub-classes are obtained from the representations of the positive sub-class.

In case $P = 0_0$ evidently all representations go over into themselves by the transformation (62). In case $P = 0_+$ and $P = 0_-$ it will turn out that for positive $\alpha$, (62) carries every representation into an equivalent one.

## C.

On account of (53) and (56), (57), the equs. (52B) and (52C) are automatically satisfied and the $Q(p,\Lambda)_{\zeta\eta}$ must be determined by (52D). This gives

$$\sum_{\eta\vartheta} Q(p,\Lambda)_{\zeta\eta} Q(\Lambda^{-1}p, I)_{\eta\vartheta}\psi(I^{-1}\Lambda^{-1}p,\vartheta)$$
$$= \pm \sum_{\vartheta} Q(p,\Lambda I)_{\zeta\vartheta}\varphi(I^{-1}\Lambda^{-1}p,\vartheta). \tag{63}$$

Since this must hold for every $\varphi$, one would conclude

$$\sum_\eta Q(p,\Lambda)_{\zeta\eta} Q(\Lambda^{-1}p, I)_{\eta\vartheta} = \pm Q(p,\Lambda I)_{\zeta\vartheta}. \tag{63a}$$

Actually, this conclusion is not justified, since two wave functions must be considered to be equal even if they are different on a set of measure zero. Thus one cannot conclude, without further consideration, that the two sides of (63a) are equal at every point $p$. On the other hand,[24] the value of $Q(p,\Lambda)_{\zeta\eta}$ can be changed on a set of measure zero and one can make it continuous in the neighborhood of every point, if the representation is continuous. This allows then, to justify (63a). It follows from (63a) that $Q(p,1)_{\zeta\eta} = \delta_{\zeta\eta}$.

Let us choose[15] now a basic $p_0$ arbitrarily. We can consider then the subgroup of all homogeneous Lorentz transformations which leave this $p_0$ unchanged. For all elements $\lambda$, $\iota$ of this "little group," we have

$$\sum_\eta Q(p_0,\lambda)_{\zeta\eta} Q(p_0,\iota)_{\eta\vartheta} = \pm Q(p_0,\lambda\iota)_{\zeta\vartheta}$$
$$q(\lambda)q(\iota) = \pm q(\lambda\iota), \tag{64}$$

where $g(\lambda)$ is the matrix $q(\lambda)_{\zeta\eta} = Q(p_0,\lambda)_{\zeta\eta}$. Because of the unitary character of $Q(\Lambda)$, the $Q(p_0,\Lambda)_{\zeta\eta}$ is unitary matrix and $q(\lambda)$ is unitary also.

If we consider, according to the last paragraph of Section 5, the group formed out of the translations and unimodular two-dimensional matrices, rather than Lorentz transformations, the $\pm$ sign in (64) can be replaced by a $+$ sign. In this case, $\lambda$ and $\iota$ are unimodular two-dimensional matrices and the little group is formed by those matrices, the corresponding Lorentz transformations $\tilde{\lambda}$, $\tilde{\iota}$ to which leave $p_0$ unchanged $\tilde{\lambda}p_0 = \tilde{\iota}p_0 = p_0$.

Adopting this interpretation of (64), one can also see, conversely, that the representation $q(\lambda)$ of the little group, together with the class and $P$ of the representation of the whole group, determines the latter representation, apart from a similarity transformation. In order to prove this, let us define for every $p$ a two-dimensional unimodular matrix $\alpha(p)$ in such a way that the corresponding Lorentz transformation

$$\tilde{\alpha}(p)p_0 = p \tag{65}$$

brings $p_0$ into $p$. The $\alpha(p)$ can be quite arbitrary except of being an almost everywhere continuous function of $p$, especially continuous for $p = p_0$ and $\alpha(p_0) = 1$. Then, we can set

$$d(\alpha(p)^{-1})\varphi(p_0,\zeta) = \varphi(p,\zeta),$$
$$d(\alpha(p))\varphi(p,\zeta) = \varphi(p_0,\zeta). \tag{66}$$

This is equivalent to setting in (58)

$$Q(p,\alpha(p)) = 1 \tag{66a}$$

and can be achieved by a similarity transformation which replaces $\varphi(p, \zeta)$ by $\sum_\eta Q(p_0, \alpha(p)^{-1})_{\zeta\eta}\varphi(p,\eta)$. As the matrix $Q(p_0, \alpha(p)^{-1})$ is unitary, this is a unitary transformation. It does not affect, furthermore, (53) since it contains $p$ only as a parameter.

Assuming this transformation to be carried out, (66) will be valid and will define, together with $d(\lambda)$, all the remaining $Q(p, \Lambda)$ uniquely. In fact, calculating $d(\Lambda)\varphi(p,\zeta)$, we can decompose $\Lambda$ into three factors

$$\Lambda = \alpha(p). \quad \alpha(p)^{-1}\Lambda\alpha(\tilde{\Lambda}^{-1}p). \quad \alpha(\tilde{\Lambda}^{-1}p)^{-1} \tag{67}$$

The second factor $\beta = \alpha(p)^{-1}\Lambda\alpha(\tilde{\Lambda}^{-1}p)$ belongs into the little group: $\tilde{\alpha}(p)^{-1}\tilde{\Lambda}\tilde{\alpha}(\tilde{\Lambda}^{-1}p)p_0 = \tilde{\alpha}(p)^{-1}\tilde{\Lambda}\cdot\tilde{\Lambda}^{-1}p = \tilde{\alpha}(p)^{-1}p = p_0$. We can write, therefore ($\tilde{\Lambda}^{-1}p = p'$)

$$\begin{aligned} d(\Lambda)\varphi(p,\zeta) &= d(\alpha(p))d(\beta)d(\alpha(p'))^{-1}\varphi(p,\zeta) \\ &= d(\beta)d(\alpha(p'))^{-1}\varphi(p_0,\zeta) \\ &= \textstyle\sum_\eta q(\beta)_{\zeta\eta}d(\alpha(p')^{-1})\varphi(p_0,\eta) = \sum_\eta q(\beta)_{\zeta\eta}\varphi(p',\eta). \end{aligned} \tag{67a}$$

This shows that all representations of the whole inhomogeneous Lorentz group are equivalent which have the same $P$ and the same representation of the little group. Further than this, the same holds even if the representations of the little group are not the same for the two representations but only equivalent to each other. Let us assume $q_1(\Lambda) = sq_2(\Lambda)s^{-1}$. Then by replacing $\varphi(p,\zeta)$ by $\sum_\eta s(\zeta,\eta)\varphi(p,\eta)$ we obtain a new form of the representation for which (53) still holds but $q_2(\beta)$ for the little group is replaced by $q_1(\beta)$. Then, by the transformation just described (Eq. (66)), we can bring $d(\Lambda)$ for both into the form (67a). The equivalence of two representations of the little group must be defined as the existence of a (*unitary*) transformation which transforms them into each other. (Only unitary transformations are used for the whole group, also).

On the other hand, if the representations of the whole group are equivalent, the representations of the little group are equivalent also: the representation of the whole group determines the representation of the little group up to a similarity transformation uniquely.

The representation of the little group was defined as the set of matrices $Q(p_0, \lambda)_{\zeta\eta}$ if the representation is so transformed that (53) and (66a) hold. Having two equivalent representations $D$ and $SDS^{-1} = D^0$ for both of which (53) and (66a) holds, the unitary transformation $S$ bringing the first into the second must leave all displacement operators invariant. Hence, it must have the form (57a), i.e., operate on the $\zeta$ only and depend on $p$ only as on a parameter.

$$S\varphi(p,\zeta) = \sum_\eta S(p)_{\zeta\eta}\varphi(p,\eta). \tag{68}$$

Denoting the matrix $Q$ for the two representations by $Q$ and $Q^0$, the condition $SD(\Lambda) = D^0(\Lambda)S$ gives that

$$\sum_\eta S(p)_{\zeta\eta} Q(p,\Lambda)_{\eta\vartheta} = \sum_\eta Q^0(p,\Lambda)_{\zeta\eta} S(\Lambda^{-1}p)_{\eta\vartheta} \tag{68a}$$

holds, for every $\Lambda$, for almost every $p$. Setting $\Lambda = \alpha(p_1)$ we can let $p$ approach $p_1$ in such a way that (68a) remains valid. Since $Q$ is a continuous function of $p$ both $Q(p,\Lambda)$ and $Q^o(p,\Lambda)$ will approach their limiting value 1. It follows that there is no domain in which

$$S(p_1) = S(\alpha(p_1)^{-1}p_1) = S(p_0) \tag{69}$$

would not hold, i.e., that (69) holds for almost every $p_1$. Since all our equations must hold only for almost every $p$, the $S(p)_{\zeta\eta}$ can be assumed to be independent of $p$ and (68a) then to hold for every $p$ also. It then follows that the representations of the little group $D$ and $D^0$ are transformed into each other by $S_{\zeta\eta}$.

The definition of the little group involved an arbitrarily chosen momentum vector $p_0$. It is clear, however, that the little groups corresponding to two different momentum vectors $p_0$ and $p$ are holomorphic. In fact they can be transformed into each other by $\alpha(p)$: If $\Lambda$ is an element of the little group leaving $p$ invariant then $\alpha(p)^{-1}\Lambda\alpha(p) = \beta$ is an element of the little group which leaves $p_0$ invariant. We can see furthermore from (67a) that if $\Lambda$ is in the little group corresponding to $p$, i.e. $\Lambda p = p$ then the representation matrix $q(\beta)$ of the little group of $p_0$, corresponding to $\beta$, is identical with the representation matrix of the little group of $p$, corresponding to $\Lambda = \alpha(p)\beta\alpha(p)^{-1}$. Thus when characterizing a representation of the whole inhomogeneous Lorentz group by $P$ and the representation of the little group, it is not necessary to say which $p_0$ is left invariant by the little group.

## D.

Lastly we shall determine the constitution of the little group in the different cases.

$1_+$. In case $1_+$ we can take for $p_0$ the vector with the components 0, 0, 0, 1. The little group which leaves this invariant obviously contains all rotations in the space of the first three coordinates. This holds for the little group of all representations of the first class.

$0_0$. In case $0_0$, the little group is the whole homogeneous Lorentz group.

$-1$. In case $P = -1$ the $p_0$ can be assumed to have the components 1, 0, 0, 0. The little group then contains all transformations which leave the form $-x_2^2 - x_3^2 + x_4^2$ invariant, i.e., is the $2+1$ dimensional homogeneous Lorentz group. The same holds for all representations with $P < 0$.

$0_+$. The determination of the little group for $P = 0_+$ is somewhat more complicated. It can be done, however, rather simply, for the group of unimodular two dimensional matrices. The Lorentz transformation corresponding to the matrix $\begin{pmatrix} a & b \\ c & d \end{pmatrix}$ with $ad - bc = 1$ brings the vector with the components $x_1, x_2, x_3, x_4,$

into the vector with the components $x_1'$, $x_2'$, $x_3'$, $x_4'$, where[18]

$$\begin{pmatrix} a & b \\ c & d \end{pmatrix} \begin{pmatrix} x_4 + x_3 & x_1 + ix_2 \\ x_1 - ix_2 & x_4 - x_3 \end{pmatrix} \begin{pmatrix} a^* & c^* \\ b^* & d^* \end{pmatrix} = \begin{pmatrix} x_4' + x_3' & x_1' + ix_2' \\ x_1' - ix_2' & x_4' - x_3' \end{pmatrix} \tag{70}$$

The condition that a null-vector $p_0$, say with the components 0, 0, 1, 1 be invariant is easily found to be $|a|^2 = 1$, $c = 0$. Hence the most general element of the little group can be written

$$\begin{pmatrix} e^{-i\beta/2} & (x+iy)e^{i\beta/2} \\ 0 & e^{i\beta/2}, \end{pmatrix} \tag{71}$$

with real $x$, $y$, $\beta$ and $0 \le \beta < 4\pi$. The general element (71) can be written as $t(x,y)\delta(\beta)$ where

$$t(x,y) = \begin{pmatrix} 1 & x+iy \\ 1 & 0 \end{pmatrix}; \quad \delta(\beta) = \begin{pmatrix} e^{-\beta/2} & 0 \\ 0 & e^{i\beta/2}. \end{pmatrix} \tag{71a}$$

The multiplication rules for these are

$$t(x,y)t(x',y') = t(x+x', y+y'), \tag{71b}$$

$$\delta(\beta)t(x,y) = t(x\cos\beta + y\sin\beta, -x\sin\beta + y\cos\beta)\delta(\beta), \tag{71c}$$

$$\delta(\beta)\delta(\beta') = \delta(\beta+\beta'). \tag{71d}$$

One could restrict the variability domain of $\beta$ in $\delta(\beta)$ from 0 to $2\pi$. As $\delta(2\pi)$ commutes with all elements of the little group, it will be a constant and from $\delta(2\pi)^2 = \delta(4\pi) = 1$ it can be $\delta(2\pi) = \pm 1$. Hence $\delta(\beta + 2\pi) = \pm\delta(\beta)$ and inserting a $\pm$ into equation (71d) one could restrict $\beta$ to $0 \le \beta < 2\pi$.

These equations are analogous to the equations (52)-(52D) and show that the little group is, in this case, isomorphic with the inhomogeneous rotation group of two dimensions, i.e. the two dimensional Euclidean group.

It may be mentioned that the Lorentz transformations corresponding to $t(x,y)$ have elementary divisors, and constitute all transformations of class e) in 4B, for which $v_e = p_0$. The transformations $\delta(\beta)$ can be considered to be rotations in the ordinary three dimensional space, about the direction of the space part of the vector $p_0$. It is possible, then, to prove equations (71) also directly.

## 7. The Representations of the Little Groups

### A. Representations of the three dimensional rotation group by unitary transformations.

The representations of the three dimensional rotation group in a space with a finite member of dimensions are well known. There is one irreducible representation with the dimensions 1, 2, 3, 4, $\cdots$ each, the representations with an odd number of dimensions are single valued, those with an even number of dimensions are two-valued. These representations will be denoted by $D^{(j)}(R)$ where the dimension is $2j+1$. Thus for single valued representations $j$ is an integer, for double valued representations a half integer. Every finite dimensional representation can be decomposed into these irreducible representations. Consequently those representations of the Lorentz group with positive $P$ in which the representation of the little group—as defined by (64)—has a finite number of dimensions, can be decomposed into such representations in which the representation of the little group is one of the well known irreducible representations of the rotation group. This result will hold for all representations of the inhomogeneous Lorentz group with positive $P$, since we shall show that even the infinite dimensional representations of the rotation group can be decomposed into the same, finite, irreducible representations.

In the following, it is more appropriate to consider the subgroup of the two dimensional unimodular group which corresponds to rotations, than the rotation group itself, as we can restrict ourselves to single valued representations in this case (cf. equations (52)). From (70), one easily sees[18] that the condition for $\begin{pmatrix} a & b \\ c & d \end{pmatrix}$ to leave the vector with the components 0, 0, 0, 1 invariant is that it shall be unitary. It is, therefore, the two dimensional unimodular unitary group the representations of which we shall consider, instead of the representations of the rotation group.

Let us introduce a discrete coordinate system in the representation space and denote the coefficients of the unitary representation by $q(R)_{k\lambda}$ where $R$ is a two dimensional unitary transformation. The condition for the unitary character of the representation $q(R)$ gives

$$\sum_k q(R)^*_{k\lambda} q(R)_{k\mu} = \delta_{\lambda\mu}; \quad \sum_\lambda q(R)^*_{k\lambda} q(R)_{\nu\lambda} = \delta_{k\nu}, \tag{72}$$

$$\textstyle\sum_k |q(R)_{k\lambda}|^2 = 1; \quad \sum_\lambda |q(R)_{k\lambda}|^2 = 1. \tag{72a}$$

This show also that $|q(R)_{k\lambda}| \leq 1$ and the $q(R)_{k\lambda}$ are therefore, as functions of $R$, square integrable:

$$\int |q(R)_{k\lambda}|^2 dR$$

exists if $\int \cdots dR$ is the well known invariant integral in group space. Since this is finite for the rotation group (or the unimodular unitary group), it can be normalized to 1. We then have

$$\sum_k \int |q(R)_{k\lambda}|^2 dR = \sum_\lambda \int |q(R)_{k\lambda}|^2 dR = 1. \tag{73}$$

The $(2j+1)^{\frac{1}{2}} D^{(j)}(R)_{kl}$ form,[26] a complete set of normalized orthogonal functions for $R$. We set

$$q(R)_{k\lambda} = \sum_{jkl} C^{k\lambda}_{jkl} D^{(j)}(R)_{kl}. \tag{74}$$

[26] H. Weyl and F. Peter, Math. Annal. *97*, 737, 1927.

We shall calculate now the integral over group space of the product of $D^{(j)}(R)^*_{kl}$ and

$$q(RS)_{k\mu} = \sum_\lambda q(R)_{k\lambda} q(S)_{\lambda\mu}. \tag{75}$$

The sum on the right converges uniformly, as for (72a)

$$\sum_{\lambda=N}^{\infty} |q(R)_{k\lambda} q(S)_{\lambda\mu}| \le \left( \sum_{\lambda=N}^{\infty} |q(R)_{k\lambda}|^2 \sum_{\lambda=N}^{\infty} |q(S)_{\lambda\mu}|^2 \right)^{\frac{1}{2}} \le \left( \sum_{\lambda=N}^{\infty} |q(S)_{\lambda\mu}|^2 \right)^{\frac{1}{2}}$$

can be made arbitrarily small by choosing an $N$, independent of $R$, making the last expression small. Hence, (75) can be integrated term by term and gives

$$\int D^{(j)}(R)^*_{kl} q(RS)_{k\mu} dR = \sum_\lambda \int D^{(j)}(R)^*_{kl} q(R)_{k\lambda} q(S)_{\lambda\mu} dR. \tag{76}$$

Substituting $\sum_m D^{(j)}(RS)_{km} D^{(j)}(S^{-1})_{ml}$ for $D^{(j)}(R)_{kl}$ one obtains

$$\begin{aligned} &\sum_m D^{(j)}(S^{-1})^*_{ml} \int D^{(j)}(RS)^*_{km} q(RS)_{k\mu} dR \\ &= \sum_\lambda q(S)_{\lambda\mu} \int D^{(j)}(R)^*_{kl} q(R)_{k\lambda} dR. \end{aligned} \tag{77}$$

In the invariant integral on the left of (77), $R$ can be substituted for $RS$ and we obtain, for (74) and the unitary character

$$\sum_m D^{(j)}(S)_{lm} C^{k\mu}_{jkm} = \sum_\lambda q(S)_{\lambda\mu} C^{k\lambda}_{jkl}. \tag{78}$$

Multiplying (78) by $D^{(h)}(S)^*_{in}$, the integration on the right side can be carried out term by term again, since the sum over $\lambda$ converges uniformly

$$\sum_{\lambda=N}^{\infty} |C^{k\lambda}_{jkl} q(S)_{\lambda\mu}| \le \left( \sum_{\lambda=N}^{\infty} |C^{k\lambda}_{jkl}|^2 \sum_{\lambda=N}^{\infty} |q(S)_{\lambda\mu}|^2 \right)^{\frac{1}{2}} \le \left( \sum_{\lambda=N}^{\infty} |C^{k\lambda}_{jkl}|^2 \right)^{\frac{1}{2}}.$$

This can be made arbitrarily small, as even $\sum_\lambda \sum_{jkl} (2j+1)^{-1} |C^{k\lambda}_{jkl}|^2$ converges, for (74) and (72a). The integration of (78) yields thus

$$\sum_\lambda C^{k\lambda}_{jkl} C^{\lambda\mu}_{hin} = \delta_{jh} \delta_{li} C^{k\mu}_{jkn}. \tag{79}$$

From $q(R)q(E) = q(R)$ follows $q(E) = 1$ and then $q(R^{-1}) = q(R)^{-1} = q(R)^\dagger$. This, with the similar equation for $D^{(j)}(R)$ gives

$$\begin{aligned} \sum_{jkl} C^{k\lambda}_{jkl} D^{(j)}(R^{-1})_{kl} &= q(R^{-1})_{k\lambda} = q(R)^*_{\lambda k} \\ &= \sum_{jkl} C^{\lambda k *}_{jlk} D^{(j)}(R)^*_{lk} \\ &= \sum_{jkl} C^{\lambda k *}_{jlk} D^{(j)}(R^{-1})_{kl}, \end{aligned} \tag{80}$$

or

$$C_{jkl}^{k\lambda} = C_{jlk}^{\lambda k*}. \tag{81}$$

On the other hand $q(E)_{k\lambda} = \delta_{k\lambda}$ yields

$$\sum_{jk} C_{jkk}^{k\lambda} = \delta_{k\lambda}. \tag{82}$$

These formulas suffice for the reduction of $q(R)$. Let us choose for every finite irreducible representation $D^{(j)}$ an index $k$, say $k = 0$. We define then, in the original space of the representation $q(R)$ vectors $v^{(kjl)}$ with the components

$$C_{jkl}^{k1},\ C_{jkl}^{k2},\ C_{jkl}^{k3},\ \ldots.$$

The vectors $v^{(kjl)}$ for different $j$ or $l$ are orthogonal, the scalar product of those with the same $j$ and $l$ is independent of $l$. This follows from (79) and (81)

$$(v^{(\mu j'l')}, v^{(kjl)}) = \sum_{\lambda} C_{j'kl'}^{\mu\lambda*} C_{jkl}^{k\lambda} = \sum_{\lambda} C_{jkl}^{k\lambda} C_{j'l'k}^{\lambda\mu} = \delta_{jj'}\delta_{ll'} C_{jkk}^{k\mu}. \tag{83}$$

The $v^{(kjl)}$ for all $k$, $j$, $l$, form a complete set of vectors. In order to show this, it is sufficient to form, for every $\nu$, linear combination from them, the $\nu$ component of which is 1, all other components 0. This linear combination is

$$\sum_{kjl} C_{jlk}^{\nu k} v^{(kjl)}. \tag{84}$$

In fact, the $\lambda$ component of (84) is, on account of (79) and (82)

$$\sum_{kjl} C_{jlk}^{\nu k} C_{jkl}^{k\lambda} = \sum_{jl} C_{jll}^{\nu\lambda} = \delta_{\nu\lambda}. \tag{85}$$

However, two $v$ with the same $j$ and $l$ but different first indices $k$ are not orthogonal. We can choose for every $j$ and $l$, say $l = 0$ and go through the vectors $v^{(1j0)}$, $v^{(2j0)}, \cdots$ and, following Schmidt's method, orthogonalize and normalize them. The vectors obtained in this way shall be denoted by

$$w^{(nj0)} = \sum_{\lambda} \alpha_{n\lambda}^{j} v^{(\lambda j0)}. \tag{86}$$

Then, since according to (83) the scalar products $(v^{(kjl)}, v^{(\lambda jl)})$ do not depend on $l$, the vectors

$$w^{(njl)} = \textstyle\sum_{\lambda} \alpha_{n\lambda}^{j} v^{(\lambda jl)} \tag{86a}$$

will be mutually orthogonal and normalized also and the vectors $w^{(njl)}$ for all $n$, $j$, $l$ will form a complete set of orthonormal vectors. The same holds for the set of the conjugate complex vectors $w^{(njl)*}$. Using these vectors as coordinate axes for the original representation $q(R)$, we shall find that $q(R)$ is completely reduced. The $\nu$ component of the vector $q(R)v^{(kjl)*}$ obtained by applying $q(R)$ on $v^{(kjl)*}$ is

$$\sum_{\mu} q(R)_{\nu\mu}(v^{(kjl)*})_{\mu} = \sum_{\mu} q(R)_{\nu\mu} C^{\mu k}_{jkl}. \tag{87}$$

The right side is uniformly convergent. Hence, its product with $(2h+1)D^{(h)}(R)^{*}_{in}$ can be integrated term by term giving

$$\sum_{\mu} \int (2h+1) D^{(h)}(R)^{*}_{in} q(R)_{\nu\mu} C^{\mu k}_{jkl} dR = \sum_{\mu} C^{\nu\mu}_{hin} C^{\mu k}_{jlk} = \delta_{hj}\delta_{ln} C^{\nu k}_{jik}. \tag{88}$$

Thus we have for almost all $R$

$$\textstyle\sum_{\mu} q(R)_{\nu\mu}(v^{(kjl)*})_{\mu} = \sum_i C^{\nu k}_{jik} D^{(j)}(R)_{il} = \sum_i D^{(j)}(R)_{il}(v^{(kji)*})_{\nu}, \tag{88a}$$

or

$$q(R)v^{(kjl)*} = \textstyle\sum_i D^{(j)}(R)_{il} v^{(kji)*}. \tag{88b}$$

Since both sides are supposed to be strongly continuous functions of $R$, (88b) holds for every $R$. In (86a), for every $n$, the summation must be carried out only over a finite number of $\lambda$. We can write therefore immediately

$$q(R)w^{(njl)*} = \sum_i D^{(j)}(R)_{il} w^{(nji)*}. \tag{89}$$

This proves that the original representation decomposes in the coordinate system of the $w$ into well known finite irreducible representations $D^{(j)}(R)$. Since the $w$ form a complete orthonormal set of vectors, the transition corresponds to a unitary transformation.

This completes the proof of the complete reducibility of all (finite and infinite dimensional) representations of the rotation group or unimodular unitary group. It is clear also that the same consideration applies for all closed groups, i.e., whenever the invariant integral $\int dR$ converges.

The result for the inhomogeneous Lorentz group is: For every positive numerical value of $P$, the representations of the little group can be, in an irreducible representation, only the $D^{(0)}$, $D^{(\frac{1}{2})}$, $D^{(1)}$, $\cdots$, both for $P_+$ and for $P_-$. All these representations have been found already by Majorana and by Dirac and for positive $P$ there are none in addition to these.

### B. Representations of the two dimensional Euclidean group

This group, as pointed out in section 6, has a great similarity with the inhomogeneous Lorentz group. It is possible, again[24], to introduce "momenta", i.e. variables $\xi$, $\eta$ and $v$ instead of $\zeta$ in such a way that

$$t(x,y)\varphi(p_0,\xi,\eta,v) = e^{i(x\xi+y\eta)}\varphi(p_0,\xi,\eta,v). \tag{90}$$

Similarly, one can define again operators $R(\beta)$

$$R(\beta)\varphi(p_0,\xi,\eta,v) = \varphi(p_0,\xi',\eta',v), \tag{91}$$

where

$$\begin{aligned} \xi' &= \xi\cos\beta - \eta\sin\beta, \\ \eta' &= \xi\sin\beta + \eta\cos\beta. \end{aligned} \tag{91a}$$

Then $\delta(\beta)R(\beta)^{-1} = S(\beta)$ will commute, on account of (71c), with $t(x,y)$ and again contain $\xi$, $\eta$ as parameter only. The equation corresponding to (57a) is

$$\delta(\beta)\varphi(p_0,\xi,\eta,v) = \sum_\omega S(\beta)_{v\omega}\varphi(p_0,\xi',\eta',\omega). \tag{92}$$

One can infer from (90) and (92) again that the variability domain of $\xi$, $\eta$ can be restricted in such a way that all pairs $\xi$, $\eta$ arise from one pair $\xi_0$, $\eta_0$ by a rotation, according (91a). We have, therefore two essentially different cases:

$$\xi^2+\eta^2 = \Xi \neq 0 \qquad (a.)$$
$$\xi^2+\eta^2 = \Xi = 0, \quad \text{i.e.} \quad \xi=\eta=0. \qquad (b.)$$

The positive definite metric in the $\xi$, $\eta$ space excludes the other possibilities of section 6 which were made possible by the Lorentzian metric for the momenta, necessitated by (55).

Case b) can be settled very easily. The "little group" is, in this case, the group of rotations in a plane and we are interested in one and two valued irreducible representations. These are all one dimensional ($e^{is\beta}$)

$$S(\beta) = e^{is\beta} \tag{93}$$

where $s$ is integer or half integer. These representations were also all found by Majorana and by Dirac. For $s=0$ we have simply the equation $\Box\varphi = 0$, for $s = \pm\frac{1}{2}$ Dirac's electron equation without mass, for $s=\pm 1$ Maxwell's electromagnetic equations, etc.

In case a) the little group consists only of the unit matrix and the matrix $\begin{pmatrix} -1 & 0 \\ 0 & -1 \end{pmatrix}$ of the two dimensional unimodular group. This group has two irreducible representations, as (1) and (−1) can correspond to the above two dimensional matrix of the little group. This gives two new representations of the whole inhomogeneous Lorentz group, corresponding to every numerical value of $\Xi$. Both these sets belong to class $0_+$ and two similar new sets belong to class $0_-$.

*The final result is thus as follows*: The representations $P_{+j}$ of the first subclass $P_+$ can be characterized by the two numbers $P$ and $j$. From these $P$ is positive, otherwise arbitrary, while $j$ is an integer or a half integer, positive, or zero. The same holds for the subclass $P_-$. There are three kinds of representations of the subclass $0_+$. Those of the first kind $0_{+s}$ can be characterized by a number $s$, which can be either an integer or a half integer, positive, negative or zero. Those of the second kind $0_+(\Xi)$ are single valued and can be characterized by an arbitrary positive number $\Xi$, those of the third kind $0'_+(\Xi)$ are double-valued and also can be characterized by a positive $\Xi$. The same holds for the subclass $0_-$. The representations of the other classes ($0_0$ and $P$ with $P < 0$) have not been determined.

## 8. Representations of the Extended Lorentz Group

### A.

As most wave equations are invariant under a wider group than the one investigated in the previous sections, and as it is very probable that the laws of physics are all invariant under this wider group, it seems appropriate to investigate now how the results of the previous sections will be modified if we go over from the "restricted Lorentz group" defined in section 4A, to the extended Lorentz group. This extended Lorentz group contains in addition to the translations all the homogeneous transformations $X$ satisfying (10)

$$XFX' = F \tag{10'}$$

while the homogeneous transformations of section 4A were restricted by two more conditions. From (10') it follows that the determinant of $X$ can be $+1$ or $-1$ only. If its $-1$, the determinant of $X_1 = XF$ is $+1$. If the four-four element of $X_1$ is negative, that of $X_2 = -X_1$ is positive. It is clear, therefore, that if $X$ is a matrix of the extended Lorentz group, one of the matrices $X$, $XF$, $-X$, $-XF$ is in the restricted Lorentz group. For $F^2 = 1$, conversely, all homogeneous transformations of the extended Lorentz group can be obtained from the homogeneous transformations of the restricted group by multiplication with one of the matrices

$$1, F, -1, -F. \tag{94}$$

The group elements corresponding to these transformations will be denoted by $E$, $F$, $I$, $IF$. The restricted group contains those elements of the extended group which can be reached continuously from the unity. It follows that the transformation of an element $L$ of the restricted group by $F$, $I$, or $IF$ gives again an element of the restricted group. This is, therefore, an invariant subgroup of the extended Lorentz group. In order to find the representations of the extended Lorentz group, we shall use again Frobenius' method.[15]

We shall denote the operators corresponding in a representation to the homogeneous transformations (94) by $d(E) = 1$, $d(F)$, $d(I)$, $d(IF)$. For deriving the equations (52) it was necessary only to assume the existence of the transformations of the restricted group, it was not necessary to assume that these are the only transformations. These equations will hold, therefore, for elements of the restricted group, in representations of the extended group also. We normalize the indeterminate factors in $d(F)$ and $d(I)$ so that their squares become unity. Then we have $d(F)d(I) = \omega d(I)d(F)$ or $d(I) = \omega d(F)d(I)d(F)$. Squaring this, one obtains $\omega^2 = \pm 1$. We can set, therefore

$$\begin{aligned} d(IF) &= d(I)d(F) = \pm d(F)d(I) \\ d(F)^2 &= d(I)^2 = 1; \quad d(IF)^2 = \pm 1. \end{aligned} \tag{95}$$

Finally, from

$$d(F)D(L_1)d(F) = \omega(L_1)D(FL_1F) \tag{96}$$

we obtain, multiplying this with the similar equation for $L_2$

$$\omega(L_1)\omega(L_2) = \omega(L_1L_2)$$

which, gives $\omega(L) = 1$ as the inhomogeneous Lorentz group (or the group used in (52B)-(52D)) has the only one dimensional representation by the unity (1). In this way, we obtain

$$d(F)D(L)d(F) = D(FLF), \tag{96a}$$
$$d(I)D(L)d(I) = D(ILI), \tag{96b}$$
$$d(IF)D(L)d(IF)^{-1} = D(IFLFI). \tag{96c}$$

## B.

Given a representation of the extended Lorentz group, one can perform the transformations described in section 6A, by considering the elements of the restricted group only. We shall consider here only such representations of the extended group, for which, after having introduced the momenta, all representations of the restricted group are either in class 1 or 2, i.e. $P \geq 0$ but not $0_0$. Following then the procedure of section 6, one can find a set of wave functions for which the operators $D(L)$ of the restricted group have one of the forms, given in section 6 as irreducible representations. We shall proceed, next to find the operator $d(F)$. For the wave functions belonging to an irreducible $D(L)$ of the restricted group, we can introduce a complete set of orthonormal functions $\psi_1(p,\zeta)$, $\psi_2(p,\zeta)$, $\cdots$. We then have

$$D(L)\psi_k(p,\zeta) = \sum_\mu D(L)_{\mu k}\psi_\mu(p,\zeta). \tag{97}$$

The infinite matrices $D(L)_{\mu k}$ defined in (97) are unitary and form a representation which is equivalent to the representation by the operators $D(L)$. The $D(L)$, $d(F)$ are, of course, operators, but the $D(L)_{\mu k}$ are components of a matrix, i.e. numbers. We can now form the wave functions $d(F)\psi_1$, $d(F)\psi_2$, $d(F)\psi_3$, $\cdots$ and apply $D(L)$ to these. For (96a) and (97) we have

$$\begin{aligned} D(L)d(F)\psi_k &= d(F)D(FLF)\psi_k \\ &= d(F)\textstyle\sum_\mu D(FLF)_{\mu k}\psi_\mu \\ &= \textstyle\sum_\mu D(FLF)_{\mu k}d(F)\psi_\mu . \end{aligned} \tag{97a}$$

The matrices $D^0(L)_{\mu k} = D(FLF)_{\mu k}$ give a representation of the restricted group ($FLF$ is an element of the restricted group, we have a new representation by an automorphism, as discussed in section 6B). We shall find out whether $D^0(L)$ is equivalent $D(L)$ or not. The translation operation $D^0$ is

$$T^0(a) = d(F)T(a)d(F) = T(Fa) \tag{98}$$

which, together with (53) shows that $D^0$ has the same $P$ as $D(L)$ itself. In fact, writing

$$U_1\varphi(p,\zeta) = \varphi(Fp,\zeta) \tag{99}$$

one has $U_1^{-1} = U_1$ and one easily calculates $U_1T^0(a)U_1 = T(a)$. Similarly for $U_1d^0(\Lambda)U_1$ one has

$$\begin{aligned} U_1d^0(\Lambda)U_1\varphi(p,\zeta) &= U_1d(F\Lambda F)U_1\varphi(p,\zeta) \\ &= d(F\Lambda F)U_1\varphi(Fp,\zeta) \\ &= \textstyle\sum_\eta Q(Fp,F\Lambda F)_{\zeta\eta}U_1\varphi(F\Lambda^{-1}p,\eta) \\ &= \textstyle\sum_\eta Q(Fp,F\Lambda F)_{\zeta\eta}\varphi(\Lambda^{-1}p,\eta). \end{aligned} \tag{99a}$$

This means that the similarity transformation with $U_1$ brings $T^0(a)$ into $T(a)$ and $d^0(\Lambda)$ into $Q(Fp,F\Lambda F)P(\Lambda)$. Thus the representation of the "little group" in $U_1d^0(\Lambda)U_1$ is

$$q^0(\lambda) = Q(Fp_0, F\lambda F).$$

For this latter matrix, one obtains from (67a)

$$\begin{aligned} q^0(\lambda) &= Q(Fp_0,F\lambda F) = q(\alpha(Fp_0)^{-1}F\lambda F\alpha(Fp_0)) \\ &= q(\lambda^0) \end{aligned} \tag{100}$$

where $\lambda^0$ is obtained from $\lambda$ by transforming it with $F\alpha(Fp_0)$.

The representations $D^0(L)$ and $D(L)$ are equivalent if the representation of $q(\lambda)$ is equivalent to the representation which coordinates $q(\lambda^0)$ to $\lambda$. The $\alpha(Fp_0)$ is a transformation of the restricted group which brings $p_0$ into $\alpha(Fp_0)p_0 = Fp_0$. (Cf. (65).) This transformation is, of course, not uniquely determined but if $\alpha(Fp_0)$ is one, the most general can be written as $\alpha(Fp_0)\iota$, where $\iota p_0 = p_0$ is in the little group. For $q(\iota^{-1}\alpha(Fp_0)^{-1}\Lambda\alpha(Fp_0)\iota) = q(\iota^{-1})q(\alpha(Fp_0)^{-1}\Lambda\alpha(Fp_0))q(\iota)$, the freedom in the choice of $\alpha(Fp_0)$ only amounts to a similarity transformation of $q^0(\lambda)$ and naturally does not change the equivalence or non equivalence of $q^0(\lambda)$ with $q(\lambda)$.

For the case $P_+$, we can choose $p_0$ in the direction of the fourth axis, with components 0, 0, 0, 1. Then $Fp_0 = p_0$ and $\alpha(Fp_0) = 1$. The little group is the group of rotations in ordinary space and $F\lambda F = \lambda$. Hence $q^0(\lambda) = q(\lambda)$ and $D^0(\Lambda)$ is equivalent to $D(\Lambda)$ in this case. The same holds for the representations of class $P_-$.

For $0_+$ we can assume that $p_0$ has the components 0, 0, 1, 1. Then the components of $Fp_0$ are 0, 0, $-1$, 1. For $\alpha(Fp_0)$ we can take a rotation by $\pi$ about the second axis and $F\alpha(Fp_0)$ will be a diagonal matrix with diagonal elements 1, $-1$, 1, 1, i.e., a reflection of the second axis. Thus if $\lambda$ is the transformation in (70), $\lambda^0 = \alpha(Fp_0)^{-1}F\lambda F\alpha(Fp_0)$ is the transformation for which

$$\lambda^0 \begin{pmatrix} x_4 + x_3 & x_1 - ix_2 \\ x_1 + ix_2 & x_4 - ix_3 \end{pmatrix} \lambda^{0\dagger} = \begin{pmatrix} x_4' + x_3' & x_1' - ix_2' \\ x_1' + ix_2' & x_4' - ix_3' \end{pmatrix} \tag{101}$$

This is, however, clearly $\lambda^0 = \lambda^*$. Thus the operators of $q^0(\lambda)$ are obtained from the operators $q(\lambda)$ by (cf. (71a))

$$\begin{aligned} t^0(x,y) &= t(x,-y) \\ \delta^0(\beta) &= \delta(-\beta). \end{aligned} \tag{101a}$$

For the representations $0_{+s}$ with discrete $s$, the $q^0(\lambda)$ and $q(\lambda)$ are clearly inequivalent as $\delta^0(\beta) = (e^{-is\beta})$ and $\delta(\beta) = (e^{is\beta})$, except for $s = 0$, when they are equivalent. For the representations $0_+(\Xi)$, $0'_+(\Xi)$, the $q^0(\lambda)$ and $q(\lambda)$ are equivalent, both in the single valued and the double valued case, as the substitution $\eta \to -\eta$ transforms them into each other. The same holds for representations of the class $0_-$. If $D^0(L)$ and $D(L)$ are equivalent

$$U^{-1}D^0(L)U = D(L), \tag{102}$$

the square of $U$ commutes with all $D(L)$. As a consequence of this, $U^2$ must be a constant matrix. Otherwise, one could form, in well known manner, [27] an idempotent which is a function of $U^2$ and thus commutes with $D(L)$ also. Such an idempotent would lead to a reduction of the representation $D(L)$ of the restricted group. As a constant is free in $U$, we can set

$$U^2 = 1 \tag{102a}$$

## C.

[27] J. von Neumann, Ann. of Math. *32*, 191, 1931; ref. 2, p. 89.

Returning now to equation (97a), if $D^0(L) = D(FLF)$ and $D(L)$ are equivalent ($P > 0$ or $0_+$, $0_-$ with continuous $\Xi$ or $s = 0$) there is a unitary matrix $U_{\mu\nu}$, corresponding to $U$, such that

$$\begin{aligned} \textstyle\sum_\mu D(FLF)_{k\mu} U_{\mu\nu} &= \textstyle\sum_\mu U_{k\mu} D(L)_{\mu\nu} \\ \textstyle\sum_\mu U_{k\mu} U_{\mu\nu} &= \delta_{k\nu}. \end{aligned} \tag{102b}$$

Let us now consider the functions

$$\varphi_\nu = \psi_\nu + \sum_\mu U_{\mu\nu} d(F) \psi_\mu. \tag{103}$$

Applying $D(L)$ to these

$$\begin{aligned} D(L)\varphi_\nu &= D(L)\psi_\nu + \textstyle\sum_\mu U_{\mu\nu} D(L) d(F) \varphi_\mu \\ &= D(L)\psi_\nu + \textstyle\sum_\mu U_{\mu\nu} d(F) D(FLF) \psi_\mu \\ &= \textstyle\sum_\mu D(L)_{\mu\nu} \psi_\mu + \textstyle\sum_{\mu k} U_{\mu\nu} d(F) D(FLF)_{k\mu} \psi_k \\ &= \textstyle\sum_\mu D(L)_{\mu\nu} \left(\psi_\mu + \textstyle\sum_k U_{k\mu} d(F) \psi_k\right) = \textstyle\sum_\mu D(L)_{\mu\nu} \varphi_\mu. \end{aligned} \tag{103a}$$

Similarly

$$\begin{aligned} d(F)\varphi_\nu &= d(F)\psi_\nu + \textstyle\sum_\mu U_{\mu\nu} \psi_\mu \\ &= \textstyle\sum_\mu U_{\mu\nu} \left(\psi_\mu + \textstyle\sum_k U_{k\mu} d(F) \psi_k\right) = \textstyle\sum_\mu U_{\mu\nu} \varphi_\mu. \end{aligned} \tag{103b}$$

Thus the wave functions $\varphi$ transform according to the representation in which $D(L)_{\mu\nu}$ corresponds to $L$ and $U_{\mu\nu}$ to $d(F)$. The same holds for the wave functions

$$\varphi'_\nu = \psi_\nu - \sum_\mu U_{\mu\nu} d(F) \psi_\mu, \tag{104}$$

except that in this case $(-U_{\mu\nu})$ corresponds to $d(F)$. The $\psi_\nu$ and $d(F)\psi_\nu$ can be expressed by the $\varphi$ and $\varphi'$. If the $\psi$ and $d(F)\psi$ were linearly independent, the $\varphi$ and $\varphi'$ will be linearly independent also. If the $d(F)\psi$ were linear combinations of the $\psi$, either the $\varphi$ or the $\varphi'$ will vanish.

If we imagine a unitary representation of the group formed by the $L$ and $FL$ in the form in which it is completely reduced out as a representation of the group of restricted transformations $L$, the above procedure will lead to a reduction of that part of the representation of the group of $L$ and $FL$, for which $D(L)$ and $D(FLF)$ are equivalent.

If $D(L)_{\mu\nu}$ and $D^0(L)_{\mu\nu}$ are inequivalent, the $\psi_k$ and $d(F)\psi_\nu = \psi'_\nu$ are orthogonal. This is again a generalization of the similar rule for finite unitary representations.[28] One can see this in the following way: Denoting $M_{k\nu} = (\psi_k, \psi'_\nu)$ one has

$$M_{k\nu} = (\psi_k, \psi'_\nu) = (D(L)\psi_k, D(L)\psi'_\nu)$$

[28]Cf. e.g. E. Wigner, ref. 4, Chapter XII.

$$= \sum_{\mu\lambda} D(L)^*_{\mu k} D^0(L)_{\lambda\nu} M_{\mu\lambda};$$

$$M = D(L)^\dagger M D^0(L).$$

Hence

$$D(L)M = MD^0(L); \quad M^\dagger D(L) = D^0(L)M^\dagger. \tag{105}$$

From these, one easily infers that $MM^\dagger$ commutes with $D(L)$, and $M^\dagger M$ commutes with $D^0(L)$. Hence both are constant matrices, and if neither of them is zero, $M$ and $M^\dagger$ are, apart from a constant, unitary. Thus $D(L)$ would be equivalent $D^0(L)$ which is contrary to supposition. Hence $MM^\dagger = 0$, $M = 0$ and the $\psi$ are orthogonal to the $d(F)\psi = \psi'$. Together, they give a representation of the group formed by the restricted Lorentz group and $F$. If they do not form a complete set, the reduction can be continued as before.

One sees, thus, that introducing the operation $F$ "doubles" the number of dimensions of the irreducible representations in which the little group was the two dimensional rotation group, while it does not increase the underlying linear manifold in the other cases. This is analogous to what happens, if one adjoins the reflection operation to the rotation groups themselves.[29]

**D.**

The operations $d(I)$ can be determined in the same manner as the $d(F)$ were found. A complete set of orthonormal functions corresponding to an irreducible representation of the group formed by the $L$ and $FL$ shall be denoted by $\psi_1$, $\psi_2$, $\cdots$. For this, we shall assume (97) again, although the $D(L)$ contained therein is now not necessarily irreducible for the restricted group alone but contains, in case of $0_{+s}$ or $0_{-s}$ and finite $s$, both $s$ and $-s$. We shall set, furthermore

$$d(F)\psi_k = \sum_\mu d(F)_{\mu k}\psi_\mu. \tag{106}$$

We can form then the functions $d(I)\psi_1$, $d(I)\psi_2$, $\cdots$. The consideration, contained in (97a) shows that these transform according to $D(ILI)_{\mu k}$ for the transformation $L$ of the restricted group:

$$D(L)d(I)\psi_k = \textstyle\sum_\mu D(ILI)_{\mu k} d(I)\psi_\mu. \tag{106a}$$

Choosing for $L$ a pure translation, a consideration analogous to that performed in (98) shows that the set of momenta in the representation $L \to D(ILI)$ has the opposite sign to the set of momenta in the representation $D(L)$. If the latter belongs to a positive subclass, the former belongs to the corresponding negative subclass

[29] I. Schur, Sitz. d. kön. Preuss. Akad. pages 189, 297, 1924.

and conversely. Thus the adjunction of the transformation $I$ always leads to a "doubling" of the number of states, the states of "negative energy" are attached to the system of possible states. One can describe all states $\psi_1, \psi_2, \cdots, d(I)\psi_1, d(I)\psi_2, \cdots$ by introducing momenta $p_1, p_2, p_3, p_4$ and restricting the variability domain of $p$ by the condition $\{p, p\} = P$ alone without stipulating a definite sign for $p_4$.

As we saw before, the $d(I)\psi_1, d(I)\psi_2$, are orthogonal to the original set of wave functions $\psi_1, \psi_2, \cdots$. The result of the application of the operations $D(L)$ and $d(F)$ to the $\psi_1, \psi_2, \cdots$ (i.e., the representation of the group formed by the $L$, $FL$) was given in part C. The $D(L)d(I)\psi_k$ are given in (106a). On account of the normalization of $d(I)$ we can set

$$d(I)d(I)\psi_k = \psi_k. \tag{106b}$$

For $d(F)d(I)\psi_k$ we have two possibilities, according to the two possibilities in (95). We can either set

$$d(F)d(I)\psi_k = d(I)d(F)\psi_k = \sum_\mu d(F)_{\mu k} d(I)\psi_\mu. \tag{107}$$

or

$$d(F) \cdot d(I)\psi_k = -d(I)d(F)\psi_k = -\textstyle\sum_\mu d(F)_{\mu k} d(I)\psi_\mu. \tag{107a}$$

Strictly speaking, we thus obtain two different representations. The system of states satisfying (107) could be distinguished from the system of states for which (107a) is valid, however, only if we could really perform the transition to a new coordinate system by the transformation $I$. As this is, in reality, impossible, the representations distinguished by (107) and (107a) are not different in the same sense as the previously described representations are different.

I am much indebted to the Wisconsin Alumni Research Foundation for their aid enabling me to complete this research.

Madison, Wis.

# A.3 Unitary Representations of the Lorentz Group

by P.A.M. Dirac

reprinted with permission from

Proceedings of the Royal Society (London). **A183**, 284-295 (1945)

# Unitary representations of the Lorentz group

By P. A. M. Dirac, F.R.S., *St John's College, Cambridge*

(*Received* 31 *May* 1944)

Certain quantities are introduced which are like tensors in space-time with an infinite enumerable number of components and with an invariant positive definite quadratic form for their squared length. Some of the main properties of these quantities are dealt with, and some applications to quantum mechanics are pointed out.

## 1. Introduction

Given any group, an important mathematical problem is to get a matrix representation of it, which means to make each element of the group correspond to a matrix in such a way that the matrix corresponding to the product of two elements is the product of the matrices corresponding to the factors. The matrices may be looked upon as linear transformations of the co-ordinates of a vector and then each element of the group corresponds to a linear transformation of a field of vectors. Of special interest are the *unitary* representations, in which the linear transformations leave invariant a positive definite quadratic form in the co-ordinates of a vector.

The Lorentz group is the group of linear transformations of four real variables $\xi_0, \xi_1, \xi_2, \xi_3$, such that $\xi_0^2-\xi_1^2-\xi_2^2-\xi_3^2$ is invariant. The finite representations of this group, i.e. those whose matrices have a finite number of rows and columns, are all well known, and are dealt with by the usual tensor analysis and its extension spinor analysis. None of them is unitary. The group has also some infinite representations which are unitary. These do not seem to have been studied much, in spite of their possible importance for physical applications.

The present paper gives a new method of attack on these representations, which was suggested by Fock's quantum theory of the harmonic oscillator. It leads to a new kind of tensor quantity in space-time, with an infinite number of components and a positive definite expression for its squared length.

## 2. Three-dimensional theory

This section will be devoted to some preliminary work applying to the rotation group of three-dimensional Euclidean space. Take an ascending power series

$$a_0+a_1\xi_1+a_2\xi_1^2+a_3\xi_1^3+\dots \tag{1}$$

in a real variable $\xi_1$ with real coefficients $a_r$. Consider these coefficients to be the co-ordinates of a vector in a certain space of an infinite number of dimensions, and define the squared length of the vector to be

$$\Sigma_0^\infty r!\,a_r^2. \tag{2}$$

The series (2) must converge for the vector to be a finite one.

Take two more similar power series $\Sigma_0^\infty b_s \xi_2^s$ and $\Sigma_0^\infty c_t \xi_3^t$ in the real variables $\xi_2$ and $\xi_3$ and consider their coefficients to be the co-ordinates of vectors in two more vector spaces, with squared lengths defined by the corresponding formula to (2). Now multiply the three vector spaces together. A general vector in the product space will be a sum of products of vectors of the three original vector spaces, and its co-ordinates $A_{rst}$ can be represented as the coefficients in a power series

$$P = \Sigma_0^\infty A_{rst} \xi_1^r \xi_2^s \xi_3^t \tag{3}$$

in the three variables $\xi_1, \xi_2, \xi_3$. The squared length of such a vector is

$$\Sigma_0^\infty r!s!t!A_{rst}^2, \tag{4}$$

and two vectors with co-ordinates $A_{rst}$ and $B_{rst}$ have a scalar product

$$\Sigma_0^\infty r!s!t!A_{rst}B_{rst}. \tag{5}$$

If the variables $\xi_1, \xi_2, \xi_3$ are subjected to a linear transformation, going over into $\xi_1', \xi_2', \xi_3'$, say, the power series (3) will go over into a power series in $\xi_1', \xi_2', \xi_3'$,

$$P = \Sigma A'_{rst} \xi_1'^r \xi_2'^s \xi_3'^t,$$

in which the coefficients $A'$ are linear functions of the previous coefficients $A$. Thus each linear transformation of the $\xi$'s generates a linear transformation of the coefficients $A$.

The theorem will now be proved: *A linear transformation of $\xi_1, \xi_2, \xi_3$ which leaves $\xi_1^2+\xi_2^2+\xi_3^2$ invariant generates a linear transformation of the coefficients $A$ which leaves the squared length* (4) *invariant.* Consider first the infinitesimal transformation

$$\xi_1 = \xi_1' + \epsilon\xi_2', \quad \xi_2 = \xi_2' - \epsilon\xi_1', \quad \xi_3 = \xi_3', \tag{6}$$

which leaves $\xi_1^2+\xi_2^2+\xi_3^2$ invariant, $\epsilon$ being a small quantity whose square is negligible. Substituting into (3), one gets

$$P = \Sigma A_{rst}(\xi_1'^r \xi_2'^s + r\epsilon \xi_1'^{r-1}\xi_2'^{s+1} - s\epsilon\xi_1'^{r+1}\xi_2'^{s-1})\,\xi_3'^t.$$

Hence
$$A'_{rst} = A_{rst} + (r+1)\epsilon A_{r+1,s-1,t} - (s+1)\epsilon A_{r-1,s+1,t},$$

in which $A_{rst}$ with a negative suffix is counted as zero. Thus

$$\Sigma r!s!t!A'^2_{rst} = \Sigma r!s!t![A^2_{rst} + 2(r+1)\epsilon A_{rst}A_{r+1,s-1,t} - 2(s+1)\epsilon A_{r-1,s+1,t}A_{rst}].$$

The last two terms in the [ ] here cancel, as may be seen by substituting $r-1$ for $r$ in the former and $s-1$ for $s$ in the latter, and hence the squared length (4) is invariant for the transformation (6). Any linear transformation of $\xi_1, \xi_2, \xi_3$ which leaves $\xi_1^2+\xi_2^2+\xi_3^2$ invariant can be built up from the infinitesimal transformation (6) and similar infinitesimal transformations with $\xi_1$, $\xi_2$ and $\xi_3$ permuted, together with possibly a reflexion $\xi_1 = -\xi_1'$, $\xi_2 = \xi_2'$, $\xi_3 = \xi_3'$, which obviously leaves the squared length (4) invariant, and hence the theorem is proved.

The group of transformations of the $\xi$'s which leave $\xi_1^2+\xi_2^2+\xi_3^2$ invariant is the rotation group in three-dimensional Euclidean space, so the transformations of the coefficients $A$ provide a representation of this rotation group. One may restrict the function $P$ to be homogeneous, of degree $u$ say, and then the representation is a finite one. The coefficients $A$ then form the components of a symmetrical tensor of rank $u$, the connexion with the usual tensor notation being effected by taking $A_{rst}$ to be $u!/r!s!t!$ times the usual tensor component with the suffix 1 occurring $r$ times, 2 occurring $s$ times and 3 occurring $t$ times, as may be seen from the invariance of expression (3) with $(\xi_1, \xi_2, \xi_3)$ transforming like a vector.

One can make a straightforward generalization of the foregoing theory by introducing other triplets of variables, $\eta_1, \eta_2, \eta_3$ and $\zeta_1, \zeta_2, \zeta_3$ say, which transform together with $\xi_1, \xi_2, \xi_3$, and setting up a power series in all the variables. The transformations of the coefficients of these more general power series will provide further representations of the three-dimensional rotation group. If such a more general power series is restricted to be homogeneous, its coefficients will form the components of an unsymmetrical tensor.

### 3. Four-dimensional theory

Take a descending power series

$$k_0/\xi_0 + k_1/\xi_0^2 + k_2/\xi_0^3 + k_3/\xi_0^4 + \dots \tag{7}$$

in a real variable $\xi_0$ with real coefficients $k_n$. Consider these coefficients to be the co-ordinates of a vector in a certain space of an infinite number of dimensions, and define the squared length of the vector to be

$$\Sigma_0^\infty k_n^2/n!. \tag{8}$$

Multiply this vector space into the vector space of the preceding section. A general vector in the product space will have co-ordinates $A_{nrst}$ which can be represented as the coefficients in a power series

$$Q = \Sigma_0^\infty A_{nrst}\xi_0^{-n-1}\xi_1^r\xi_2^s\xi_3^t \tag{9}$$

in the four variables $\xi_0, \xi_1, \xi_2, \xi_3$. The squared length of such a vector is

$$\Sigma_0^\infty n!^{-1}r!s!t!A_{nrst}^2, \tag{10}$$

and two vectors with co-ordinates $A_{nrst}$ and $B_{nrst}$ have a scalar product

$$\Sigma_0^\infty n!^{-1}r!s!t!A_{nrst}B_{nrst}. \tag{11}$$

The series (7) may be extended backwards to include some terms with non-negative powers of $\xi_0$, so that coefficients $k_n$ occur with negative $n$-values, leading to coefficients $A_{nrst}$ with negative $n$-values. Since $n!$ is infinite for $n$ negative, these new coefficients do not contribute to the squared length of a vector or the scalar product of two vectors. Thus the terms with non-negative powers of $\xi_0$ should be counted as corresponding to the vector zero, and whether they are present in an expansion or not does not matter.

Now apply a Lorentz transformation to the $\xi$'s,

$$\xi_\mu = \alpha^\nu_\mu \xi'_\nu, \tag{12}$$

the $\alpha$'s satisfying certain conditions so that $\xi_0^2-\xi_1^2-\xi_2^2-\xi_3^2$ is invariant. This makes $\xi_1^r\xi_2^s\xi_3^t$ go over into a finite polynomial in $\xi'_0, \xi'_1, \xi'_2, \xi'_3$, and $\xi_0^{-n-1}$ go over into

$$(\alpha^0_0\xi'_0+\alpha^1_0\xi'_1+\alpha^2_0\xi'_2+\alpha^3_0\xi'_3)^{-n-1}, \tag{13}$$

which may be expanded in ascending powers of $\xi'_1, \xi'_2, \xi'_3$ and descending powers of $\xi'_0$. (The question of the convergence of this expansion is left to the next section, so as not to break the main argument here.) The power series (9) then goes over into a series

$$Q = \Sigma A'_{nrst}\xi_0'^{-n-1}\xi_1'^r\xi_2'^s\xi_3'^t \tag{14}$$

in ascending powers of $\xi_1, \xi_2, \xi_3$ and descending powers of $\xi_0$, with coefficients $A'$ which are linear functions of the previous coefficients $A$. There may be terms in (14) with non-negative powers of $\xi_0$, but on account of what was said above these can be discarded. If such terms are present in the original series (9), they will not affect any of the coefficients with non-negative $n$-values in (14).

The Lorentz transformation thus generates a linear transformation of the coefficients $A_{nrst}$ with non-negative $n$-values. The theorem will now be proved: *The transformation of the coefficients A leaves the squared length* (10) *invariant.*

Consider first the infinitesimal Lorentz transformation

$$\xi_0 = \xi'_0+\epsilon\xi'_1, \quad \xi_1 = \xi'_1+\epsilon\xi'_0, \quad \xi_2 = \xi'_2, \quad \xi_3 = \xi'_3. \tag{15}$$

Substituting into (9), one gets

$$Q = \Sigma A_{nrst}[\xi_0'^{-n-1}\xi_1'^r-(n+1)\epsilon\xi_0'^{-n-2}\xi_1'^{r+1}+r\epsilon\xi_0'^{-n}\xi_1'^{r-1}]\,\xi_2'^s\xi_3'^t.$$

Hence
$$A'_{nrst} = A_{nrst}-n\epsilon A_{n-1,r-1,s,t}+(r+1)\epsilon A_{n+1,r+1,s,t},$$

in which $A_{nrst}$ with a negative value for $r$, $s$ or $t$ is counted as zero. Thus

$$\Sigma n!^{-1}r!\,s!\,t!\,A'^2_{nrst}$$
$$= \Sigma n!^{-1}r!\,s!\,t!\,[A^2_{nrst}-2n\epsilon A_{nrst}A_{n-1,r-1,s,t}+2(r+1)\epsilon A_{n+1,r+1,s,t}A_{nrst}].$$

The last two terms in the [ ] here cancel, as may be seen by substituting $n+1$ for $n$ in the former and $r-1$ for $r$ in the latter, and hence the squared length (10) is invariant for the transformation (15). Any Lorentz transformation can be built up from infinitesimal transformations like (15) and three-dimensional rotations like those considered in the preceding section, together with possibly a reflexion, which obviously leaves the squared length (10) invariant, and hence the theorem is proved.

The transformations of the coefficients $A$ thus provide a unitary representation of the Lorentz group. The coefficients themselves form the components of a new kind of tensor quantity in space-time. I propose for it the name *expansor*, because of its connexion with binomial expansions. One may restrict the function (9) to be homogeneous and one then gets a simpler kind of expansor, which may be called a *homogeneous expansor*. The analogy with the three-dimensional case suggests that one should look upon a homogeneous expansor as a symmetrical tensor in space-time with the suffix 0 occurring in its components a negative number of times.

The foregoing theory can be generalized, like the three-dimensional theory, by the introduction of other quadruplets of variables, $\eta_0, \eta_1, \eta_2, \eta_3$ and $\zeta_0, \zeta_1, \zeta_2, \zeta_3$ say, which transform together with $\xi_0, \xi_1, \xi_2, \xi_3$. One can then set up a power series in all the variables, ascending in those variables with suffixes 1, 2 and 3, and descending in those variables with suffix 0. The transformations of the coefficients of these more general power series will provide further unitary representations of the Lorentz group, and the coefficients themselves will form the components of more general expansors.

There is another generalization which may readily be made in the theory, namely, to take the values of the index $n$ in (9) to be not integers, but any set of real numbers $n_0, n_0+1, n_0+2, \ldots$ extending to infinity. In the formula (10) for the squared length $n!$ is then to be interpreted as $\Gamma(n+1)$. The terms with negative $n$-values can no longer be discarded. The expression for the squared length is still positive definite if the minimum value of $n$ is greater than $-1$, which is the case if the function (9) is homogeneous and its degree is negative. The resulting representation is then still unitary. If, however, the function (9) is homogeneous and its degree is positive, there will be a finite number of negative terms in the expression for the squared length. The resulting kind of representation may be called *nearly unitary*.

## 4. Some theorems on convergence

If the series (2) is convergent, then (1) is convergent for all values of $\xi_1$. Similarly, if (4) is convergent, then (3) is convergent for all values of $\xi_1, \xi_2, \xi_3$. On the other hand, if (8) is convergent, (7) need not be convergent for any value of $\xi_0$, in which case, of course, it does not define a function of $\xi_0$. Similarly, if (10) is convergent, (9) need not be convergent for any values for the $\xi$'s. Thus, corresponding to a general expansor $A_{nrst}$, there need not exist any function $Q$ of the $\xi$'s. However, it will now be proved that *if the series* (9) *is homogeneous and* (10) *is convergent, then* (9) *is absolutely convergent for all values of the $\xi$'s satisfying*

$$\xi_0^2-\xi_1^2-\xi_2^2-\xi_3^2>0. \tag{16}$$

If the series (9) is homogeneous of degree $u-1$, it may be written

$$\Sigma_n \xi_0^{-n-1}\Sigma_g A_{nrst}\xi_1^r\xi_2^s\xi_3^t, \tag{17}$$

where $\Sigma_g$ means a sum over all values of $r$, $s$ and $t$ satisfying

$$r+s+t=n+u. \tag{18}$$

With this notation the series (10) may be written

$$\Sigma_n n!^{-1}\Sigma_g r!\,s!\,t!\,A_{nrst}^2. \tag{19}$$

Now apply Cauchy's inequality

$$(x_1y_1+x_2y_2+x_3y_3+\dots)^2 \leqslant (x_1^2+x_2^2+x_3^2+\dots)(y_1^2+y_2^2+y_3^2+\dots) \tag{20}$$

in the following way. Take $(r+s+t)!/r!s!t!$ of the $x$'s in (20) to be each equal to $\xi_1^r\xi_2^s\xi_3^t$ and the corresponding $y$'s to be each equal to $A_{nrst}r!s!t!/(r+s+t)!$, and do this for all $r, s, t$ subject to (18) for a fixed value of $n$. Then (20) becomes

$$(\Sigma_S A_{nrst}\xi_1^r\xi_2^s\xi_3^t)^2 \leqslant (\xi_1^2+\xi_2^2+\xi_3^2)^{n+u}(n+u)!^{-1}\Sigma_S r!s!t!A_{nrst}^2. \tag{21}$$

If the sum with respect to $n$ in (19) converges, there must be some number $\kappa$ such that

$$n!^{-1}\Sigma_S r!s!t!A_{nrst}^2<\kappa$$

for all $n$. Then from (21)

$$|\xi_0^{-n-1}\Sigma_S A_{nrst}\xi_1^r\xi_2^s\xi_3^t| < \kappa^{\frac{1}{2}}|\xi_0|^{u-1}\left\{\frac{n!}{(n+u)!}\right\}^{\frac{1}{2}}\left\{\frac{\xi_1^2+\xi_2^2+\xi_3^2}{\xi_0^2}\right\}^{\frac{1}{2}(n+u)},$$

which shows that (17) is convergent when (16) is satisfied. One may take $\xi_0, \xi_1, \xi_2, \xi_3$ and the $A$'s to be all positive without disturbing the argument, so (17), considered as a quadruple series in $n, r, s, t$, is absolutely convergent.

Thus for any homogeneous expansor of finite length, there exists a function $Q$ given by (9) defined within the light-cone (16). In transforming this function with a Lorentz transformation (12), it is legitimate to use the expansion of (13) in ascending powers of $\xi_1', \xi_2', \xi_3'$ and descending powers of $\xi_0'$, for the following reason. Suppose $\xi_\mu$ is within the light-cone and take for definiteness $\xi_0>0$. Then

$$\alpha_0^0\xi_0'+\alpha_0^1\xi_1'+\alpha_0^2\xi_2'+\alpha_0^3\xi_3'>0. \tag{22}$$

One can change the sign of any of the co-ordinates $\xi_1', \xi_2', \xi_3'$, leaving $\xi_0'$ unchanged, and $\xi_\mu$ will still lie within the light-cone with $\xi_0>0$, so (22) must still be satisfied. Hence

$$\alpha_0^0\xi_0'-|\alpha_0^1\xi_1'|-|\alpha_0^2\xi_2'|-|\alpha_0^3\xi_3'|>0,$$

which shows that the expansion of (13) in the required manner is absolutely convergent. The use of this expansion in the preceding section is thus justified for the case when (9) is homogeneous, the new coefficients $A$ being determined by the transformed function $Q$ within the light-cone. The justification for (9) not homogeneous then follows since terms of different degree do not interfere. It should be noted that all the foregoing arguments are valid also when the values of $n$ are not integers.

There are some expansors which are invariant under all Lorentz transformations, namely those whose components are the coefficients of $(\xi_0^2-\xi_1^2-\xi_2^2-\xi_3^2)^{-l}$ expanded in ascending powers of $\xi_1, \xi_2, \xi_3$ and descending powers of $\xi_0$. For such an expansion to be possible, $l$ must not be a negative integer or zero, but it can be any other real number. Again, there are expansors which transform like ordinary tensors under Lorentz transformations, namely those whose components are the coefficients in the expansion of

$$f(\xi_0,\xi_1,\xi_2,\xi_3)(\xi_0^2-\xi_1^2-\xi_2^2-\xi_3^2)^{-l}, \tag{23}$$

where $f$ is any homogeneous integral polynomial in $\xi_0, \xi_1, \xi_2, \xi_3$, and $l$ is restricted as before. (23) transforms like the polynomial $f$, and thus like a tensor of order equal to the degree of $f$. One would expect all these expansors to be of infinite length, as otherwise one could set up a positive definite form for the squared length of a tensor in space-time. A formal proof that they are is as follows.

Suppose the $f$ in (23) is of degree $u$ and express it as

$$f = g_u + \xi_0 g_{u-1} + (\xi_0^2 - \xi^2) g_{u-2} + \xi_0(\xi_0^2 - \xi^2) g_{u-3} + (\xi_0^2 - \xi^2)^2 g_{u-4} + \dots,$$

where $\xi^2 = \xi_1^2 + \xi_2^2 + \xi_3^2$ and each of the $g$'s is a polynomial in $\xi_1, \xi_2, \xi_3$ only, of degree indicated by the suffix. The successive terms here contribute to (23) the amounts

$$g_u(\xi_0^2 - \xi^2)^{-l} \qquad = \Sigma_{m=0}^{\infty} \frac{l(l+1)(l+2)\dots(l+m-1)}{m!} \frac{g_u \xi^{2m}}{\xi_0^{2(m+l)}},$$

$$g_{u-1}\xi_0(\xi_0^2 - \xi^2)^{-l} \quad = \Sigma_{m=0}^{\infty} \frac{l(l+1)(l+2)\dots(l+m-1)}{m!} \frac{g_{u-1}\xi^{2m}}{\xi_0^{2(m+l)-1}},$$

$$g_{u-2}(\xi_0^2 - \xi^2)^{-l+1} \quad = \Sigma_{m=-1}^{\infty} \frac{(l-1)l(l+1)\dots(l+m-1)}{(m+1)!} \frac{g_{u-2}\xi^{2(m+1)}}{\xi_0^{2(m+l)}},$$

$$g_{u-3}\xi_0(\xi_0^2 - \xi^2)^{-l+1} = \Sigma_{m=-1}^{\infty} \frac{(l-1)l(l+1)\dots(l+m-1)}{(m+1)!} \frac{g_{u-3}\xi^{2(m+1)}}{\xi_0^{2(m+l)-1}},$$

and so on. These expansions show that, for large values of $m$, the terms arising from $g_{u-2}, g_{u-4}, \dots$ are of smaller order than the corresponding terms arising from $g_u$, and the terms arising from $g_{u-3}, g_{u-5}, \dots$ are of smaller order than the corresponding terms arising from $g_{u-1}$, so that in testing for the convergence of the series which gives the squared length, only the $g_u$ and $g_{u-1}$ terms need be taken into account. (It will be found that the convergence conditions are not sufficiently critical to be affected by this neglect.) Now express $g_u$ and $g_{u-1}$ as

$$g_u = S_u + \xi^2 S_{u-2} + \xi^4 S_{u-4} + \dots, \quad g_{u-1} = S_{u-1} + \xi^2 S_{u-3} + \xi^4 S_{u-5} + \dots,$$

where the $S$'s are solid harmonic functions of $\xi_1, \xi_2, \xi_3$, of degrees indicated by the suffixes. Each of them gives a contribution to (23) of the form

$$S_{u-2r}\xi^{2r}(\xi_0^2 - \xi^2)^{-l} = S_{u-2r}\Sigma_{m=0}^{\infty} \frac{(m+l-1)!}{m!(l-1)!} \frac{\xi^{2(m+r)}}{\xi_0^{2(m+l)}} \tag{24}$$

or

$$S_{u-2r-1}\xi_0\xi^{2r}(\xi_0^2 - \xi^2)^{-l} = S_{u-2r-1}\Sigma_{m=0}^{\infty} \frac{(m+l-1)!}{m!(l-1)!} \frac{\xi^{2(m+r)}}{\xi_0^{2(m+l)-1}}. \tag{25}$$

Using the result (41) of the appendix, one finds for the squared lengths of the expansors whose components are the coefficients of (24) and (25), series of the form

$$c\Sigma_m \frac{(m+l-1)!^2}{m!^2(l-1)!^2} \frac{4^{m+r}(m+r)!(m+u-r+\frac{1}{2})!}{(2m+2l-1)!}$$

and

$$c\Sigma_m \frac{(m+l-1)!^2}{m!^2(l-1)!^2} \frac{4^{m+r}(m+r)!(m+u-r-\frac{1}{2})!}{(2m+2l-2)!},$$

respectively. In both these series the ratio of the $(m+1)$th term to the $m$th is $1+u/m$ for large $m$, and since $u$ is necessarily positive or zero, the series diverge and the expansors are of infinite length. From the result (40) of the appendix, expansors associated with solid harmonics of different degrees are orthogonal to one another, and hence the total expansor with the coefficients of (23) as components must also be of infinite length.

## 5. A TRANSFORMATION OF VARIABLES

Up to the present, expansors have always been considered in connexion with certain $\xi$-functions, the components of the expansor being the coefficients of the $\xi$-function. In the case of integral $n$-values, one can make a transformation of variables of the kind which is familiar in quantum mechanics, and get the expansors connected with some other functions, which serves to clarify some of their properties.

Introduce the four operators $x_\mu$, defined by

$$2^{\frac{1}{2}}x_0 = \xi_0 - \partial/\partial\xi_0, \quad 2^{\frac{1}{2}}x_r = \xi_r + \partial/\partial\xi_r \quad (r = 1, 2, 3). \tag{26}$$

Under Lorentz transformations they transform like the components of a 4-vector. The operators $\partial/\partial x_\mu$ may be taken to be

$$2^{\frac{1}{2}}\,\partial/\partial x_0 = \xi_0 + \partial/\partial\xi_0, \quad 2^{\frac{1}{2}}\,\partial/\partial x_r = -\xi_r + \partial/\partial\xi_r, \tag{27}$$

as this gives the correct commutation relations between all the operators. One can now represent $\xi$-functions with integral $n$-values by functions of the $x$'s according to the following scheme.

Take as starting point the $\xi$-function $\xi_0^{-1}$. It vanishes when operated on by $\partial/\partial\xi_1$, $\partial/\partial\xi_2$, or $\partial/\partial\xi_3$, and also effectively vanishes when multiplied by $\xi_0$, since the result of the multiplication can be discarded. It must therefore be represented by a function of the $x$'s which vanishes when operated on by $x_r + \partial/\partial x_r$ with $r = 1, 2$ or $3$, or by $x_0 + \partial/\partial x_0$, and thus by a multiple of $e^{-\frac{1}{2}(x_0^2+\mathbf{x}^2)}$, where $\mathbf{x}^2 = x_1^2 + x_2^2 + x_3^2$. Take

$$\xi_0^{-1} \equiv \pi^{-1} e^{-\frac{1}{2}(x_0^2+\mathbf{x}^2)}, \tag{28}$$

where the sign $\equiv$ means 'represented by'. Multiply the left-hand side of (28) by the operator $(-\partial/\partial\xi_0)^n\,\xi_1^r\xi_2^s\xi_3^t$ and the right-hand side by the operator equal to it according to (26) and (27). The result is, after dividing through by $n!$,

$$\begin{aligned}\xi_0^{-n-1}\xi_1^r\xi_2^s\xi_3^t &\equiv \pi^{-1} n!^{-1}\, 2^{-\frac{1}{2}(n+r+s+t)} \left(x_0 - \frac{\partial}{\partial x_0}\right)^n \left(x_1 - \frac{\partial}{\partial x_1}\right)^r \left(x_2 - \frac{\partial}{\partial x_2}\right)^s \left(x_3 - \frac{\partial}{\partial x_3}\right)^t e^{-\frac{1}{2}(x_0^2+\mathbf{x}^2)} \\ &= F_{nrst}(x_0, x_1, x_2, x_3),\end{aligned} \tag{29}$$

say. This gives the function $F$ of the $x$'s which represents $\xi_0^{-n-1}\xi_1^r\xi_2^s\xi_3^t$. A general $\xi$-function is now represented thus,

$$\Sigma A_{nrst}\,\xi_0^{-n-1}\xi_1^r\xi_2^s\xi_3^t \equiv \Sigma A_{nrst} F_{nrst}(x_0, x_1, x_2, x_3). \tag{30}$$

In this way a general expansor $A_{nrst}$ gets connected with a function of the $x$'s.

The chief interest of this connexion is that the law for the scalar product of two expansors becomes very simple when expressed in terms of the functions of the $x$'s. *The scalar product is the integral of the product of the two functions of the $x$'s over all $x_0$, $x_1$, $x_2$ and $x_3$.* To prove this, first evaluate the integral

$$\int_{-\infty}^{\infty}\left(z-\frac{d}{dz}\right)^{m} e^{-\frac{1}{2}z^2}.\left(z-\frac{d}{dz}\right)^{m'} e^{-\frac{1}{2}z^2}\,dz, \tag{31}$$

where the dot has the meaning that operators to the left of it do not operate on functions to the right of it. For $m > 0$, (31) goes over by partial integration into

$$\int_{-\infty}^{\infty}\left(z-\frac{d}{dz}\right)^{m-1} e^{-\frac{1}{2}z^2}.\left(z+\frac{d}{dz}\right)\left(z-\frac{d}{dz}\right)^{m'} e^{-\frac{1}{2}z^2}\,dz$$

$$=\int_{-\infty}^{\infty}\left(z-\frac{d}{dz}\right)^{m-1} e^{-\frac{1}{2}z^2}.\left\{\left(z-\frac{d}{dz}\right)^{m'}\left(z+\frac{d}{dz}\right)+2m'\left(z-\frac{d}{dz}\right)^{m'-1}\right\} e^{-\frac{1}{2}z^2}\,dz$$

$$=2m'\int_{-\infty}^{\infty}\left(z-\frac{d}{dz}\right)^{m-1} e^{-\frac{1}{2}z^2}.\left(z-\frac{d}{dz}\right)^{m'-1} e^{-\frac{1}{2}z^2}\,dz.$$

Applying this procedure $m$ times, one gets zero if $m' < m$ and $2^m m!\,\pi^{\frac{1}{2}}$ if $m' = m$. Since there is symmetry between $m$ and $m'$, the result is

$$\int_{-\infty}^{\infty}\left(z-\frac{d}{dz}\right)^{m} e^{-\frac{1}{2}z^2}.\left(z-\frac{d}{dz}\right)^{m'} e^{-\frac{1}{2}z^2}\,dz = 2^m m!\,\pi^{\frac{1}{2}}\delta_{mm'}.$$

Now substitute for $z$ each of the variables $x_0$, $x_1$, $x_2$, $x_3$ in turn, with $m$ equal to $n, r, s, t$ and $m'$ equal to $n', r', s', t'$ respectively, and multiply the four equations so obtained. The result, after dividing through by $\pi^2 n!^2\, 2^{n+r+s+t}$, is

$$\iiiint F_{nrst}F_{n'r's't'}\,dx_0\,dx_1\,dx_2\,dx_3 = n!^{-1}\,r!\,s!\,t!\,\delta_{nn'}\delta_{rr'}\delta_{ss'}\delta_{tt'}.$$

This proves the theorem for the case when the expansors each have only one non-vanishing component, which is sufficient to prove it generally.

One can obviously get a unitary representation of the Lorentz group by considering the transformations of a set of vectors in a space of an infinite number of dimensions, where each vector corresponds to a function of four Lorentz variables $x_0$, $x_1$, $x_2$, $x_3$, and where the squared length of a vector is defined as the integral of the square of the function over all $x_0$, $x_1$, $x_2$, $x_3$. The above work shows how this obvious unitary representation is connected with the expansor theory.

## 6. Applications to quantum mechanics

The four $x$'s of the preceding section may be looked upon as the co-ordinates of a four-dimensional harmonic oscillator, the four operators $i\partial/\partial x_\mu$ being the conjugate momenta $p^\mu$, and the energy of the oscillator may be taken to be the Lorentz invariant

$$\tfrac{1}{2}[x_1^2+x_2^2+x_3^2-x_0^2+(p^1)^2+(p^2)^2+(p^3)^2-(p^0)^2]. \tag{32}$$

The 1, 2, 3 components of the oscillator thus have positive energies and the 0 component negative energy.

The state of the oscillator for which the 0, 1, 2, 3 components are in the $n$th, $r$th, $s$th, $t$th quantum states respectively is then represented by the function $F_{nrst}$ defined by (29), with a suitable normalizing factor. This representative may be transformed to the $\xi$-representation and becomes $\xi_0^{-n-1}\xi_1^r\xi_2^s\xi_3^t$. Thus a state of the oscillator for which each of its components is in a quantum state corresponds to an expansor with one non-vanishing component. A general state of the oscillator therefore corresponds to a general expansor with integral $n$-values. A stationary state of the oscillator corresponds to a homogeneous expansor, the degree of the expansor giving the energy of the state with neglect of zero-point energy.

Four-dimensional harmonic oscillators of the above type occur in the theory of the electromagnetic field. Each Fourier component of the field, specified by a particular frequency and a particular direction of motion of the waves, provides one such oscillator, its four components coming from the four electromagnetic potentials. Thus a state of the electromagnetic field in quantum mechanics is described by a number of expansors, one for each Fourier component. By using the electromagnetic equation which gives the value of the divergence of the potentials, one can eliminate in a non-relativistic way the 0 component and one other component of each of the four-dimensional oscillators, so that only two-dimensional oscillators are left. This circumstance has made it possible for people to develop quantum electrodynamics without using expansors.

Another possible application of expansors is to the spins of particles. The wave function describing a particle may be a function of the four co-ordinates $x_\mu$ of the particle in space-time and also of the four variables $\xi_\mu$ whose coefficients are the components of an expansor. As a simple example of relativistic wave equations for such a particle in the absence of external forces, one may consider

$$\left(\hbar^2 \frac{\partial^2}{\partial x_\mu \partial x^\mu} + m^2\right)\psi = 0, \quad \xi_\mu \frac{\partial}{\partial \xi_\mu}\psi = -\psi, \quad \xi_\mu \frac{\partial}{\partial x_\mu}\psi = 0. \tag{33}$$

The first of these is the usual equation for the motion of the particle as a whole. The second shows that at each point in space-time the wave function $\psi$ is homogeneous in the $\xi$'s of degree $-1$. The third shows that the state for which the momentum-energy four-vector of the particle has the value $p_\mu$ is represented by the wave function

$$\psi = \frac{m}{\xi_0 p_0 - (\xi\mathbf{p})} e^{-i(p_0 x_0 - (\mathbf{px}))/\hbar}$$

in three-dimensional vector notation. This may be expanded as

$$\psi = m e^{-i(p_0 x_0 - (\mathbf{px}))/\hbar} \sum_{n=0}^{\infty} \frac{(\xi\mathbf{p})^n}{\xi_0^{n+1} p_0^{n+1}}. \tag{34}$$

For the state for which the particle is at rest, $p_1 = p_2 = p_3 = 0$, $p_0 = m$, and $\psi$ reduces to

$$\psi = e^{-imx_0/\hbar}\xi_0^{-1}.$$

This $\psi$ is spherically symmetrical, showing that when the particle is at rest it has no spin. But when the particle is moving, it is represented by the general $\psi$ (34) and has a finite probability of a non-zero spin. In fact, taking for simplicity $p_2 = p_3 = 0$, the particle has a probability $(mp_1^n/p_0^{n+1})^2$ of being in a state of spin corresponding to the transformations of $\xi_1^n$ under three-dimensional rotations.

This example shows there is a possibility of a particle having no spin when at rest but acquiring a spin when moving, a state of affairs which was not allowed by previous theory. It is desirable that the new spin possibilities opened up by the present theory should be investigated to see whether they could in some cases give an improved description of Nature. The present theory of expansors applies, of course, only to integral spins, but probably it will be possible to set up a corresponding theory of two-valued representations of the Lorentz group, which will apply to half odd integral spins.

## Appendix

The rules (5), (11) for forming scalar products are not always convenient for direct use. There are various ways of transforming them and making them more suitable for practical application. One such way has been given (Dirac 1942, equation 3·22) for the case of a single $\xi$ with ascending power series. Another way, applicable to the case of homogeneous functions of $\xi_1$, $\xi_2$, $\xi_3$, is provided by the following.

By partial integration with respect to $\xi'$, one gets, for $m > 0$,

$$\iint_{-\infty}^{\infty} \xi^m \xi'^n e^{i\xi\xi'} d\xi d\xi' = \int_{-\infty}^{\infty} \xi^{m-1} d\xi \left\{ \Big[ -i\xi'^n e^{i\xi\xi'} \Big]_{\xi'=-\infty}^{\xi'=\infty} + in \int_{-\infty}^{\infty} \xi'^{n-1} e^{i\xi\xi'} d\xi' \right\}.$$

If the integrals are made precise in the sense of Cesaro, which means neglecting oscillating terms like $\xi'^n e^{i\xi\xi'}$ for $\xi'$ infinite, this gives

$$\iint_{-\infty}^{\infty} \xi^m \xi'^n e^{i\xi\xi'} d\xi d\xi' = in \iint_{-\infty}^{\infty} \xi^{m-1} \xi'^{n-1} e^{i\xi\xi'} d\xi d\xi'.$$

Taking $m \geqslant n$ and applying the partial integration process $n$ times, one gets

$$\begin{aligned} \iint_{-\infty}^{\infty} \xi^m \xi'^n e^{i\xi\xi'} d\xi d\xi' &= i^n n! \iint_{-\infty}^{\infty} \xi^{m-n} e^{i\xi\xi'} d\xi d\xi' \\ &= 2\pi i^n n! \int_{-\infty}^{\infty} \xi^{m-n} \delta(\xi) d\xi \\ &= 2\pi i^n n! \, \delta_{mn}. \end{aligned} \tag{35}$$

It follows that if $A$ and $B$ are homogeneous functions of $\xi_1$, $\xi_2$, $\xi_3$ of degree $u$, their scalar product according to (5) is

$$(AB) = (2\pi)^{-3} i^{-u} \iint_{\dots\dots} A(\xi_1 \xi_2 \xi_3) \, B(\xi_1' \xi_2' \xi_3') \, e^{i(\xi_1\xi_1' + \xi_2\xi_2' + \xi_3\xi_3')} d\xi_1 d\xi_1' d\xi_2 d\xi_2' d\xi_3 d\xi_3'. \tag{36}$$

As an application of this rule, take

$$A = (\xi_1^2+\xi_2^2+\xi_3^2)^r S_{u-2r}, \quad B = (\xi_1^2+\xi_2^2+\xi_3^2)^s S_{u-2s},$$

where the $S$'s are solid harmonic functions. Then, using three-dimensional vector notation, (36) gives

$$(AB) = (2\pi)^{-3} i^{-u} \iint_{\cdots\cdots} \xi^{2r} \xi'^{2s} S_{u-2r}(\xi)\, S_{u-2s}(\xi')\, e^{i(\xi\xi')}\, d\xi_1 \ldots d\xi_3'. \tag{37}$$

From Green's theorem

$$\iiint \left[ e^{i(\xi\xi')} \left( \frac{\partial^2}{\partial \xi_1^2} + \frac{\partial^2}{\partial \xi_2^2} + \frac{\partial^2}{\partial \xi_3^2} \right) \left\{ \xi^{2r} S_{u-2r}(\xi) \right\} - \xi^{2r} S_{u-2r}(\xi) \left( \frac{\partial^2}{\partial \xi_1^2} + \frac{\partial^2}{\partial \xi_2^2} + \frac{\partial^2}{\partial \xi_3^2} \right) e^{i(\xi\xi')} \right] d\xi_1 d\xi_2 d\xi_3$$

equals a surface integral of an oscillating kind which is to be counted as vanishing at infinity. This result reduces to

$$4r(u-r+\tfrac{1}{2}) \iiint_{-\infty}^{\infty} \xi^{2(r-1)} S_{u-2r}(\xi)\, e^{i(\xi\xi')}\, d\xi_1 d\xi_2 d\xi_3 + \xi'^2 \iiint_{-\infty}^{\infty} \xi^{2r} S_{u-2r}(\xi)\, e^{i(\xi\xi')}\, d\xi_1 d\xi_2 d\xi_3 = 0.$$

so (37) becomes, for $r, s > 0$,

$$(AB) = (2\pi)^{-3} i^{-u+2}\, 4r(u-r+\tfrac{1}{2}) \iint_{\cdots\cdots} \xi^{2(r-1)} \xi'^{2(s-1)} S_{u-2r}(\xi)\, S_{u-2s}(\xi')\, e^{i(\xi\xi')}\, d\xi_1 \ldots d\xi_3'. \tag{38}$$

Now suppose $s \geqslant r$ and apply the procedure by which (37) was changed to (38) $r$ times. The result is

$$(AB) = (2\pi)^{-3} i^{-u+2r}\, 4^r r!(u-r+\tfrac{1}{2})!\,(u-2r+\tfrac{1}{2})!^{-1}$$
$$\times \iint_{\cdots\cdots} \xi'^{2(s-r)} S_{u-2r}(\xi)\, S_{u-2s}(\xi')\, e^{i(\xi\xi')}\, d\xi_1 \ldots d\xi_3'. \tag{39}$$

where $n!$ means $\Gamma(n+1)$ for $n$ not an integer. If $s > r$, the procedure can be applied once more, and then shows that

$$(AB) = 0 \quad \text{for} \quad r \neq s. \tag{40}$$

If $s = r$, (39) shows that
$$(AB) = c4^r r!(u-r+\tfrac{1}{2})!, \tag{41}$$

where $c$ depends only on $u-2r$ and on the two $S$ functions.

## Reference

Dirac, P. A. M. 1942 *Proc. Roy. Soc.* A, **180**, 1–40.

# A.4 A Remarkable Representation of the 3 + 2 de Sitter Group

by P.A.M. Dirac

reprinted with permission from

Journal of Mathematical Physics **4**, 901-909 (1963).

JOURNAL OF MATHEMATICAL PHYSICS VOLUME 4, NUMBER 7 JULY 1963

# A Remarkable Representation of the 3 + 2 de Sitter Group

P. A. M. DIRAC
*Cambridge University, Cambridge, England* and
*Institute for Advanced Study, Princeton, New Jersey**
(Received 20 February 1963)

Among the infinitesimal operators of the 3 + 2 de Sitter group, there are four independent cyclic ones, one of which is separate from the other three. A representation is obtained for which this one has integral eigenvalues while the other three have half-odd eigenvalues, or vice versa. The representation is of a specially simple kind, with the wavefunctions involving only two variables.

## INTRODUCTION

We consider the group of rotations of five real variables $x_1$, $x_2$, $x_3$, $x_4$, $x_5$ which leave the quadratic form

$$x_1^2 + x_2^2 + x_3^2 - x_4^2 - x_5^2 \tag{1}$$

invariant. The infinitesimal operators of the group are $m_{ab} = -m_{ba}(a, b = 1, 2, 3, 4, 5)$. There are ten independent ones. They satisfy the following commutation relations, in the notation $[\xi, \eta] = \xi\eta - \eta\xi$, with $a, b, c, d$ all different:

$$[m_{ab}, m_{cd}] = 0. \tag{2}$$

With $a$, $b$, $c$, all different and no summation over $a$,

$$\begin{aligned}[m_{ab}, m_{ac}] &= m_{bc} \quad \text{for} \quad a = 4, 5, \\ &= -m_{bc} \quad \text{for} \quad a = 1, 2, 3.\end{aligned} \tag{3}$$

We shall consider only unitary representations, so that $im_{ab}$ has real eigenvalues, for all $a$, $b$.

Some of the basic rotations $m_{ab}$ are cyclic and some are hyperbolic. The cyclic ones are $m_{12}$, $m_{23}$, $m_{31}$, $m_{45}$, and the hyperbolic ones are $m_{14}$, $m_{24}$, $m_{34}$, $m_{15}$, $m_{25}$, $m_{35}$. For $m_{ab}$ hyperbolic, $im_{ab}$ has a continuous range of eigenvalues and for $m_{ab}$ cyclic, $im_{ab}$ has discrete eigenvalues, which may be either integers or half-odd integers.

If we set up wave equations in the 3 + 2 de Sitter space,

$$x_1^2 + x_2^2 + x_3^2 - x_4^2 - x_5^2 = -R^2, \tag{4}$$

with various tensor or spinor kinds of wavefunctions, we get various representations of the de Sitter group, but they are such that, for all the cyclic operators, $m_{ab}$, $im_{ab}$ has integral eigenvalues, or else for all of them it has half-odd integral eigenvalues. For these straightforward representations, a mixing of integral and half-odd integral eigenvalues does not occur.

The present paper is concerned with a more primitive representation, for which $im_{12}$, $im_{23}$, $im_{31}$, have half-odd integral eigenvalues while $im_{45}$ has integral eigenvalues. There is nothing inconsistent in such a mixing, because the $m_{45}$ rotation is completely detached from the $m_{12}$, $m_{23}$, $m_{31}$ rotations. In fact, if one goes over to the covering group of the 3 + 2 de Sitter group, the detachment of the $m_{45}$ rotation allows $im_{45}$ to have any real eigenvalues, independently of what eigenvalues the other rotations have. A general theory of the representations of this covering group has been given by Ehrman.[1]

## THE γ MATRICES

We consider 4 × 4 matrices whose elements are all real. There are 16 independent ones. We choose them in a certain way and call them γ matrices. One of them is the unit matrix $\gamma_0$. The other 15 are chosen to be all symmetric or skew matrices, with the square of each symmetric one equal to $\gamma_0$ and the square of each skew one equal to $-\gamma_0$, and with the product of any two equal to ± a third one. These 15 may be written $\gamma_{AB} = -\gamma_{BA}$, where $A$ and $B$ are two different suffixes going from 1 to 6.

The rules for multiplying the $\gamma_{AB}$ are as follows: We use the notation $\gamma_{AB}\gamma_{CD} = \gamma_{ABCD}$, and so on for products of more than two factors. Thus any product appears as a γ with an even number of suffixes. There are two general rules: (i) Any two different suffixes may be interchanged, if one brings in the factor −1. (ii) A suffix $A$ occurring in two consecutive positions may be suppressed but one must then bring in the factor −1 for $A$ = 4, 5 or 6. Thus for example, with $a, b = 1, \cdots 5$,

$$\gamma_{a6}\gamma_{b6} = \gamma_{a6b6} = -\gamma_{ab66} = \gamma_{ab}. \tag{5}$$

The suppression of all the suffixes yields $\gamma_0$.

As a consequence of these rules, we find $\gamma_{AB}^2 = \gamma_0$ if one of the suffixes $A$, $B$ is in the set 1, 2, 3, and

* The author's stay in Princeton was supported by the National Science Foundation.

[1] J. B. Ehrman, Proc. Cambridge Phil. Soc. 53, 290 (1957).

the other in the set 4, 5, 6, so $\gamma_{AB}$ is then symmetric. Otherwise $\gamma_{AB}^2 = -\gamma_0$ and $\gamma_{AB}$ is skew. We find also that $\gamma_{AB}$ and $\gamma_{CD}$ commute if $A$, $B$, $C$, $D$ are all different and anticommute if one of the suffixes $A$, $B$ is the same as one of the suffixes $C$, $D$. Thus a set of five $\gamma_{AB}$'s with one suffix in common, all anticommute. Further we find that

$$\gamma_{123456}^2 = \gamma_0,$$

so that

$$\gamma_{123456} = \pm\gamma_0.$$

We arrange that

$$\gamma_{123456} = \gamma_0. \tag{6}$$

This equation enables us to reduce $\gamma_{ABCD}$ to a two-suffix $\gamma$ when $A$, $B$, $C$, $D$ are all different. For example, multiplying (6) by $-\gamma_{56}$, we get

$$\gamma_{1234} = -\gamma_{56}.$$

The reduction of $\gamma_{ABCD}$ to a two-suffix $\gamma$ when two of the suffixes $A$, $B$, $C$, $D$ are the same is given directly by the rules (i) and (ii), as is illustrated by the example (5).

The result of interchanging rows and columns in any matrix $\alpha$ will be written $\alpha^\dagger$. We have $\gamma_{AB}^\dagger = \pm\gamma_{AB}$, the $+$ sign occurring when $\gamma_{AB}^2 = \gamma_0$ and the $-$ sign when $\gamma_{AB}^2 = -\gamma_0$.

Each $\gamma_{AB}$ has the diagonal sum zero;

$$\langle\gamma_{AB}\rangle = 0. \tag{7}$$

To prove this, let $\sigma$ be any $4 \times 4$ matrix with a reciprocal. Then

$$\langle\gamma_{AB}\rangle = \langle\gamma_{AB}\sigma\sigma^{-1}\rangle = \langle\sigma^{-1}\gamma_{AB}\sigma\rangle.$$

By taking $\sigma$ to be one of the $\gamma$ matrices that anticommutes with $\gamma_{AB}$, we get

$$\langle\gamma_{AB}\rangle = -\langle\gamma_{AB}\rangle.$$

## A REPRESENTATION OF THE CONFORMAL GROUP

Introduce four variables $u_1$, $u_2$, $u_3$, $u_4$ to correspond to the four rows of the $\gamma$ matrices. They may be written as a column matrix and we then denote them by the symbol $u$. Alternatively, they may be written as a row matrix and are then denoted by $u^\dagger$.

Let $v_1$, $v_2$, $v_3$, $v_4$ be a second such set of four variables. With $\alpha$ any $4 \times 4$ matrix, we may form, by matrix multiplication, $v^\dagger\alpha u$, which is a $1 \times 1$ matrix or a number. We have evidently

$$v^\dagger\alpha u = u^\dagger\alpha^\dagger v. \tag{8}$$

Introduce the four differential operators $\partial_n = \partial/\partial u_n (n = 1, 2, 3, 4)$. They satisfy the commutation relations

$$[\partial_n, u_{n'}] = \delta_{nn'}. \tag{9}$$

We may write them as a column matrix $\partial$ or as a row matrix $\partial^\dagger$. If we substitute $\partial$ for $v$ in (8), the equation is no longer valid, on account of the lack of commutation of the $\partial$'s and $u$'s. One easily finds from (9),

$$\partial^\dagger\alpha u = u^\dagger\alpha^\dagger\partial + \langle\alpha\rangle. \tag{10}$$

From (8) and (9) we find also, for $\alpha$ and $\beta$ any two $4 \times 4$ matrices,

$$[u^\dagger\alpha\partial, u^\dagger\beta\partial] = u^\dagger[\alpha, \beta]\partial. \tag{11}$$

For each matrix $\gamma_{AB}$, define the differential operator

$$\chi_{AB} = \tfrac{1}{2}u^\dagger\gamma_{AB}\partial. \tag{12}$$

From (11) we get

$$[\chi_{AB}, \chi_{CD}] = \tfrac{1}{4}u^\dagger(\gamma_{ABCD} - \gamma_{CDAB})\partial.$$

It follows that, for $A$, $B$, $C$, $D$ all different,

$$[\chi_{AB}, \chi_{CD}] = 0, \tag{13}$$

and for $A$, $B$, $C$ all different, with no summation over $A$,

$$\begin{aligned}[\chi_{AB}, \chi_{AC}] &= \chi_{BC} \quad \text{for } A = 4, 5, 6, \\ &= -\chi_{BC} \quad \text{for } A = 1, 2, 3.\end{aligned} \tag{14}$$

The commutation relations (13), (14) are like the commutation relations (2), (3), with the difference that the suffixes can take on six values instead of five. They show that the $\chi_{AB}$ may be looked upon as infinitesimal operators of the group of rotations of six real variables $x_1 \cdots x_6$ which leave

$$x_1^2 + x_2^2 + x_3^2 - x_4^2 - x_5^2 - x_6^2 \tag{15}$$

invariant. This is the conformal group of four-dimensional space. It contains the 3 + 2 de Sitter group as a subgroup.

With the $u$'s restricted to be real variables, the $\chi_{AB}$ provide a representation of the conformal group for which the wavefunctions are functions $\psi(u_1, u_2, u_3, u_4)$ of the four real $u$'s. The representation is unitary, because the adjoint of $\chi_{AB}$ is

$$-\tfrac{1}{2}\partial^\dagger\gamma_{AB}^\dagger u = -\chi_{AB},$$

from (10) and (7). If we choose $\psi$ so that $\int |\psi|^2 d^4u$ converges, the integral is left invariant by the application of any infinitesimal rotation to $\psi$.

Corresponding to (12), we define

$$\chi_0 = \tfrac{1}{2}u^\dagger\gamma_0\partial = \tfrac{1}{2}u^\dagger\partial.$$

We see from (11) that $\chi_0$ commutes with every $\chi_{AB}$. Thus we may put $\chi_0$ equal to a number, $a$, say. The condition that $\chi_0 = a$ means that the wavefunction $\psi$ is homogeneous in the $u$'s of degree $2a$. Let us introduce polar variables $\rho$, $\theta_1$, $\theta_2$, $\theta_3$ in the $u$ space, with $\rho = (u^\dagger u)^{\frac{1}{2}}$, and $\theta_1$, $\theta_2$, $\theta_3$, three independent functions of the ratios of the $u$'s. Then $\psi$ is of the form

$$\psi = \rho^{2a}\psi_1(\theta_1, \theta_2, \theta_3). \qquad (16)$$

To secure the unitary character of the representation, we require

$$i(u\partial^\dagger + \partial^\dagger u) = \text{real number}.$$

Now,

$$u^\dagger\partial + \partial^\dagger u = 2u^\dagger\partial + 4 = 4(\chi_0 + 1).$$

Thus,

$$\chi_0 = -1 + ib, \qquad (17)$$

where $b$ is a real number. Equation (16) becomes

$$\psi = \rho^{-2+2ib}\psi_1(\theta). \qquad (18)$$

The wavefunction $\psi$ is now not square integrable, being an eigenfunction of an operator $\chi_0$ with a continuous range of eigenvalues. We have

$$\int |\psi|^2\, d^4u = \int |\psi|^2 \rho^3\, d\Omega\, d\rho,$$

where $d\Omega$ is an element of 3-dimensional solid angle in the $u$ space and is some multiple of $d\theta_1\, d\theta_2\, d\theta_3$. Substituting the form (18) for $\psi$, we get

$$\int_{-\infty}^{\infty} |\psi|^2\, d^4u = \int_0^\infty \rho^{-1}\, d\rho \int |\psi_1|^2\, d\Omega. \qquad (19)$$

We may now drop the variable $\rho$ from the representation, which involves working with the wave function $\psi_1(\theta)$ and dropping the infinite factor $\int \rho^{-1}\, d\rho$ from the formula (19) for the squared length of a wavefunction.

Going back to general wavefunctions $\psi$ that are not eigenfunctions of $\chi_0$, let us consider wavefunctions of the form

$$\psi = P(u_1, u_2, u_3, u_4)f(\rho), \qquad (20)$$

where $P$ is a power series in the $u$'s and $f(\rho)$ is chosen so that $\psi$ is square integrable. One finds readily

$$[\chi_{AB}, \rho^2] = u^\dagger\gamma_{AB}u. \qquad (21)$$

If the rotation $\chi_{AB}$ is cyclic, $\gamma_{AB}$ is skew, and so $\chi_{AB}$ commutes with $\rho$. Thus a cyclic rotation applied to the wavefunction (20) affects only the power series $P$ and leaves the factor $f(\rho)$ invariant.

Take $P$ to be linear in the $u$'s and suppose it is an eigenfunction of a particular cyclic $\chi_{AB}$. The eigenvalue is then an eigenvalue of $\frac{1}{2}\gamma_{AB}$, and is thus $\pm\frac{1}{2}i$. One can readily infer that, if $P$ is homogeneous of the $n$th degree in the $u$'s and is an eigenfunction of a particular cyclic $\chi_{AB}$, the eigenvalue of $i\chi_{AB}$ is an integer or half-odd integer according to whether $n$ is even or odd.

The parity of the power series $P$ is invariant under all the rotations, hyperbolic as well as cyclic. This parity determines the integral or half-odd integral character of the eigenvalues of each of the cyclic rotations. It follows that in the present representation there is no mixing of integral and half-odd integral eigenvalues, such as occurs in the remarkable representation we are seeking.

### SOME PROPERTIES OF THE OPERATORS $\chi$

*Lemma.* For any symmetrical $4 \times 4$ matrix $S$,

$$\gamma_{12}S\gamma_{12} + \gamma_{23}S\gamma_{23} + \gamma_{31}S\gamma_{31} = S - \langle S\rangle\gamma_0, \qquad (22)$$

and similarly,

$$\gamma_{45}S\gamma_{45} + \gamma_{56}S\gamma_{56} + \gamma_{64}S\gamma_{64} = S - \langle S\rangle\gamma_0. \qquad (23)$$

To prove (22), we note that any symmetric $\gamma_{AB}$ has one of its suffixes equal to 1, 2, or 3 and the other equal to 4, 5, or 6 and therefore it commutes with one of the three quantities $\gamma_{12}$, $\gamma_{23}$, $\gamma_{31}$ and anticommutes with the other two. Thus if such a $\gamma_{AB}$ is substituted for $S$ in (22), two of the terms on the left become $S$ and the third one $-S$, and the $\langle S\rangle$ on the right is zero. Also (22) holds with $S = \gamma_0$, so it holds for any symmetric $S$. The proof of (23) is similar.

The lemma enables one to deduce a number of quandratic relations between the $\chi$'s. From (10),

$$2\chi_{AB} = u^\dagger\gamma_{AB}\partial = \partial^\dagger\gamma_{AB}^\dagger u.$$

Hence

$$-4(\chi_{12}^2 + \chi_{23}^2 + \chi_{31}^2) = \partial^\dagger\gamma_{12}uu^\dagger\gamma_{12}\partial + \partial^\dagger\gamma_{23}uu^\dagger\gamma_{23}\partial + \partial^\dagger\gamma_{31}uu^\dagger\gamma_{31}\partial.$$

The multiplication is associative, so we can pick out the $uu^\dagger$ in the middle of each term on the right and consider it as a $4 \times 4$ matrix. It is symmetric, so we can take it to be the $S$ of (22) and get

$$-4(\chi_{12}^2 + \chi_{23}^2 + \chi_{31}^2) = \partial^\dagger uu^\dagger\partial - \partial^\dagger\langle uu^\dagger\rangle\partial = (u^\dagger\partial + 4)u^\dagger\partial - u^\dagger u\partial^\dagger\partial - 2u^\dagger\partial.$$

Thus,

$$\chi_{12}^2 + \chi_{23}^2 + \chi_{31}^2 = -\chi_0(\chi_0 + 1) + \tfrac{1}{4}u^\dagger u\partial^\dagger\partial. \qquad (24)$$

Similarly,

$$\chi_{45}^2 + \chi_{56}^2 + \chi_{64}^2 = -\chi_0(\chi_0 + 1) + \tfrac{1}{4}u^\dagger u\partial^\dagger\partial. \tag{25}$$

Hence,

$$\chi_{12}^2 + \chi_{23}^2 + \chi_{31}^2 = \chi_{45}^2 + \chi_{56}^2 + \chi_{64}^2. \tag{26}$$

A more general result may be obtained as follows: Let $\gamma_P$ and $\gamma_Q$ be any two $\gamma$ matrices, i.e., each of them is either $\gamma_0$ or a $\gamma_{AB}$. Putting $S = uu^\dagger$ in (22), and multiplying by $\partial^\dagger\gamma_P^\dagger$ on the left and $\gamma_Q\partial$ on the right, we get

$$\begin{aligned}\partial^\dagger\gamma_P^\dagger\gamma_{12}uu^\dagger\gamma_{12}\gamma_Q\partial &+ \partial^\dagger\gamma_P^\dagger\gamma_{23}uu^\dagger\gamma_{23}\gamma_Q\partial \\ &+ \partial^\dagger\gamma_P^\dagger\gamma_{31}uu^\dagger\gamma_{31}\gamma_Q\partial - \partial^\dagger\gamma_P^\dagger uu^\dagger\gamma_Q\partial \\ &= -\partial^\dagger\gamma_P^\dagger\langle uu^\dagger\rangle\gamma_Q\partial \\ &= -u^\dagger u\partial^\dagger\gamma_{P\dagger Q}\partial - 2u^\dagger\gamma_{P\dagger Q}\partial,\end{aligned}$$

where $\gamma_{P\dagger Q}$ denotes $\gamma_P^\dagger\gamma_Q$. With the help of (10) and (12), this becomes

$$\begin{aligned}&\{\chi_{12P} + \tfrac{1}{2}\langle\gamma_{12P}\rangle\}\chi_{12Q} + \{\chi_{23P} + \tfrac{1}{2}\langle\gamma_{23P}\rangle\}\chi_{23Q} \\ &\quad + \{\chi_{31P} + \tfrac{1}{2}\langle\gamma_{31P}\rangle\}\chi_{31Q} + \{\chi_P + \tfrac{1}{2}\langle\gamma_P\rangle\}\chi_Q \\ &= \tfrac{1}{4}u^\dagger u\partial^\dagger\gamma_{P\dagger Q}\partial + \chi_{P\dagger Q}.\end{aligned} \tag{27}$$

Now $\langle\gamma_{12P}\rangle$ vanishes unless $\gamma_{12P} = \pm\gamma_0$, in which case $\gamma_{12} = \mp\gamma_P = \pm\gamma_P^\dagger$. Thus,

$$\langle\gamma_{12P}\rangle\chi_{12Q} = |\langle\gamma_{12P}\rangle|\,\chi_{P\dagger Q}.$$

Treating the other $\langle\gamma\rangle$ terms in (27) in the same way, we get finally

$$\begin{aligned}&\chi_{12P}\chi_{12Q} + \chi_{23P}\chi_{23Q} + \chi_{31P}\chi_{31Q} + \chi_P\chi_Q \\ &= \{1 - \tfrac{1}{2}\,|\langle\gamma_{12P} + \gamma_{23P} + \gamma_{31P} + \gamma_P\rangle|\}\chi_{P\dagger Q} \\ &\quad + \tfrac{1}{4}u^\dagger u\partial^\dagger\gamma_{P\dagger Q}\partial.\end{aligned} \tag{28}$$

Similarly, it may be shown that

$$\begin{aligned}&\chi_{45P}\chi_{45Q} + \chi_{56P}\chi_{56Q} + \chi_{64P}\chi_{64Q} + \chi_P\chi_Q \\ &= \{1 - \tfrac{1}{2}\,|\langle\gamma_{45P} + \gamma_{56P} + \gamma_{64P} + \gamma_P\rangle|\}\chi_{P\dagger Q} \\ &\quad + \tfrac{1}{4}u^\dagger u\partial^\dagger\gamma_{P\dagger Q}\partial.\end{aligned} \tag{29}$$

It should be noted that the last term in (28) or (29) vanishes if $\gamma_{PQ}^\dagger$ or $\gamma_{PQ}$ is skew.

There are many applications of the general formulas (28) and (29). With $\gamma_P = \gamma_Q = \gamma_0$, we get back to (24) and (25). With $\gamma_P = \gamma_0$, $\gamma_Q = \gamma_{56}$ in (28), we get

$$\chi_{12}\chi_{34} + \chi_{23}\chi_{14} + \chi_{31}\chi_{24} = -(1 + \chi_0)\chi_{56}. \tag{30}$$

With $\gamma_P = \gamma_{12}$, $\gamma_Q = \gamma_{45}$ in (28), we get

$$-\chi_0\chi_{36} + \chi_{31}\chi_{16} + \chi_{32}\chi_{26} + \chi_{12}\chi_{45} = \chi_{36} - \tfrac{1}{4}u^\dagger u\partial^\dagger\gamma_{36}\partial. \tag{31}$$

With $\gamma_P = \gamma_{31}$, $\gamma_Q = \gamma_{16}$ in (29), we get

$$\chi_{26}\chi_{32} + \chi_{24}\chi_{51} + \chi_{25}\chi_{14} + \chi_{31}\chi_{16} = -\chi_{36} - \tfrac{1}{4}u^\dagger u\partial^\dagger\gamma_{36}\partial. \tag{32}$$

Subtracting (32) from (31), we get (taking into account the noncommutativity of $\chi_{26}$ and $\chi_{32}$).

$$\chi_{12}\chi_{45} + \chi_{24}\chi_{15} + \chi_{41}\chi_{25} = (1 + \chi_0)\chi_{36}. \tag{33}$$

Equations (30) and (33) and similar equations lead to the general formula, for $A$, $B$, $C$, $D$ all different:

$$\chi_{AB}\chi_{CD} + \chi_{BC}\chi_{AD} + \chi_{CA}\chi_{BD} = (1 + \chi_0)\chi_{ABCD}. \tag{34}$$

With $\gamma_P = \gamma_Q = \gamma_{45}$ in (28), we get

$$\chi_{36}^2 + \chi_{16}^2 + \chi_{26}^2 + \chi_{45}^2 = \chi_0 + \tfrac{1}{4}u^\dagger u\partial^\dagger\partial. \tag{35}$$

Subtracting (25) from (35), we get

$$\chi_{16}^2 + \chi_{26}^2 + \chi_{36}^2 - \chi_{46}^2 - \chi_{56}^2 = \chi_0(\chi_0 + 2). \tag{36}$$

More generally, we have for each value of $B$,

$$\sum_{A\neq B}\chi_{AB}\chi^{AB} = -\chi_0(\chi_0 + 2), \tag{37}$$

where the raising of a suffix involves a minus sign if the suffix is 4, 5, or 6. Summing for all $B$, we get

$$\sum_A\sum_{B<A}\chi_{AB}\chi^{AB} = -3\chi_0(\chi_0 + 2). \tag{38}$$

### THE REMARKABLE REPRESENTATION

With $\mu, \nu = 1, 2, 3, 4$, define

$$\begin{aligned}m_{\mu\nu} &= \chi_{\mu\nu}, \\ m_{\mu 5} &= C(\chi_{\mu 5} - i\chi_{\mu 6}),\end{aligned} \tag{39}$$

where $C$ is the operator which converts numbers into their conjugate complexes. It has the algebraic properties

$$C^2 = 1, \qquad Ci = -iC, \tag{40}$$

and $C$ commutes with all the $\partial$'s, $u$'s and $\gamma$'s. Thus it commutes with all the $\chi$'s.

We must check whether the $m_{ab}$ defined by (39) satisfy the commutation relations (2), (3). It is evident from (14) that all these relations are satisfied except the ones for $[m_{\mu 5}, m_{\nu 5}]$ with $\mu \neq \nu$. We have

$$\begin{aligned}[m_{\mu 5}, m_{\nu 5}] &= C(\chi_{\mu 5} - i\chi_{\mu 6})C(\chi_{\nu 5} - i\chi_{\nu 6}) \\ &\quad - C(\chi_{\nu 5} - i\chi_{\nu 6})C(\chi_{\mu 5} - i\chi_{\mu 6}) \\ &= (\chi_{\mu 5} + i\chi_{\mu 6})(\chi_{\nu 5} - i\chi_{\nu 6}) \\ &\quad - (\chi_{\nu 5} + i\chi_{\nu 6})(\chi_{\mu 5} - i\chi_{\mu 6}) \\ &= [\chi_{\mu 5}, \chi_{\nu 5}] + [\chi_{\mu 6}, \chi_{\nu 6}] \\ &\quad + 2i(\chi_{\mu 6}\chi_{\nu 5} - \chi_{\mu 5}\chi_{\nu 6}) \\ &= 2\chi_{\mu\nu} + 2i\{-\chi_{\mu\nu}\chi_{56} + (1 + \chi_0)\chi_{\mu\nu 56}\}\end{aligned}$$

from (34), with $A, B, C, D = \mu, \nu, 5, 6$. Thus,

$$[m_{\mu 5}, m_{\nu 5}] = \chi_{\mu\nu} - 2i\chi_{\mu\nu}(\chi_{56} + \tfrac{1}{2}i) + 2i\chi_{\mu\nu 56}(1 + \chi_0). \quad (41)$$

Consider wavefunctions $\psi$ that satisfy the two supplementary conditions

$$(\chi_0 + 1)\psi = 0, \quad (42)$$

$$(\chi_{56} + \tfrac{1}{2}i)\psi = 0. \quad (43)$$

The conditions are consistent since $\chi_0$ and $\chi_{56}$ commute. The operators $\chi_0 + 1$ and $\chi_{56} + \frac{1}{2}i$ occurring in these conditions commute with all the $m_{ab}$. This is obvious for the operator $\chi_0 + 1$. It is also obvious that $\chi_{56} + \frac{1}{2}i$ commutes with $m_{\mu\nu}$, and it can be proved to commute with $m_{\mu 5}$ as follows. We have

$$\begin{aligned} m_{\mu 5}(\chi_{56} + \tfrac{1}{2}i) &= C(\chi_{\mu 5} - i\chi_{\mu 6})(\chi_{56} + \tfrac{1}{2}i) \\ &= C\{\chi_{56}(\chi_{\mu 5} - i\chi_{\mu 6}) - \chi_{\mu 6} - i\chi_{\mu 5} + \tfrac{1}{2}i(\chi_{\mu 5} - i\chi_{\mu 6})\} \\ &= C(\chi_{56} - \tfrac{1}{2}i)(\chi_{\mu 5} - i\chi_{\mu 6}) \\ &= (\chi_{56} + \tfrac{1}{2}i)m_{\mu 5}. \end{aligned}$$

The supplementary conditions thus pick out a set of $\psi$'s that is invariant under all the operations $m_{ab}$.

For this set of $\psi$'s, (41) reduces to

$$[m_{\mu 5}, m_{\nu 5}] = m_{\mu\nu},$$

so the $m_{ab}$ satisfy all the commutation relations (2), (3). They thus provide a representation of the 3 + 2 de Sitter group. The representation is unitary, because the $m_{ab}$ are all equal to minus their adjoints, as follows in the case of $m_{\mu 5}$ from $C$ and $Ci$ being self-adjoint.

Some properties of the representation will be worked out. We have, from (30) and (42),

$$m_{12}m_{34} + m_{23}m_{14} + m_{31}m_{24} = 0. \quad (44)$$

The corresponding relations involving other sets of four of the suffixes 1, $\cdots$5 also hold.

The condition (43) means that $i\chi_{56}$ has just the one eigenvalue $\frac{1}{2}$. It follows from the parity discussion in connection with the form (20) that $i$ times any cyclic $\chi_{AB}$ has half-odd integral eigenvalues. Thus $im_{12}$, $im_{23}$, $im_{31}$ have half-odd integral eigenvalues. It follows then that

$$m_{12}^2 + m_{23}^2 + m_{31}^2 = -k(k + 1), \quad (45)$$

where $k$ has half-odd integral eigenvalues.

We have, from (39) and (14),

$$\begin{aligned} m_{45}^2 &= (\chi_{45} + i\chi_{46})(\chi_{45} - i\chi_{46}) \\ &= \chi_{45}^2 + \chi_{46}^2 - i\chi_{56} \quad (46) \\ &= \chi_{12}^2 + \chi_{23}^2 + \chi_{31}^2 - \chi_{56}^2 - i\chi_{56}, \end{aligned}$$

with the help of (26). Thus, with the supplementary condition (43),

$$m_{45}^2 = m_{12}^2 + m_{23}^2 + m_{31}^2 - \tfrac{1}{4}, \quad (47)$$

$$= -(k + \tfrac{1}{2})^2, \quad (48)$$

with the help of (45). It follows that $im_{45}$ has integral eigenvalues. Thus the representation gives a mixing of integral and half-odd integral eigenvalues for the cyclic rotations.

The $k$ introduced by Eq. (45) may be restricted to have only positive eigenvalues. It follows from (48) that $m_{45}$ does not have the eigenvalue zero.

The absence of the zero eigenvalue has the consequence that the representation can be reduced to two component representations, involving the eigenfunctions of $im_{45}$ with positive and negative eigenvalues, respectively. To prove the result, we note that $m_{12}$, $m_{23}$, or $m_{31}$ applied to an eigenfunction of $im_{45}$ does not change the eigenvalue, while $m_{r4} (r = 1, 2, 3)$ *applied to it changes the eigenvalue* by $\pm 1$, on account of the commutation relation

$$[[m_{r4}, m_{45}], m_{4s}] = -m_{r4}. \quad (49)$$

Thus, any application of operators $m_{\mu\nu}$ to an eigenfunction with a positive eigenvalue produces a linear combination *of eigenfunctions* with positive eigenvalues, as they cannot make the jump from 1 to $-1$. One can easily check that the two components are irreducible.

The eigenfunctions of $im_{45}$ may be obtained in the following way: We have, from (46) and (25),

$$\begin{aligned} m_{45}^2\psi &= \{-\chi_{56}^2 - i\chi_{56} - \chi_0(\chi_0 + 1) + \tfrac{1}{4}u'u\partial'\partial\}\psi \\ &= \tfrac{1}{4}(-1 + u'u\partial'\partial)\psi. \quad (50) \end{aligned}$$

Now suppose $\psi$ is of the form

$$\psi = (u'u)^{-\frac{1}{2}n-1}\phi, \quad (51)$$

*where $\phi$ is homogeneous of degree $n$. This $\psi$ satisfies* (42). Suppose further that $\phi$ is harmonic, i.e.,

$$\partial'\partial\phi = 0.$$

Equation (50) now leads to

$$m_{45}^2\psi = -\tfrac{1}{4}(n + 1)^2\psi. \quad (52)$$

The condition (43) leads to

$$(\chi_{56} + \tfrac{1}{2}i)\phi = 0. \quad (53)$$

Thus any harmonic $\phi$ satisfying (53) provides, according to (51), a $\psi$ that is a linear combination *of the eigenfunctions of $im_{45}$ with eigenvalues* $\pm\frac{1}{2}(n + 1)$. Of course $n$ has to be an odd integer.

One can infer this directly by noting that (51) is of the form (20) with $\phi$ for $P$ and the condition (43) requires the function $\phi$ to be of odd parity.

### INHOMOGENEOUS FORM OF THE REPRESENTATION

The representation given by (39) involves the four $u$'s in a homogeneous manner, each $m_{\mu\nu}$ being of degree zero in the $u$'s. One may use the two supplementary conditions (42), (43) to eliminate two of the variables in the wavefunctions, so that there are only two left, which then appear in a nonhomogeneous manner. The resulting form of the representation is more convenient for problems requiring detailed calculation.

Let us first pass from the four real $u$'s to two complex variables. Put

$$v_1 = u_1 + iu_2, \qquad v_2 = u_3 + iu_4.$$

The conjugate complex equations are

$$v_1^* = u_1 - iu_2, \qquad v_2^* = u_3 - iu_4.$$

The operators $\partial/\partial v_1$, $\partial/\partial v_2$ will be written $d_1$, $d_2$ for brevity, and they have the values

$$d_1 = \tfrac{1}{2}(\partial_1 - i\partial_2), \qquad d_2 = \tfrac{1}{2}(\partial_3 - i\partial_4).$$

Their conjugates are

$$d_1^* = \tfrac{1}{2}(\partial_1 + i\partial_2), \qquad d_2^* = \tfrac{1}{2}(\partial_3 + i\partial_4).$$

We may choose the $\gamma$ matrices so that, with the $\chi$'s given by (12),

$$\begin{aligned} \chi_{14} - i\chi_{23} &= -v_2 d_1 - v_1 d_2, \\ \chi_{15} - i\chi_{16} &= v_1^* d_1 - v_2^* d_2, \\ \chi_{24} - i\chi_{31} &= -i(v_2 d_1 - v_1 d_2), \\ \chi_{25} - i\chi_{26} &= i(v_1^* d_1 + v_2^* d_2), \\ \chi_{34} - i\chi_{12} &= v_1 d_1 - v_2 d_2, \\ \chi_{35} - i\chi_{36} &= v_2^* d_1 + v_1^* d_2, \\ \chi_0 + i\chi_{56} &= v_1 d_1 + v_2 d_2, \\ \chi_{45} - i\chi_{46} &= v_2^* d_1 - v_1^* d_2. \end{aligned} \tag{54}$$

These eight complex equations give sixteen real equations which determine the sixteen $\gamma$'s, and one finds that the $\gamma$'s so determined have all the desired properties.

The supplementary conditions (42), (43) may be written

$$\begin{aligned} (\chi_0 + i\chi_{56})\psi &= -\tfrac{1}{2}\psi, \\ (\chi_0 - i\chi_{56})\psi &= -\tfrac{3}{2}\psi. \end{aligned} \tag{55}$$

In the new notation they become

$$\begin{aligned} (v_1 d_1 + v_2 d_2)\psi &= -\tfrac{1}{2}\psi, \\ (v_1^* d_1^* + v_2^* d_2^*)\psi &= -\tfrac{3}{2}\psi. \end{aligned} \tag{56}$$

They show that $\psi$ is homogeneous of degree $-\tfrac{1}{2}$ in $v_1$, $v_2$ and homogeneous of degree $-\tfrac{3}{2}$ in $v_1^*$, $v_2^*$. Thus $\psi$ is of the form

$$\psi = v_2^{-\frac{1}{2}} v_2^{*-\frac{3}{2}} f(z, z^*), \tag{57}$$

where

$$z = v_1/v_2.$$

Resolving $z$ into its real and pure imaginary parts,

$$z = x_1 + ix_2,$$

we have

$$\psi = v_2^{-\frac{1}{2}} v_2^{*-\frac{3}{2}} g(x_1, x_2). \tag{58}$$

The wavefunction $\psi$ is determined by the function $f$ or $g$ of two variables. Thus we may take $f$ or $g$ to be the wavefunction in a new form of the representation.

Let us see what is the squared length of a wavefunction in the new form. The condition (42) shows that $\psi$ is of the form (18) with $b = 0$, so we have again a squared length like (19) and have to drop out the infinite factor $\int \rho^{-1}\, d\rho$. We may define the variables $\theta_1$, $\theta_2$, $\theta_3$ by means of the equations

$$\begin{aligned} v_1 &= |v_1| e^{i\theta_1} = \rho \sin\theta_3 e^{i\theta_1}, \\ v_2 &= |v_2| e^{i\theta_2} = \rho \cos\theta_3 e^{i\theta_2}. \end{aligned}$$

Thus,

$$|z| = |v_1/v_2| = \tan\theta_3.$$

We now have

$$\begin{aligned} d^4u &= |v_1|\, d|v_1|\, d\theta_1\, |v_2|\, d|v_2|\, d\theta_2 \\ &= |v_1|\, |v_2|\, \rho\, d\rho\, d\theta_3\, d\theta_1\, d\theta_2 \\ &= \rho^3\, d\rho \sin\theta_3 \cos\theta_3\, d\theta_1\, d\theta_2\, d\theta_3 \\ &= \rho^3\, d\rho \cos^4\theta_3\, d\theta_1\, d\theta_2\, |z|\, d|z|. \end{aligned}$$

Hence, with $\psi$ of the form (58),

$$\int |\psi|^2\, d^4u = \int \rho^{-1}\, d\rho \int |g|^2\, d\theta_1\, d\theta_2\, |z|\, d|z|.$$

The function $g$ involves only $|z|$ and $\theta_1 - \theta_2$, so we may pass from the variables $\theta_1$, $\theta_2$ to $\theta_1 + \theta_2$ and $\theta_1 - \theta_2$, and then carry out the integration with respect to $\theta_1 + \theta_2$. The result is, with omission of the infinite factor $\int \rho^{-1}\, d\rho$,

$$\begin{aligned} \tfrac{1}{2} \int |g|^2\, d(\theta_1 + \theta_2)\, d(\theta_1 - \theta_2)\, |z|\, d|z| \\ = \pi \int |g|^2\, dx_1\, dx_2. \end{aligned} \tag{59}$$

Thus we must take the squared length of the new wavefunctions in the usual way, with weight factor unity.

Define the operators

$$D = v_2 d_1, \qquad D^* = v_2^* d_1^*, \qquad C_1 = C v_2^*/v_2. \tag{60}$$

Note that $C_1^2 = 1$,

$$C_1 z C_1 = z^*, \qquad C_1 D C_1 = D^*. \tag{61}$$

With $\psi$ of the form (57), we have

$$\begin{aligned} D\psi &= v_2^{-\frac{1}{2}} v_2^{*-\frac{3}{2}}\, \partial f/\partial z, \\ D^*\psi &= v_2^{-\frac{1}{2}} v_2^{*-\frac{3}{2}}\, \partial f/\partial z^*, \\ C_1\psi &= v_2^{-\frac{1}{2}} v_2^{*-\frac{3}{2}} f^*. \end{aligned} \tag{62}$$

Thus the operators $D$, $D^*$, $C_1$ applied to a $\psi$ of the form (57) give other $\psi$'s of this form, and are equivalent to the operators $\partial/\partial z$, $\partial/\partial z^*$, and taking the conjugate complex, applied to $f$. So these three operators occur in the expressions for the $m_{ab}$ in the new form of the representation.

One finds from (54), with the help of (56),

$$\begin{aligned} m_{14} - im_{23} &= (z^2 - 1)D - z(v_1 d_1 + v_2 d_2) \\ &= (z^2 - 1)D + \tfrac{1}{2}z, \\ m_{14} + im_{23} &= (z^{*2} - 1)D^* - z^*(v_1^* d_1^* + v_2^* d_2^*) \\ &= (z^{*2} - 1)D^* + \tfrac{3}{2}z^*, \\ m_{24} - im_{31} &= -i(z^2 + 1)D + iz(v_1 d_1 + v_2 d_2) \\ &= -i(z^2 + 1)D - \tfrac{1}{2}iz, \\ m_{24} + im_{31} &= i(z^{*2} + 1)D^* - iz^*(v_1^* d_1^* + v_2^* d_2^*) \\ &= i(z^{*2} + 1)D^* + \tfrac{3}{2}iz^*, \\ m_{34} - im_{12} &= 2zD - v_1 d_1 - v_2 d_2 = 2zD + \tfrac{1}{2}, \\ m_{34} + im_{12} &= 2z^* D^* - v_1^* d_1^* - v_2^* d_2^* \\ &= 2z^* D^* + \tfrac{3}{2}, \\ m_{15} &= C_1\{(z + z^*)D - v_1 d_1 - v_2 d_2\} \\ &= C_1\{(z + z^*)D + \tfrac{1}{2}\}, \\ m_{25} &= -C_1 i\{(z - z^*)D - v_1 d_1 - v_2 d_2\} \\ &= -C_1 i\{(z - z^*)D + \tfrac{1}{2}\}, \\ m_{35} &= C_1\{(1 - zz^*)D + z^*(v_1 d_1 + v_2 d_2)\} \\ &= C_1\{(1 - zz^*)D - \tfrac{1}{2}z^*\}, \\ m_{45} &= C_1\{(1 + zz^*)D - z^*(v_1 d_1 + v_2 d_2)\} \\ &= C_1\{(1 + zz^*)D + \tfrac{1}{2}z^*\}. \end{aligned} \tag{63}$$

These equations give the $m_{ab}$ in the new form. One can easily check that they are equal to minus their adjoints, as they have to be with the form (59) for the squared length.

## TRANSFORMATION OF VARIABLES

We shall make a transformation of the operators $z$, $z^*$, $D$, $D^*$ occurring in the expressions (63) for the $m_{ab}$ and get a new representation. Define the operators $\zeta$, $\zeta^*$, $\Delta$, $\Delta^*$ according to

$$\zeta^2 = 2D, \qquad \zeta^{*2} = -2D^*, \tag{64}$$

$$\Delta = -z\zeta, \qquad \Delta^* = \zeta^* z^*. \tag{65}$$

$\zeta$ and $\zeta^*$ commute, and will be taken as the basic variable in terms of which the new wavefunctions are expressed. They are adjoint operators, so they form a pair of conjugate complex variables. We have

$$[\Delta, \zeta^2] = -2[z\zeta, D] = 2\zeta,$$

so

$$[\Delta, \zeta] = 1.$$

Also,

$$[\Delta, \zeta^*] = 0.$$

Thus, $\Delta$ may be considered as the operator $\partial/\partial\zeta$. $\Delta^*$ is minus the adjoint of $\Delta$, so it may be considered as $\partial/\partial\zeta^*$.

From (61),

$$\zeta^{*2} = -C_1\zeta^2 C_1 = -(C_1\zeta C_1)^2.$$

We may assume

$$\zeta^* = iC_1\zeta C_1, \tag{66}$$

this being consistent with the other equations. Equations (65) now give

$$\begin{aligned} \Delta^* &= iC_1\zeta z C_1 \\ &= -iC_1\zeta\Delta\zeta^{-1}C_1. \end{aligned}$$

Since $\Delta^*$ commutes with $\zeta$, we get

$$\begin{aligned} \Delta^* &= -i\zeta^{-1}C_1\zeta\Delta\zeta^{-1}C_1\zeta \\ &= -iC_2\Delta C_2, \end{aligned} \tag{67}$$

where

$$C_2 = \zeta^{-1}C_1\zeta. \tag{68}$$

$C_2$ has the properties

$$C_2^2 = 1, \qquad C_2 i = -iC_2. \tag{69}$$

From (67),

$$C_2\Delta C_2 = i\Delta^*, \qquad C_2\Delta^* C_2 = i\Delta. \tag{70}$$

From (66),

$$C_2\zeta C_2 = -i\zeta^*, \qquad C_2\zeta^* C_2 = -i\zeta. \tag{71}$$

From (69), (70), (71), it follows that $C_2$ commutes with each of the following:

$$i\Delta\Delta^*, \quad i\zeta\zeta^*, \quad i(\zeta^*\Delta - \zeta\Delta^*), \quad \zeta^*\Delta + \zeta\Delta^*. \tag{72}$$

From (66) again,

$$C_1\zeta^* = -i\zeta C_1 = \zeta C_1 i,$$

or

$$\zeta^{-1}C_1 = C_1 i\zeta^{*-1}.$$

Thus an alternative form for (68) is

$$C_2 = C_1 i\zeta\zeta^{*-1}. \tag{73}$$

In terms of the new operators $\zeta$, $\zeta^*$, $\Delta$, $\Delta^*$, $C_2$, Eqs. (63) become

$$\begin{aligned} m_{14} - im_{23} &= \tfrac{1}{2}\{(\Delta\zeta^{-1})^2 - 1\}\zeta^2 - \tfrac{1}{2}\Delta\zeta^{-1} \\ &= \tfrac{1}{2}(\Delta^2 - \zeta^2), \\ m_{24} - im_{31} &= -\tfrac{1}{2}i(\Delta^2 + \zeta^2), \\ m_{34} - im_{12} &= -\zeta\Delta - \tfrac{1}{2}, \end{aligned}$$

and the three adjoint equations. Also,

$$\begin{aligned} m_{15} &= \tfrac{1}{2}C_1\{(-\Delta\zeta^{-1} + \zeta^{*-1}\Delta^*)\zeta^2 + 1\} \\ &= \tfrac{1}{2}C_2 i(\zeta^*\Delta - \zeta\Delta^*), \end{aligned}$$

with the help of (73), and similarly,

$$\begin{aligned} m_{25} &= \tfrac{1}{2}C_2(\zeta^*\Delta + \zeta\Delta^*), \\ m_{35} &= -\tfrac{1}{2}C_2 i(\Delta\Delta^* + \zeta\zeta^*), \\ m_{45} &= \tfrac{1}{2}C_2 i(\Delta\Delta^* - \zeta\zeta^*). \end{aligned}$$

Since $C_2$ commutes with the expressions (72), it commutes with the $m_{\mu 5}$. It must therefore also commute with the $m_{\mu\nu}$, which are just the commutators of the $m_{\mu 5}$. Let us define new infinitesimal operators $m'_{ab}$ by

$$\begin{aligned} m'_{\mu\nu} &= m_{\mu\nu}, \\ m'_{\mu 5} &= C_2 m_{\mu 5}. \end{aligned}$$

One sees at once that the $m'_{ab}$ also satisfy the commutation relations (2), (3) and thus provide a representation of the 3 + 2 de Sitter group. The $C_2$ operator does not occur in the $m'_{ab}$, each of them being merely a quadratic function of the variables $\zeta$, $\zeta^*$, $\Delta$, $\Delta^*$.

The new representation can be better understood if we refer it to real variables instead of the complex $\zeta$, $\zeta^*$. Put

$$\zeta = 2^{-\frac{1}{2}}(q_1 + iq_2), \qquad \zeta^* = 2^{-\frac{1}{2}}(q_1 - iq_2),$$

$$-i\Delta = 2^{-\frac{1}{2}}(p_1 - ip_2), \qquad -i\Delta^* = 2^{-\frac{1}{2}}(p_1 + ip_2).$$

Thus $q_1$, $q_2$, $p_1$, $p_2$ are Hermitian operators satisfying

$$[q_1, p_1] = [q_2, p_2] = i,$$

with their other commutators vanishing, so they are like canonical variables in quantum mechanics. We now get

$$\begin{aligned} im'_{12} &= \tfrac{1}{2}(q_1p_2 - q_2p_1), \\ im'_{23} &= \tfrac{1}{4}(p_1^2 + q_1^2 - p_2^2 - q_2^2), \\ im'_{31} &= -\tfrac{1}{2}(q_1q_2 + p_1p_2), \\ im'_{14} &= \tfrac{1}{2}(q_1q_2 - p_1p_2), \\ im'_{24} &= \tfrac{1}{4}(q_1^2 - p_1^2 - q_2^2 + p_2^2), \\ im'_{34} &= \tfrac{1}{2}(q_1p_1 + p_2q_2), \\ im'_{15} &= -\tfrac{1}{2}(q_1p_2 + q_2p_1), \\ im'_{25} &= -\tfrac{1}{2}(q_1p_1 - q_2p_2), \\ im'_{35} &= \tfrac{1}{4}(q_1^2 + q_2^2 - p_1^2 - p_2^2), \\ im'_{45} &= \tfrac{1}{4}(p_1^2 + q_1^2 + p_2^2 + q_2^2). \end{aligned}$$

There are ten independent quadratic functions of the four variables $q_1$, $q_2$, $p_1$, $p_2$, and suitable linear combinations of these ten provide the expressions for the $m'_{ab}$.

The algebraic connection between the operators in the new representation and those of the previous representation (39) or (63) ensures that the new operators satisfy the same algebraic relations as before, e.g., the commutation relations (2), (3) and the various quadratic relations such as (47). However, the new representation is not equivalent to the previous one.

The new representation can be reduced to two component representations, for which the wave-functions $\psi(q_1, q_2)$ are even or odd functions of the $q$'s, respectively. One can easily check that these components are irreducible.

One sees that $im'_{12}$ has integral or half-odd integral eigenvalues according to whether $\psi$ is even or odd. Further, the expression for $im'_{45}$ is like half the energy of two harmonic oscillators. Its eigenvalues are all positive and the lowest eigenvalue is $\frac{1}{2}$, coming from the zero-point energy of each oscillator. Even wavefunctions lead to half-odd integral eigenvalues for $im'_{45}$, and odd wavefunctions lead to integral eigenvalues. Thus for each of the component representations, one of the operators $im'_{12}$, $im'_{45}$ has integral eigenvalues, and the other half-odd integral eigenvalues. The component with $im'_{12}$ half-odd and

$im'_{45}$ integral is equivalent to the component of (39) or (63) for which $im_{45}$ has positive eigenvalues.

The new representation is thus more fundamental than the previous one in two respects—it separates the negative eigenvalues of $im_{45}$ away from the positive ones and it allows both the alternatives of integral and half-odd eigenvalues for $im_{45}$. The greater power of the new representation comes from the introduction, by (64), of new variables equal to the square roots of previous variables.

The quadratic functions of the four variables $q_1, q_2, p_1, p_2$ are the infinitesimal generators of the group of linear transformations of these variables that leave their commutation relations invariant. It has been pointed out to me by R. Jost that this group is just the 4-dimensional simplectic group, which is equivalent to the 3 + 2 de Sitter group. These results are easily established by the following argument:

For convenience, put $p_1 = q_3$, $p_2 = q_4$. We then have to consider linear transformations of the four $q$'s which leave the commutators $[q_m, q_n]$ invariant $(m, n = 1, 2, 3, 4)$. Let $a_m$ be a set of four numbers which transform contravariantly to the four $q$'s, so that $a_m q_m$ is invariant, and let $b_m$ be another such set of four numbers. Put

$$W_{mn} = a_m b_n - a_n b_m.$$

Thus, each linear transformation of the $q$'s gives rise to a linear transformation of the $a$'s and $b$'s, which in turn gives rise to a linear transformation of the $W$'s.

We have

$$[a_m q_m, b_n q_n] = a_m b_n [q_m, q_n] = 2i(W_{13} + W_{24}).$$

This is invariant. Also,

$$W_{12}W_{34} + W_{23}W_{14} + W_{31}W_{24} = 0.$$

Put

$$W_{12} = x_1 + x_4, \quad W_{23} = x_2 + x_5, \quad W_{31} = x_3 + x_6,$$
$$W_{34} = x_1 - x_4, \quad W_{14} = x_2 - x_5, \quad W_{24} = x_3 - x_6.$$

Then the six $x$'s are subject to linear transformations which leave $x_6$ invariant, and which preserve the quadratic equation

$$x_1^2 + x_2^2 + x_3^2 - x_4^2 - x_5^2 - x_6^2 = 0.$$

This is just the 3 + 2 de Sitter group.

# REFERENCES


Abbot, L.F., Atwood, W.B., and Barnett, R.M., 1980, *Phys. Rev.* D **22**, 582-593.

Abraham, R. and Marsden, J.E., 1978, *Foundations of Mechanics* 2nd Ed. (Benjamin/Cummings. Reading, MA).

Agarwal, G.S., 1987, *J. Mod. Opt.* **34**, 909-921.

Ali, S.T. and Emch, G.G., 1974, *J. Math. Phys.* **15**, 176-182.

—, 1985, *Riv. Nuovo Cimento* **8** (No. 11), 1-128.

Allasia, D. *et al.*, 1984, *Phys. Lett.* **135B**, 231-236.

Almeida, A.M.O., 1983, *Ann. Phys.* **145**, 100-115.

Alonso, J.D., 1985, *Ann. Phys.* (New York) **160**, 1-53.

Altarelli, G. and Parisi, G., 1977, *Nucl. Phys.* **B126**, 298-318.

Aravind, P.K., 1989, *Am. J. Phys.* **57**, 309-411.

—, 1990, *Phys. Rev.* A**42**, 4077-4084.

Arecchi, F.T., 1965, *Phys. Rev. Lett.* **15**, 912-916.

Arnold, V.I., 1978, *Mathematical Methods of Classical Mechanics* (Springer-Verlag, New York, NY).

Aubert, J.J. *et al.*, 1983, *Phys. Lett.* **123B**, 123-126.

Balazs, N.L. and Zippel, G.G., 1973, *Ann. Phys.* (New York) **77**, 139-156.

—, 1980, *Physica* **102A**, 236-254.

—, 1981, *Physica* **109A**, 317-327.

— and Jennings, B.K., 1984, *Phys. Rep.* **104**, 347-391.

Bargmann, V. and Wigner, E.P., 1948, *Proc. Nat. Acad. Scie.* (USA) **34**, 211-223.

Barocchi, F., Neumann, M., and Zoppi, M., 1985, *Phys. Rev.* A**31**, 4015-4017.

Bartlett, M.S. and Moyal, J.E., 1949, *Proc. Cambridge Philo. Soc.* **45**, 545-553.

Beg, M.B., Lee, B.W., and Pais, A., 1964, *Phys. Rev. Lett.* **13**, 514-517.

Berry, M.V., 1977, *Phil. Trans. Roy. Soc.* (London) A**287**, 237-271.

Bertrand, P., Doremus, J.P., Izrar, B., Ngyuen, V.T., and Feix, M.R., 1983, *Phys. Lett.* **94A**, 415-417.

Bjorken, J.D., 1969, *Phys. Rev.* **179**, 1546-1553.

— and Paschos, E.A., 1969, *Phys. Rev.* **185**, 1975-1982.

Blanchard, C.H., 1982, *Am. J. Phys.* **50**, 642-645.

Blum, K., 1981, *Density Matrix Theory and Applications* (Plenum, New York, NY).

Bodek, A. *et al.*, 1973, *Phys. Rev. Lett.* **30**, 1087-1091.

Bohm, D. and Hiley, B.J., 1981, *Found. of Phys.* **11**, 179-203.

Bondourant, R.S. and Shapiro, J.H., 1984, *Phys. Rev.* D**30**, 2584-2556.

Boya, L.J. and de Azcarraga, J., 1967, *Ann. R. Soc. Esp. Fis. Quim.* A**63**, 143-155.

Buras, A.J., 1980, *Rev. Mod. Phys.* **52**, 199-276.

Capri, A.Z. and Chiang, C.C., 1976, *Nuovo Cimento* **36A**, 331-353.

— and Chiang, C.C., 1977, *Nuovo Cimento* **38A**, 191-208.

— and Chiang, C.C., 1978, *Prog. Theor. Phys.* **59**, 996-1008.

Carruthers, P. and Zachariasen, F., 1976, *Phys. Rev.* D**13**, 950-960.

— and Shih, C.C., 1983, *Phys. Lett.* **127B**, 242-250.

— and Zachariasen, F., 1983, *Rev. Mod. Phys.* **55**, 245-285.

Caves, C.M., Thorne, K.S., Drever, R.W.P., Sandberg, V.D., and Zimmermann, M., 1980, *Rev. Mod. Phys.* **52**, 341-392.

—, 1981, *Phys. Rev.* D**23**, 1693-1708.

— and Schumaker, B.L., 1985, *Phys. Rev.* A**31**, 3068-3092.

Chakrabarti, A., 1964, *J. Math. Phys.* **5**, 1747-1755.

Chang, S.J. and Shi, K.J., 1985, *Phys. Rev.* A**34**, 7-22.

Chiao, R.Y. and Jordan, T.F., 1988, *Phys. Lett.* A**132**, 77-81.

—, 1989, in the *Proceedings of the International Symposium on Spacetime Symmetries in Commemoration of the* 50$^{th}$ *Anniversary of Eugene Paul Wigner's Fundamental Paper on the Inhomogeneous Lorentz Group*, College Park, Maryland, 1988, Eds. Y.S. Kim and W.W. Zachary (North-Holland, Amsterdam), pp.327-223.

Chou, K.C. and Zastavenco, L.G., 1958, *Zhur. Exptl. i Teoret. Fiz.* **35**, 1417-1245 or *Soviet Phys. JETP* **8** (1959), 990-995.

Cohen, L., 1966, *J. Math. Phys.* **7**, 781-786.

—, 1984, *J. Chem. Phys.* **80**, 4277-4279.

Cole, J.B. *et al.*, 1988, *Phys. Rev.* D**37**, 1105-1112.

Collet, M.J. and Walls, D.F., 1985, *Phys. Rev.* A**32**, 2887-2892.

Conner, A. and Li, Y., 1985, *Appl. Opt.* **24**, 3825-3829.

Critchfield, C.L., 1976, *J. Math. Phys.* **17**, 261-266.

Dahl, J.P., 1982, *Physica Scripta* **25**, 499-503.

— and Springborg, M., 1982, *Molec. Phys.* **47**, 1001-1019.

Das, A., 1983, *Prog. Theor. Phys.* **70**, 1666-1671.

—, 1988, *Z. Phys.* C**41**, 505-512.

Davies, R.W. and Davies, K.T.R., 1975, *Ann. Phys.* (New York) **89**, 261-271.

Davis, M.J. and Heller, E.J., 1979, *J. Chem. Phys.* **71**, 3383-3395.

De, T, Kim, Y.S., and Noz, M.E., 1973, *Nuovo Cimento* **13A**, 1089-1101.

Dickman, R. and O'Connell, R.F., 1985, *Phys. Rev.* B**32**, 471-473.

Dirac, P.A.M., 1927, *Proc. Roy. Soc.* (London) **A114**, 234-265 and 710-728.

—, 1930, *The Principle of Quantum Mechanics* (Oxford University Press, Lon-

don).
—, 1945, *Proc. Roy. Soc.* (London) **A183**, 284-295.
—, 1949, *Rev. Mod. Phys.* **21**, 392-399.
—, 1958, *The Principles of Quantum Mechanics* 4th Ed. (Oxford University Press, London).
—, 1963, *J. Math. Phys.* **4**, 901-909.
—, 1970, *Physics Today* **23**, No. 4 (April) 29-31.
Dirl, R., Kasperkovitz, P., and Moshinsky, M., 1988, *J. Phys.* **A21**, 1835-1846.
Drell, S. D. and Yan, T. M., 1971, *Ann. Phys.* (New York) **66**, 578-623.
Duke, D.W., and Owens, J.F., 1984, *Phys. Rev.* **D30**, 49-54.
Durand, M., Schuck, M., and Kunz, J., 1985, *Nucl. Phys.* **A439**, 263-288.
Easton, R.L., Ticknor, A.J., and Barrett, H.H., 1984, *Opt. Engineering* **23**, 738-744.
Ekert, A.K. and Knight, P.L., 1989, *Am. J. Phys.* **57**, 692-697.
Emch, G.G., 1982, *J. Math. Phys.* **23**, 1785-1791.
Eu, B.C., 1983, *J. Chem. Phys.* **78**, 409-419.
Feingold, M. and Peres, A., 1985, *Phys. Rev.* **A31**, 2472-2476.
Fetter, A.L. and Walecka, J.D., 1971, *Quantum Theory of Many-Particle Systems* (McGraw-Hill, New York, NY).
Feynman, R.P., 1969, in *High Energy Collisions, Proceedings of the third International Conference*, Stony Brook, New York, Eds. C.N. Yang *et al.* (Gordon and Breach, New York, NY).
—, Kislinger, M., and Ravndal, F., 1971, *Phys. Rev.* **D3**, 2706-2732.
—, 1973, *Statistical Mechanics* (Benjamin/Cummings, Reading, MA)
Fisher, R.A., Nieto, M.M., and Sandberg, V.D., 1984, *Phys. Rev.* **D29**, 1107-1110.
Franca, H.M. and Thomas, M.T., 1985, *Phys. Rev.* **D31**, 1337-1340.
Frazer, W.E. and J. Fulco, 1960, *Phys. Rev. Lett.* **2**, 365-368.
—, 1966, *Elementary Particles* (Prentice-Hall, Englewood Cliffs, NJ).
Friedmann, A., 1922, *Z. Phys.* **10**, 377-386.
—, 1924, *Z. Phys.* **21**, 326-332.
Fujimura, K., Kobayashi, T., and Namiki, M., 1970, *Prog. Theor. Phys.* **43**, 73-79.
Fujita, S., 1966, *Introduction to Non-Equilibrium Statistical Mechanics* (Saunders, Philadelphia).
Gell-Mann, M., 1964, *Phys. Lett.* **8**, 214-215.
Gilmore, R., 1974, *Lie Groups, Lie Algebras, and Some of Their Applications* (Wiley-Interscience, New York, NY).
Ginzburg, V.L. and Man'ko, V.I., 1965, *Nucl. Phys.* **74**, 577-588.
Glauber, R.J., 1963, *Phys. Rev.* **130**, 2529-2539.
Gluck, M., Reya, E., and Hoffmann, E., 1982, *Z. Physik* **C13**, 119-130.
Goldin, E., 1982, *Waves and Photons* (Wiley, New York, NY).
Goldstein, H., 1980, *Classical Mechanics*, 2nd Ed. (Addison-Wesley, Reading, MA)
Gracia-Bondia, J.M., 1984, *Phys. Rev.* **A30**, 691-697.

Gradshteyn, I.S. and Ryzhik, I.M., 1965, *Tables of Integrals, Series, and Products* (Academic Press, New York, NY).
Greenberg, O.W., 1964, *Phys. Rev. Lett.* **13**, 598-602.
— and Resnikoff, M., 1967, *Phys. Rev.* **163**, 1844-1851.
— and Nelson, C.A., 1977, *Phys. Reports* **32**, 69-121.
—, 1982, Am. J. Phys. **50**, 1074-1089.
Groenewold, H.J., 1946, *Physica* **12**, 405-460.
Guillemin, V. and Sternberg, S., 1984, *Symplectic Techniques in Physics* (Cambridge Univ. Press, Cambridge).
Haberman, M.L., 1984, *Phys. Rev.* **D29**, 1412-1416.
— and Hussar, P.E., 1988, *Z. Phys.* **C40**, 153-162.
Hakim, R., 1978, *Riv. Nuovo Cimento* **1** (No. 6), 1-52.
Han, D. and Kim, Y.S., and Noz, M.E., 1981, *Found. of Phys.* **11**, 895-905.
—, Kim, Y.S., and Son D., 1982, *Phys. Rev.* **D26**, 3717-3725.
—, Kim, Y.S., and Son D., 1983, *Phys. Lett.* **131B**, 327-330.
—, Kim, Y.S., and Son D., 1986, *J. Math. Phys.* **27**, 2228-2235.
—, Kim, Y.S., and Son D., 1987a, *J. Math. Phys.* **28**, 2373-2378.
—, Kim, Y.S., and Son D., 1987b, *Class. Quantum Grav.* **4**, 1777-1783.
—, Kim, Y.S., and Noz, M.E., 1987c, *Phys. Rev.* **A35**, 1682-1691.
—, Kim, Y.S., and Noz, M.E., 1988, *Phys. Rev.* **A37**, 807-814.
—, Hardekopf, E.E., and Kim, Y.S., 1989, *Phys. Rev.* **A39**, 1269-1276.
—, Kim, Y.S., and Noz, M.E., 1989, *Phys. Rev.* **A40**, 902-912.
—, Kim, Y.S., and Noz, M.E., 1990a, *Phys. Lett.* **144A**, 111-115.
—, Kim, Y.S., and Noz, M.E., 1990b, *Phys. Rev.* **A41**, 6233-6244.
Hanlon, J. *et al.*, 1980, *Phys. Rev. Lett.* **45**, 1817-1821.
Hartle, J., 1990, in the *Proceedings of the Aharonov-Bohm Symposium*, 1989, Columbia, South Carolina, Ed. J. Anandan (World Scientific, Singapore).
Heisenberg, W., 1927, *Z. Phys.* **43**, 172-198.
Heitler, W., 1954, *The Quantum Theory of Radiation*, 3rd Ed. (Oxford University Press, London).
Heller, E.J., 1976, *J. Chem. Phys.* **65**, 1289-1298.
—, 1977, *J. Chem. Phys.* **67**, 3339-3353.
Henriques, A.B., Kellet, B.H., and Moorehouse, R.G., 1975, *Ann. Phys.* (New York) **93**, 125-151.
Henry, R.W. and Glotzer, S.C., 1988, *Am. J. Phys.* **56**, 318-328.
Hillery, M., O'Connell, R.F., Scully, M.O., and Wigner, E.P., 1984, *Phys. Rep.* **106**, 121-167.
Ho, S.T., Kumar, P., and Shapiro, J.H., 1986, *Phys. Rev.* **A34**, 293-303.
—, Kumar, P., and Shapiro, J.H., 1987, *Phys. Rev.* **A35**, 3982-3985.
Hofstadter, R. and McAllister, R.W., 1955, *Phys. Rev.* **98**, 183-184.
—, 1963, *Nuclear and Nucleon Structure* (Benjamin, New York).

Huang, K., 1982, ***Quarks, Leptons and Gauge Fields*** (World Scientific, Singapore).

Hussar, P.E., Kim, Y.S., and Noz, M.E., 1980, ***Am. J. Phys.*** **48**, 1043-1049.

—, 1981, ***Phys. Rev.*** D**23**, 2781-2783.

—, Kim, Y.S., and Noz, M.E., 1985, ***Am. J. Phys.*** **53**, 142-147.

Hwa, R. C., 1980, ***Phys. Rev.*** D**22**, 759-764.

—, and Zahir, M.S., 1981, ***Phys. Rev.*** D**23**, 2539-2553.

Imre, K., Ozimir, E., Rosenbaum, M., and Zweifel, P.F., 1967, *J.* ***Math. Phys.*** **8**, 1097-1108.

Inonu, E. and Wigner, E.P., 1953, ***Proc. Nat. Acad. Scie.*** (USA) **39**, 510-524.

Isgur, N. and Karl, G., 1979, ***Phys. Rev.*** D**19**, 2653-2677.

Ishida, S., 1971, ***Prog. Theor. Phys.*** **46**, 1570-1586 and 1905-1923.

—, Matsuda, A., Namiki, M., 1977, ***Prog. Theor. Phys.*** **57**, 210-228.

—, Takeuchi, K., Tsuruta, S., Watanabe, M., and Oda, M., 1979, ***Phys. Rev.*** D**20**, 2906-2922.

Itzykson, C. and Zuber, J.B., 1984, ***Quantum Field Theory*** (McGraw-Hill, New York, NY).

Jacob, M. and Wick, G.C., 1959, ***Ann. Phys.*** (New York) **7**, 404-428.

Janssen, A.J.E.M., 1985, *J.* ***Math. Phys.*** **26**, 1986-1994.

Jose, J.V., 1984, ***Phys. Rev.*** B**29**, 2836-2838.

Kadanoff, L.P. and Baym, G., 1962, ***Quantum Statistical Mechanics*** (Benjamin, New York, NY).

Karr, T.J., 1976, ***Field Theory of Extended Hadrons based on Covariant Harmonic Oscillators***, Ph.D. Thesis (Univ. of Maryland).

Kijowski, J., 1974, ***Rep. Math. Phys.*** **6**, 361-386.

Kim, K.J., 1986, ***Nucl. Instr.*** A**246**, 71-76.

Kim, Y.S. and Noz, M.E., 1973, ***Phys. Rev.*** D**8**, 3521-3527.

—, 1976, ***Phys. Rev.*** D**14**, 273-279.

— and Noz, M.E., 1977, ***Phys. Rev.*** D**15**, 335-358.

—, Noz, M.E., and Oh, S.H., 1979a, *J.* ***Math. Phys.*** **20**, 1341-1344.

—, Noz, M.E., and Oh, S.H., 1979b, ***Found. of Phys.*** **9**, 947-954.

— and Noz, M.E., 1983, ***Am. J. Phys.*** **51**, 368-374.

— and Noz, M.E., 1986, ***Theory and Applications of the Poincaré Group*** (Reidel, Dordrecht, The Netherlands).

— and Wigner, E.P., 1987a, *J.* ***Math. Phys.*** **28**, 1175-1179.

— and Wigner, E.P., 1987b, ***Phys. Rev.*** A**36**, 1293-1297.

— and Wigner, E.P., 1988, ***Phys. Rev.*** A**38**, 1159-1167.

— and Wigner, E.P., 1989, ***Phys. Rev.*** A**39**, 2829-2834.

—, 1989, ***Phys. Rev. Lett.*** **63**, 348-351.

— and Li, M., 1989, ***Phys. Lett.*** A**139**, 445-448.

— and Wigner, E.P., 1990a, *J.* ***Math. Phys.*** **31**, 55-60.

— and Wigner, E.P., 1990b, ***Am. J. Phys.*** **58**, 439-448.

— and Wigner, E.P., 1990c, *Phys. Lett.* **A147**, 343-347.

— and Noz, M. E., 1991, in *Proceedings of the 18$^{th}$ International Colloquium on Group Theoretical Methods in Physics*, Moscow, Eds. V. V. Dodonov and V. I. Man'ko.

Kitazoe, T. and Hama, S., 1979, *Phys. Rev.* **D19**, 2006-2017.

— and Morii, T., 1980a, *Phys. Rev.* **D21**, 685-694.

— and Morii, T., 1980b, *Nucl. Phys.* **B164**, 76-102.

Klauder, J.R. and Sudarshan E.C.G., 1968, *Fundamentals of Quantum Optics* (Benjamin, New York, NY).

— and Skagerstam, B.S., 1985, *Coherent States* (World Scientific, Singapore).

—, McCall, S.L. and Yurke, B., 1986, *Phys. Rev.* **A33**, 3204-3209.

Kokkedee, J.J.J., 1969, *The Quark Model* (Benjamin, New York).

Korsch, J.J. and Berry, M.V., 1981, *Physica* **D3**, 627-636

Kumar, V.B.K.V. and Carroll, C.W., 1984, *Opt. Engineering* **23**, 732-737.

Kupersztych, J., 1976, *Nuovo Cimento* **31B**, 1-11.

Kuratsuji, H., 1981, *Prog. Theor. Phys.* **65**, 224-240.

Lee, H.W. and Scully, M.O., 1980, *J. Chem. Phys.* **73**, 2238-2242.

—, 1982, *Am. J. Phys.* **50**, 438-440.

— and Scully, M.O., 1982, *J. Chem. Phys.* **77**, 4604-4610.

— and Scully, M.O., 1983, *Found. of Phys.* **13**, 61-72.

Lesche, B., 1984, *Phys. Rev.* **D29**, 2270-2274.

— and Seligman, T.H., 1986, *J. Phys.* **A19**, 91-105.

Le Yaouanc, A.L., Oliver, L., Pene, O., and Raynal, J.C., 1978, *Phys. Rev.* **D18**, 1733-1736.

Lichtenberg, D.B., 1970, *Unitary Symmetry and Elementary Particles* (Academic Press New York).

Lipes, R.G., 1972, *Phys. Rev.* **D5**, 2849-2863.

Littlejohn, R.G., 1986, *Phys. Rep.* **138**, 193-291.

Loudon, R., 1973, *The Quantum Theory of Light* (Oxford University Press, London).

Louisell, W.H., 1963, *Phys. Lett.* **7**, 60-61.

—, 1973, *Quantum Statistical Properties of Radiation* (Wiley, New York, NY).

Machida, S. and Yamamoto, Y., 1988, *Phys. Rev. Lett.* **60** 792-793.

Maeda, M.W., Kumar, P., and Shapiro, J.H., 1987, *J. Opt. Soc. Am.* **B4**, 1501-1513.

Magnus, W. and Oberhettinger, F., 1949, *Formulas and Theorems for the Functions of Mathematical Physics* (Chelsea, New York, NY).

Maltman, K. and Isgur, N., 1984, *Phys. Rev.* **D29**, 952-977.

Mann, A. and Revzen, M., 1989, *Phys. Lett.* **A134**, 273-275.

Marciano, W. and Pagels, H., 1978, *Phys. Rep.* **36**, 138-276.

Markov, M., 1956, *Suppl. Nuovo Cimento* **3**, 760-772.

Maslov, V.P. and Fedoriuk, M.V., 1981, *Semiclassical Approximation in Quantum*

*Mechanics* (Reidel, Dordrecht, The Netherlands).
McDonald, S.W. and Kaufman, A.N., 1985, *Phys. Rev.* **A32**, 1708-1713.
Merzbacher, E., 1970, *Quantum Mechanics*, 2nd Ed. (Wiley, New York, NY).
Miller, W., 1972, *Symmetry Groups and Their Applications* (Academic Press, New York, NY).
Misra, S.P. and Maharana, J., 1976, *Phys. Rev.* **D14**, 133-139.
Mita, K., 1978, *Phys. Rev.* **D18**, 4545-4550.
Molzahn, F.H. and Osborn, T.A., 1986, *J. Math. Phys.* **27**, 88-99.
Moshinsky, M., 1987, in the *Proceedings of the First International Conference on the Physics of Phase Space*, 1986, College Park, Maryland, Eds. Y.S. Kim and W.W. Zachary (Springer-Verlag, Heidelberg).
Moyal, J.E., 1949, *Proc. Cambridge Philo. Soc.* **45**, 99-124.
Narcowich, F.J. and O'Connell, R.F., 1986, *Phys. Rev.* **A34**, 1-6.
Newton, T.D. and Wigner, E.P., 1949, *Rev. Mod. Phys.* **21**, 400-406.
Nix, J.R., 1969, *Nucl. Phys.* **A130**, 241-292.
Noz, M.E. and Kim, Y.S., 1988, *Special Relativity and Quantum Theory* (Kluwer, Dordrecht, The Netherlands)
O'Connell, R.F. and Wigner, E.P., 1981a, *Phys. Lett.* **83A**, 145-148.
— and Wigner, E.P., 1981b, *Phys. Lett.* **85A**, 121-126.
— and Rajagopal, A.K., 1982, *Phys. Rev. Lett.* **48**, 525-526.
—, 1983, *Found. of Phys.* **13**, 83-92.
– and Wang, L., 1985, *Phys. Rev.* **A31**, 1707-1711.
—, 1987, in the *Proceedings of the First International Conference on the Physics of Phase Space*, 1986, College Park, Maryland, Eds. Y.S. Kim and W.W. Zachary (Springer-Verlag, Heidelberg).
Oneda, S., 1985, *Hadron Spectroscopy, AIP Conference Proceedings* No. 132 (American Institute of Physics, New York, NY).
Osborn, T.A. and Molzahn, F.H., 1986, *Phys. Rev.* **A34**, 1696-1707.
Ou, Z.Y., Hong, C.K., and Mandel, L., 1987, *Phys. Rev.* **A36**, 192-196.
Pantell, R.H. and Puthoff, H.E., 1969, *Foundations of Quantum Electronics*, (Wiley, New York, NY).
Pathria, P.K., 1972, *Statistical Mechanics* (Pergamon Press, Elmsford, NY).
Perelemov, A., 1986, *Generalized Coherent States and Their Applications* (Springer-Verlag, Berlin).
Perina, J., 1971, *Coherence of Light* (Van Nostrand Reinhold, New York, NY).
Petroni, N.C., Gueret, P., Vigier, J.P., and Kyprianidis, A., 1985, *Phys. Rev.* **D31**, 3157-3161.
Pocsik, G., 1978, *Acta Physica Austriaca* **49**, 47-52.
Preparata, G. and Craigie, N.S., 1976, *Nucl. Phys.* **B102**, 478-524.
Procida, R. and Lee, H.W., 1984, *Opt. Comm.* **49**, 201-204.
Rana, L. and Kim, Y.S., 1989, *Univ. of Maryland Physics Paper* **89-067**.

Ree, F.H., 1983, *J. Chem. Phys.* **78**, 409-415.

Rai, J. and Mehta, C.L., 1988, *Phys. Rev.* **A37**, 4497-4499.

Reid, M.D. and Walls, D.F., 1985, *Phys. Rev.* **A31**, 1622-1635.

Remler, R.A. and Sathe, A.P., 1975, *Ann. Phys.* (New York) **91**, 295-324.

—, 1975, *Ann. Phys.* (New York) **95**, 455-495.

—, 1977, *Phys. Rev.* **D16**, 3464-3473.

— and Sathe, A.P., 1978, *Phys. Rev.* **C18**, 2293-2315.

Ritus, V.I., 1961, *Soviet Phys. JETP* **13** 240-248.

Robertson, H. P., 1933, *Rev. Mod. Phys.* **5**, 62-90.

Ross, J. and Kirkwood, L., 1954, *J. Chem. Phys.* **22**, 1094-1103.

Rotbart, F.C., 1981, *Phys. Rev.* **D23**, 3078-3080.

Royer, A., 1985, *Phys. Rev.* **A32**, 1729-1743.

—, 1987, in the *Proceedings of the First International Conference on the Physics of Phase Space*, 1986, College Park, Maryland, Eds. Y.S. Kim and W.W. Zachary (Springer-Verlag, Heidelberg).

Ruiz, M.J., 1974, *Phys. Rev.* **D10**, 4306-4307.

—, 1975, *Phys. Rev.* **D12**, 2922-4924.

Sargent III, M., Scully, M., and Lamb, Jr. W.E., 1974, *Laser Physics* (Addison-Wesley, Reading, MA).

Schiff, L.I., 1968, *Quantum Mechanics*, 3rd Ed. (McGraw-Hill, New York, NY).

Schleich, W., Walls, D.C., and Wheeler J.A., 1988, *Phys. Rev.* **A38**, 1177-1186.

Schumaker, B.L., 1986, *Phys. Rep.* **135**, 317-408.

— and Caves C.M., 1985, *Phys. Rev.* **A31**, 3093-3111.

Serima, O.T., Javanainen, J., and Varro, S., 1986, *Phys. Rev.* **A33**, 2913-2927.

Shelby, R.M., Levenson, M.D., Perlmutter, S.H., DeVoe, R.G., and Walls, D.F., 1986, *Phys. Rev. Lett.* **57**, 691-694.

Shirley, J.H., 1965, *Phys. Rev.* **138** B979-987.

Shlomo, S., 1983, *J. Phys.* **A16**, 3463-3469.

—, 1985, *Nuovo Cimento* **87A**, 211-223.

— and Prakash, M., 1981, *Nucl. Phys.* **A357**, 157-170.

Sirugue, M., Sirugue-Collin, M., and Truman, A., 1984, *Ann. Inst. Henri Poincaré* **41**, 429-442.

Slusher, R.E., Hollenberg, L.W., Yurke, B., Mertz, J.C., and Valley, J.F., 1985, *Phys. Rev. Lett.* **55**, 2409-2412.

Sogami, I., 1973, *Prog. Theor. Phys.* **50**, 1729-1747.

Stanley, D.P. and Robson, D., 1982, *Phys. Rev.* **D26**, 223-232.

Stoler, D., 1970, *Phys. Rev.* **D1**, 3217-3219.

Subotic, N.S. and Saleh, B.E.A., 1984, *Opt. Comm.* **52**, 259-264.

Sudarshan, E.C.G., 1963, *Phys. Rev. Lett.* **10**, 277-279

Sumi, H., 1983, *Phys. Rev.* **B27**, 2374-2386.

—, 1984, *Phys. Rev.* **B29**, 4616-4630.

Sundberg, R.L., Abramson, A., Kinsey, J.L., and Fied, R.W., 1985, *J. Chem. Phys.* **83**, 466-475.

Susskind, L. and Gologower, J., 1964, *Physics* **1**, 49-61.

Szu, H., 1987, in the *Proceedings of the First International Conference on the Physics of Phase Space*, 1986, College Park, Maryland, Eds. Y.S. Kim and W.W. Zachary (Springer-Verlag, Heidelberg).

Takabayasi, T., 1954, *Prog. Theor. Phys.* **11**, 341-473.

—, 1964, *Nuovo Cimento* **33**, 668-672.

—, 1979, *Prog. Theor. Phys. Suppl.* **67**, 1-68.

Takatsuka, K. and Nakamura, H., 1985, *J. Chem. Phys.* **82**, 2573-2589.

Thies, M., 1979, *Ann. Phys.* (New York) **123**, 411-441.

Thomas, L.H., 1927, *Phil. Mag.* **3**, 1-22.

Teich, M.C. and Saleh, B.E.A., 1985, *J. Opt. Soc. Am.* B**2**, 275-282.

— and Saleh, B.E.A., 1989, *Quantum Opt.* **1**, 153-191.

— and Saleh, B.E.A., 1990, *Physics Today* **26**, No. 6 (June) 26-34.

Truax, D.R., 1985, *Phys. Rev.* D**31**, 1988-1991.

Umezawa, H., Matsumoto, H., and Tachiki, M., 1982, *Thermo Field Dynamics and Condensed States* (North-Holland, Amsterdam).

Van Dam, H. and Wigner, E.P., 1965, *Phys. Rev.* **138**, B1576-1582.

Van Royen, R. and Weisskopf, V., 1967, *Nuovo Cimento* **50A**, 617-645.

Vasak, D., Gyulassy, M. and Elze, H.T., 1987, *Ann. Phys.* (New York) **173**, 462-492.

Vassiliadis, D.V., 1989, *J. Math. Phys.* **30**, 2177-2180.

Von Neumann, J., 1927, Nachriten Göttingen **1927**, 245-272 and 276-291.

—, 1955, *Mathematical Foundation of Quantum Mechanics* (Princeton University Press, Princeton, NJ).

Walls, D.F., 1983, *Nature* (London) **306**, 141-146.

Weinberg, S., 1964, *Phys. Rev.* **135**, B1049-1056.

Weisskopf, V. and Wigner, E.P., 1930a, *Z. Phys.* **63**, 54-73.

— and Wigner, E.P., 1930b, *Z. Phys.* **65**, 18-29.

Werner, R., 1984, *J. Math. Phys.* **25**, 1404-1411.

Weyl, H., 1946, *Classical Groups* (Princeton Univ. Press, Princeton, NJ).

Whitenton, J.B., Durand, B., and Durand, L., 1983, *Phys. Rev.* D**28**, 597-623.

Wigner, E.P., 1932a, *Phys. Rev.* **40**, 749-759.

—, 1932b, *Z. Phys. Chem.* **B19**, 203-216.

—, 1938, *Trans. Faraday Soc.* **34**, 29-41.

—, 1939, *Ann. Math.* **40**, 149-204.

—, 1957, *Rev. Mod. Phys.* **29**, 255-268.

—, 1959, *Group Theory and Its Applications to the Quantum Mechanics of Atomic Spectra* (Academic Press, New York, NY)

— and Yanase, M., 1963, *Proc. Nat. Acad. Scie.* (USA) **49**, 910-918.

—, 1971, in *Perspectives in Quantum Theory*, Eds. W. Yourgrau and A. van der Merwe (MIT Press, Cambridge, MA).

—, 1972 in *Aspects of Quantum Theory, in Honour of P.A.M. Dirac's* $70^{th}$ *Birthday*, Eds. A. Salam and E.P. Wigner (Cambridge Univ. Press, London).

—, 1987, in the *Proceedings of the First International Conference on the Physics of Phase Space*, 1986, College Park, Maryland, Eds. Y.S. Kim and W.W. Zachary (Springer-Verlag, Heidelberg).

Winter, J., 1985, *Phys. Rev.* D**32**, 1871-1888.

Wu, L.A., Kimble, H.J., Hall, J.L., and Wu, H., 1986, *Phys. Rev. Lett.* 20, 2520-2523.

Yuen, H.P., 1976, *Phys. Rev.* A**13**, 2226-2243.

— and Shapiro, 1979, *Optics Lett.* **4**, 334-336.

Yukawa, H., 1953, *Phys. Rev.* **91**, 415-416.

Yurke, B., 1985, *Phys. Rev.* A**32**, 300-310 and 311-323.

—, McCall, S., and Klauder, J.R., 1986, *Phys. Rev.* A**33**, 4033-4054.

— and Potasek, M., 1987, *Phys. Rev.* A**36**, 3464-3466.

Zachariasen, F., 1987, in the *Proceedings of the First International Conference on the Physics of Phase Space*, 1986, College Park, Maryland, Eds. Y.S. Kim and W.W. Zachary (Springer-Verlag, Heidelberg).

Zhang, W.M., Feng, D.H., and Gilmore, R., 1990, *Rev. Mod. Phys.* to be published.

Zuk, J.A., 1985, *Phys. Rev.* D**32**, 2653-2658.

# Index

www.ingramcontent.com/pod-product-compliance
Lightning Source LLC
Chambersburg PA
CBHW061622220326
41598CB00026BA/3846
* 9 7 8 9 8 1 0 2 0 3 6 1 0 *